Pyrylium Salts:
Syntheses, Reactions,
and Physical Properties

Advances in Heterocyclic Chemistry

Edited by

A. R. Katritzky

PYRYLIUM SALTS: SYNTHESES, REACTIONS, AND PHYSICAL PROPERTIES

Advances in Heterocyclic Chemistry
Supplement 2

ALEXANDRU T. BALABAN

*Institute of Atomic Physics
and Department of Organic
Chemistry
The Polytechnic
Bucharest, Rumania*

GERHARD W. FISCHER

*Academy of Sciences of the German
Democratic Republic
Research Centre of Chemical
Toxicology
Leipzig, German Democratic
Republic*

ANTONIE DINCULESCU

*Institute of Chemical-Pharmaceutical
Research
Bucharest, Rumania*

ALLA V. KOBLIK

*Scientific Research Institute of
Physical and Organic Chemistry
Rostov-on-Don State University
Rostov-on-Don, USSR*

GENADII N. DOROFEENKO

*Scientific Research Institute of
Physical and Organic Chemistry
Rostov-on-Don State University
Rostov-on-Don, USSR*

VALERII V. MEZHERITSKII

*Scientific Research Institute of
Physical and Organic Chemistry
Rostov-on-Don State University
Rostov-on-Don, USSR*

WERNER SCHROTH

*Department of Chemistry
Martin Luther University, Halle-
Wittenberg
Halle, German Democratic Republic*

1982

ACADEMIC PRESS

A Subsidiary of Harcourt Brace Jovanovich, Publishers

New York • London
Paris • San Diego • San Francisco • São Paulo • Sydney • Toronto

ACADEMIC PRESS, INC.
111 Fifth Avenue, New York, New York 10003

United Kingdom Edition published by
ACADEMIC PRESS, INC. (LONDON) LTD.
24/28 Oval Road, London NW1 7DX

LIBRARY OF CONGRESS CATALOG CARD NUMBER: 62–13037

ISBN 0–12–020652–8

PRINTED IN THE UNITED STATES OF AMERICA

82 83 84 85   9 8 7 6 5 4 3 2 1

Contents

Preface

The pyrylium salts are of very considerable practical and theoretical interest. For the first time we now have a monograph that—together with Part I which appeared in Volume 10 of *Advances in Heterocyclic Chemistry*—summarizes completely their preparation and their chemical and physical properties. Their many-fold reactivity enables the attainment of a wide variety of synthetic objectives. Their peculiar stability has attracted the attention of theoretically inclined chemists for many generations. The present work is indeed an international effort written jointly by teams from Rumania, the Soviet Union, and the German Democratic Republic. All the authors have been actively engaged in the field themselves and speak with very considerable authority. It is with great regret that we learned of the death of Professor Dorofeenko just a few months after the final manuscript had been completed. We believe that the present volume will be both a worthy epitaph and a signpost to the many scientists who will utilize pyrylium chemistry.

A. R. KATRITZKY

Detailed Table of Contents

Pyrylium Salts: Syntheses, Reactions, and Physical Properties

I. Introduction

Pyrylium salts represent, perhaps more than any other heterocyclic system, a nodal point for many synthetic routes; they can function as intermediates for an extraordinary variety of syntheses. They owe their key role both to a high formation tendency and to a high reactivity toward nucleophiles.

In the framework of the present "renaissance" of organic chemical synthesis, syntheses with or through heterocycles play an important part; the chemistry of pyrylium salts represents within this part a representative and convincing example of versatility and variety.

At the same time, pyrylium salts are interesting objects of study in themselves because they represent the extreme case of a single perturbation introduced by a heteroatom into a benzene ring: the replacement of CH in benzene by O^+ modifies the electron distribution much more than any other common heteroatom (the electronegativity increases in the order $CR < N < NR^+ < O^+$), or than any substituent R in CR or NR^+. Thus, pyrylium salts give no electrophilic substitution, but only addition of nucleophiles (as primary reaction step). Since the resonance energy in pyrylium is smaller than in benzene or pyridine, unlike these ring systems, the pyrylium ring is as easily opened as it is formed. Such ring-opening reactions are only encountered under more drastic conditions (e.g., temperature, high pH) in pyridinium salts with electronegative substituents R like CN, SO_3^-, NO_2, polynitrophenyl, or 4-pyridyl; in

1

these cases NR$^+$ approaches the electronegativity of O$^+$, but does not reach it.

Finally, pyrylium and pyrone rings as well as benzo derivatives of these systems appear in many natural products so that the study of the reactions and properties of the parent system is also of interest for natural product chemistry.

It is therefore understandable why papers on properties and reactions of pyrylium salts have multiplied so fast in recent years, adding new information to established research directions or opening new vistas. The evolution of the literature on pyrylium chemistry is depicted in Fig. 1, where in each five-year period the number of papers in Chemical Abstracts for that period was counted. The inauguration of pyrylium chemistry is due to Baeyer, the peak around 1920 to discoveries by Dilthey and Schneider, and the upward trend beginning in the fifties to Dimroth, Praill, Nenitzescu, and Balaban.

Several nonexhaustive reviews covering certain limited aspects of the chemistry of pyrylium salts have appeared since the publication of Part I,[1] containing important information relevant to the contents of the pres-

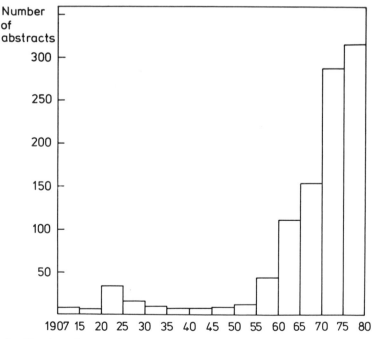

FIG. 1.   Number of papers from Chemical Abstracts dealing with pyrylium salts during 1907–1980.

ent Part II: Dimroth's reviews on the formation of aromatic compounds from pyrylium salts[2-4]; Dorofeenko's books on perchloric acid in organic chemistry,[5] on preparative aspects of pyrylium salt chemistry,[6] on heterocycles in organic synthesis[7]; the reviews on steroids with fused pyrylium rings[8] (all of them in Russian), and on pyrylium side-chain reactions[9] encompass certain areas of pyrylium chemistry; Van der Plas's book "Ring Transformations of Heterocycles"[10] discusses in a more general framework these reactions which convert pyrylium into another ring; Meyers's book "Heterocycles in Organic Synthesis"[11] describes important examples of the synthetic potential of pyrylium salts in connection with the general development of heterocyclic chemistry; Balaban's review "The Pyrylium Cation as a Synthon in Organic Chemistry"[12] deals with the pathways leading to the formation of the C_5 chain in pyrylium, and presents the various modes in which this chain is incorporated totally or partly into other acyclic, carbocyclic, or heterocyclic systems; Katritzky's report[13] on the uses of N-R-2,4,6-triarylpyridinium salts for obtaining products derived from R^+ cations after elimination of 2,4,6-triarylpyridine as the leaving group. Brief general reviews on pyrylium salts are included in wider coverage books or treatises[14,15]; a review on pyrylium chemistry has appeared in Japanese[16]; thiopyrans and thiopyrylium salts were also reviewed (in Russian).[17]

The present review (Part II) has a more ambitious aim, namely to treat systematically and exhaustively all known reactions and properties of pyrylium salts, so as to constitute (together with Part I[1]) a complete review of pyrylium salt chemistry: the present Part II takes much more space than the first part, which appeared in 1969 and described the syntheses of pyrylium salts. It starts with an updating of progress in syntheses of pyrylium salts (Section II) in the period 1968–1978, and then discusses the reactions and properties of pyrylium salts. The literature coverage includes 1979, and the newer literature up to 1981 was considered as often as possible. This review is confined to monocyclic pyrylium salts as defined in Section I,A of Part I[1]; benzo derivatives have been included exceptionally, when the subject made it necessary. The nomenclature follows the rules outlined in Section I,B in the first part; positions 2 or 6 may be denoted by α, 3 or 5 by β, and 4 by γ. The anion is left out in the formula pictures if it has no special influence on the chemical or physical properties of the pyrylium cation.

The present treatment of numerous literature references aims at a synoptical and critical discussion according to systematic classification criteria. It would be impossible to discuss in a limited space all compounds prepared from pyrylium salts, especially in cases when one reaction scheme leads to a host of analogous compounds (e.g., the conversion of pyrylium salts to pyridines, Section III,C). In such cases it was preferred to indicate the synthetic application for characteristic representative compounds, but to quote, however, all relevant papers.

From a graph-theoretical viewpoint, the synthesis of the five-carbon chain C_5 of pyrylium cations may be analyzed in two manners: (i) construction of C_5 from one, two, or three synthons leading to rather simple synthon graphs[18] (cf. Fig. 2; the cases with two synthons are simpler, being subgraphs of the above graphs. The pyrylium ring π shown in Fig. 2 is formed by cyclization of the five-carbon chain symbolized by a-b-a or a-a-b involving three synthons a, a, and b, each of which has 1–3 carbon atoms); (ii) analysis of the various modes for fragmenting the C_5 chain. We shall briefly elaborate on the latter approach, which leads to power graphs $G(r^s)$.[19] In the present case, the power graph is $G(2^4)$: there are four C—C bonds which may be broken or formed, hence the exponent 4. The graph is a four-dimensional hypercube shown in two representations in Fig. 3. If one denotes by letters A–P all 16 possible modes for the bond formation (fragmentation) each of these modes represents a point (vertex) in the power graph $G(2^4)$; two vertices are connected by a line (edge) if they differ by only one bond being formed or fragmented. The resulting graph (Fig. 3) is regular of degree four, i.e., four lines meet at each vertex.

For practical applications, only a portion of this power graph is of interest (Fig. 4), namely that part involving vertices A–E and G–J. Indeed, only in these cases is it possible to have at most two different reagents, even in the three-synthon approaches G–J, because only then two out of the three synthons can be identical. The requirement for at most two different reagents results from the desire to obtain tractable reaction mixtures with significant product yields and few side products.

Figure 5 illustrates the key role played by pyrylium salts in many syntheses: for obtaining pyrylium cations there exist many methods (Part I[1] and Section II of the present part). These methods can be classified

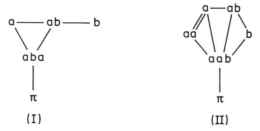

FIG. 2.   Examples of synthon graphs for syntheses of pyrylium salts from three synthons: (I) of type G or J; (II) of type H or I, where capital letters for types are as in Figs. 3 and 4; π stands for a pyrylium ring, **a** and **b** are synthons.

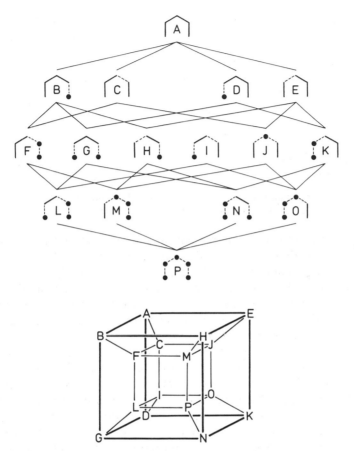

FIG. 3. Two equivalent representations of the theoretical graph for the synthesis of the C_5-chain in pyrylium salts; power graph G (2^4).

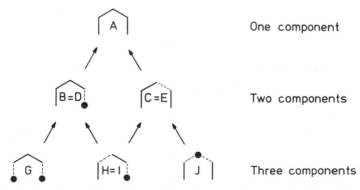

FIG. 4. Practical graph (partial graph from the preceding one) for the synthesis of the C_5-chain in pyrylium salts.

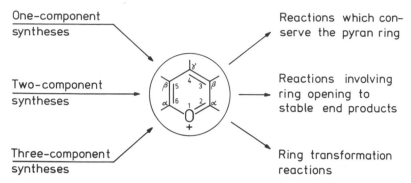

FIG. 5.   The pyryliun ring as a turning plate for the synthesis of acyclic, carbocyclic and heterocyclic systems starting from simple acyclic reagents.

in one-component, two-component, and three-component syntheses. In turn, some of these may afford several combinations ($C_1 + C_4$ and $C_2 + C_3$ for two-component syntheses, $C_2 + C_1 + C_2$, $C_1 + C_2 + C_2$, and $C_1 + C_3 + C_1$ for three-component syntheses), as indicated in Fig. 4. During the synthesis which is not a synchronous process the three-component syntheses go over into two-component syntheses, and these in turn into one-component syntheses.

As general conclusions about syntheses of pyrylium rings, it is certain that (i) by appropriate choice of the synthetic pathway and of the starting materials one can build up pyrylium rings with a desired substitution pattern from simple building blocks,* that (ii) in practically all cases at least one carbonyl derivative (aldehyde, ketone or carboxylic acid derivative) must be present, and that (iii) according to availability, price, and convenience, in most cases alternative synthetically equivalent reagents may be employed: methyl ketones and acetylenes ($RCOCH_3 \rightleftharpoons RC{\equiv}CH + H_2O$); 1,3-diketones and their enols, β-halovinyl ketones, enaminoketones, acetylenic ketones, tertiary carbinols, tertiary halides, branked alkenes, etc.

$$RCOCH_2COR' \rightleftharpoons \begin{array}{l} RC{\equiv}CCOR' + H_2O \\ RCCl{=}CHCOR' + H_2O - HCl \\ Alk_2NCR{=}CHCOR' + H_2O - HNAlk_2 \end{array}$$

The high formation tendency of the pyrylium ring is due (i) to the well-known easy closure of a six-membered ring, and (ii) to aromatic stabi-

_____

* With limitations imposed by the pathway, e.g., in three-component syntheses, at least two of the substituents will be identical, namely the α-substituents in the pathways G and J (Fig. 4), the β-substituents in pathway J, and in pathway H = I substituents in positions 2 and 4, as well as 3 and 5, pairwise.

lization by the six π-electrons within the ring. The high electronegativity of the oxygen heteroatom leads to charge localization and to a lower resonance energy than in benzene or pyridine; the conjugation energy of pyrylium is, however, high enough to make pyrylium salts stable in acid or neutral aqueous solutions (unlike simple oxonium salts), but low enough to enable pyrylium cations to react with many nucleophiles under ring opening.

As indicated by limiting structures **1b–d**, nucleophilic attack may occur in positions 2, 4, or 6, where the positive charge appears. Most reactions occur through a primary nucleophilic attack in positions 2 or 6 (α-positions) which, as will be discussed in Section IV, have the highest electron deficit (evident from ^{1}H- and ^{13}C-NMR spectra and from theoretical calculations); the reaction then usually proceeds through a thermally allowed electrocyclic ring opening of the resulting α-pyran, which valence isomerizes to a 2,4-pentadien-1-one derivative. In the case of pyrylium salts without γ-substituents or of very strong nucleophiles like Grignard reagents or borohydride anions, nucleophilic attack at position γ (4) competes with attack at α-positions leading to a γ-pyran. As a bisenol ether, this 4*H*-pyran can also undergo ring opening under solvolytic conditions to stable 1,5-pentanedione derivatives (Section III,B).

(1a)   (1b)   (1c)   (1d)

On the other hand, the ring-opened products (2,4-pentadien-1-ones) usually recyclize regiospecifically leading to new carbocyclic or heterocyclic systems, mostly aromatic, with 5-, 6-, or 7-membered rings (Section III,C). Both the ring-opening and the ring-transformation reactions may be accompanied or followed by chain fission so that the C_5 chain and the corresponding substituents of the pyrylium ring are only partly incorporated into the final product in these cases. Finally, by suitable modification of the substituents bonded to an already existing pyrylium ring, one can obtain new structures (Section III,A) which then may be subjected to ring opening or to ring transformation reactions.

The limiting structures for a pyrylium cation make it analogous either to benzene (**1a** is an oxoniabenzene, oxygen acts as π-equivalent of a methine group) or to tropylium (in **1b–d** oxygen acts as π-equivalent of a C=C double bond), and these analogies help in understanding some of the pyrylium reactions.

On the other hand, from the main pathway for ring transformation reactions of pyrylium salts, namely primary nucleophilic attack at the

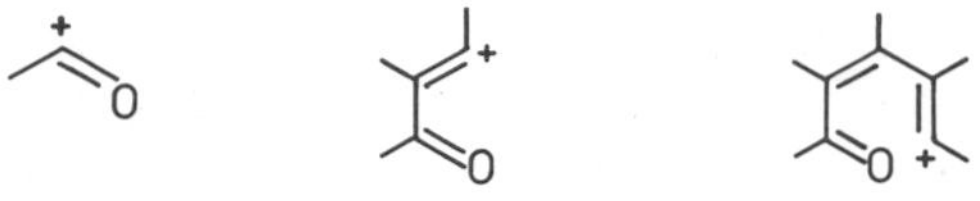

(2)   (3)   (4)

FIG. 6. Vinylogous cations which may act as nucleophilic synthons N^1, N^3, and N^5, respectively.

α-position followed by an electrocyclic ring opening to a pentadienone, one can draw another analogy with other building blocks in organic syntheses: the pyrylium cation is found after ring opening as if it had possessed the structure 4, an electrophilic N^5 cation of pentadienone.* Thus, pyrylium appears as the third member of a vinylogous series (Fig. 6) starting with acyl cations 2 (N^1 synthon, carboxylic acid derivatives such as acyl chlorides), followed by acylvinyl cations 3 (N^3 synthon, such as β-halovinyl ketones). These considerations and analogies lead to important conclusions for designing organic syntheses.

II. Syntheses (Supplement to Part I)†

A. ONE-COMPONENT SYNTHESES.

1. *From Compounds Containing the Pyran Ring*

Reactions leading to pyrylium cations and proceeding with retention of the pyrylium system will not be described in this section. They are discussed in detail in Section III,A, e.g., anion exchange reactions (Section III,A,1), various condensation reactions at the α- and γ-methyl(ene) groups of pyrylium salts (Section III,A,2), and other transformations of substituents (Sections III,A,3–5).

a. *Alkylation, Protonation, and Acylation of Pyrones.* As described in detail in Part I,[1] protonation, alkylation and acylation of 4-pyrones 5 leading to 4-hydroxy-, 4-alkoxy-, and 4-acyloxypyrylium salts 6 (R = H, Alkyl, Acyl) represents one of the oldest syntheses for pyrylium salts. Analogous reactions were investigated with 2-pyrones 7 and with benzo derivatives of 5 and 7.

2-Pyrones are weaker bases than 4-pyrones. This is why the latter form crystalline 4-hydroxypyrylium salts with strong acids in aprotic

---

* Formula 4 may be regarded as a "no-bond" resonance structure of a pyrylium cation.
† Mainly newer data will be described in this section, supplementing Part I[1] and maintaining the arrangement of that part.

solvents, whereas with 2-pyrones crystalline salts cannot be isolated, though the 2-pyrones also undergo protonation with acids.[20-22] The same is true for benzopyrones. Thus, benzo-4-pyrones (chromones) are stronger bases than benzo-2-pyrones (coumarins). This difference is advantageous for separating the compounds. Bubbling of dry hydrogen chloride in ether solutions containing both chromone and coumarin results in the precipitation of 4-hydroxychromylium chloride only, whereas coumarin remains in solution.[23]

(5)   (6)   (7)   (8)

The capability of 4-pyrone derivatives to undergo protonation and alkylation decreases along the series 4-pyrones > chromones > xanthones. This observation is consistent with the basicity data.[24] π-Electron charges on carbonyl oxygen atoms were calculated for 4-pyrone and its benzo derivatives.[25] With xanthones, related hydroxyxanthylium salts could be isolated only as perchlorates, which are unstable crystalline products and which rapidly undergo hydrolysis in the air.[26] With weaker mineral acids (e.g., H_2SO_4, HCl), the salts formed can only exist in solution.[26]

The difference in reactivities between 4-pyrone and its benzo derivatives is even more manifest in O-alkylation reactions. Thus, 2,6-dimethyl- and 2,6-diphenyl-4-pyrones undergo alkylation with dimethyl sulfate and methyl iodide,[27-29] though only with difficulty. Dimethyl sulfate and methyl iodide cannot be used in the synthesis of 4-alkoxybenzo[b]pyrylium salts. Stronger alkylating agents should be applied in the latter case, such as esters of nitrobenzenesulfonic acid.[30] Even these reagents, however, fail to give salt-like products with xanthones.[30]

2-Alkoxypyrylium salts 8 (R = alkyl) can be prepared only by alkylation of 2-pyrones with such strong alkylating agents as methyl fluorosulfonate[31] or trialkyloxonium fluoborates.[32] The latter were also used for converting coumarins into 2-alkoxybenzo[b]pyrylium salts.[33]

The reaction of 4-pyrones 5 (or flavones) with phosgene leads via intermediate acyl derivatives of type 9 to 4-chloropyrylium salts 10.[34,35] Similarly, on treating 4-pyrones with halo derivatives of phosphorus such as PCl_5, PBr_5, or $POCl_3$, 4-halopyrylium salts are formed.[36] These may be used for preparing 4-alkoxy-, 4-aryloxy-, 4-acyloxy-, and other γ-substituted pyrylium salts (cf. Section III,A,5,b).

b. *Nucleophilic Displacements at Pyrones.* As shown in Part I,[1] 2-
and 4-pyrones react with numerous compounds possessing active methylene groups to give methylene pyrans. Protonation of the latter yields
pyrylium salts. A more recent example of such a reaction is the condensation of disulfene with 2,6-dimethyl-4-pyrone (**11**) in acetic anhydride
leading to 2,6-dimethyl-4-pyranylidenedisulfene (**12**).[37] The chemical shift
of the methylene proton resonance signal in the ^{1}H-NMR spectrum is
indicative of a contribution from the dipolar resonance hybrid to the
ground state of the molecule.

Even as weak an electrophile as acetic anhydride can change the
direction of 4-pyrone condensation with active methylene compounds
due to the formation of 4-acetoxypyrylium acetate (**13**) in low equilibrium
concentration. Thus, the reaction with 1,3-dicyanoacetone in acetic anhydride leads to the bis-condensation product **14**, whereas in the absence
of acetic anhydride only the mono-condensation product **15** is formed.[38]

In the presence of phosphorus oxychloride 4-pyrones react, probably via 4-chloropyrylium salts, with 2-phenyloxazol-5-one to give γ-pyran-ylideneoxazolones.[39] Also in the reaction of 2,6-dimethyl-4-pyrone with N,N-dialkylanilines, resorcinol, N-alkylindoles or N-alkylpyrroles in the presence of POCl$_3$ resulting in 4-aryl- or 4-hetaryl-substituted pyrylium salts like **16**, **17**, **18**, and **19**,[40,41] 4-chloropyrylium salts are assumed as reactive intermediates because the latter react in the same manner (cf. Section III,A,5,b). Similar "pyrylation" reactions of aromatic compounds are possible using γ-unsubstituted pyrylium salts (cf. Section II,A,1,c) or γ-alkoxypyrylium salts (cf. Section III,A,5,a).

Van Allan et al.[42] suggested a simple and convenient route to 4-ami-nopyrylium salts (e.g., **21**) involving the reaction between 4-pyrones (or flavones) with activated isocyanates RCNO (R = COCCl$_3$, SO$_2$Cl, SO$_2$C$_6$H$_4$CH$_3$, COPh) followed by treatment of the primarily formed pyroneimines **20** with mineral acid. The yields from both steps are nearly quantitative.

According to McKinnon,[43] reduction of pyrones and thiopyrones with LiAlH$_4$ followed by treatment of the pyranol and thiopyranol pseudo-bases with mercury(II) acetate in glacial acetic acid provides a preparative route to pyrylium and thiopyrylium salts.

The reaction of pyrones with organomagnesium compounds is an important method of synthesis of pyrylium salts described by Baeyer and Piccard[44] long ago. This technique makes it possible to introduce various substitutents in positions 2 and 4 of the pyrylium ring. Reynolds and Van Allan[45] treated 2,6-di-t-butyl-4-pyrone with t-butylmagnesium chloride to obtain 2,4,6-tri-*t*-butylpyrylium perchlorate. The use of organolithium derivatives of heterocyclic compounds in this reaction type provided a means for isolating 4-pyranol intermediates **22** and for synthesizing 4-hetaryl derivatives **23** of pyrylium[46–48] (and furochromylium[49]) in high yields. Similarly, ferrocenyllithium reacts with 2,6-dimethyl-4-pyrone to give a 4-ferrocenylpyrylium salt.[50,51]

Reformatski's reaction between 4-pyrones or their benzo derivatives with methyl bromoacetate or methyl γ-bromocrotonate in the presence of activated zinc metal followed by acidification of the reaction mixture with perchloric acid gives 4-carbomethoxymethyl- or 4-(3-carbomethoxy-2-propenyl)-substituted pyrylium perchlorates.[52] Reagent **24** suggested by Ivanov and prepared from phenylacetic acid reacts with 4-pyrones to give 4-carboxymethyl-substituted pyrylium salts **25** in good yields (51–92%).[53]

Pirkle and Dines[32] obtained 3-nitro-2-pyrone (**27**) on heating 2-pyrone with nitronium fluoborate; they assume that a nitroxypyrylium salt **26** is formed intermediately and that the product is obtained by a rearrangement of the nitro group from the exocyclic oxygen to the ring.

c. *Dehydrogenation of Pyrans.* The methods of dehydrogenation of pyrans to pyrylium salts have become widely applicable during recent years: 2- and 4-unsubstituted pyrylium salts readily give 2- and 4H-pyrans under the action of nucleophiles (cf. Section III,A,3). The latter may easily be converted to new pyrylium cations by oxidative dehydrogenation, especially by hydride transfer reactions.

As hydride acceptor, the original γ-unsubstituted pyrylium salt itself may function (cf. Section III,A,6,f). Thus, Wizinger et al.[54,55] and Kröhnke and Dickoré[56] observed that pyrylium systems of type 28 (X = O, S) react with activated aromatics like N,N-dimethylaniline via 4H-pyrans 29 to give γ-arylated pyrylium salts 30. Starting from γ-unsubstituted pyrylium salts with various 2,6-substituents this principle was used by Krivun, Dorofeenko et al. to "pyrylate" a variety of aromatic, heteroaromatic, azulenoid and other compounds (e.g., N,N-dialkylanilines[57,58]), polyfunctional phenols, and their alkoxy derivatives,[58] indoles,[57,58] pyrroles,[57,58] azulene,[59] 1,2-diphenylbenzo[b]cyclopenta[e]pyran (a pseudoazulene),[60] and 4-arylmethylene pyrans.[61,62] In the latter case pyrylocyanines are formed. As will be shown in Section III,A,5 the same pyrylation products may be obtained without hydride transfer from the corresponding 4-alkoxy- or 4-chloropyrylium salts by nucleophilic replacement of the alkoxy or chloro substituent.

(28)   (29)   (30)

Pyranyl-4-phosphonic acids,[63–65] 4-cyano- and 4-carboxypyrans,[66] as well as indeno[2,1-b]pyrans[67,68] were shown to undergo oxidative dehydrogenation with triphenylmethyl perchlorate to give pyrylium salts of the structures 31, 32, 33, and 34, respectively.

(31)   (32)   (33)   (34)

The syntheses by Undheim et al.[69,70] of pyrylium salts containing strong electron acceptors (such as COOH, COOR, CN, COMe) in positions 2 and 6 of the pyrylium ring are of considerable interest. These products were likewise obtained by dehydrogenation of the corresponding pyrans with triphenylmethyl perchlorate.

2,6-Diphenylpyrylium (35) reacts with lithium phenylacetylide to give the product 37, whose structure was deduced from IR and ^{1}H-NMR evidence.[49,71] The formation of 37 was explained by a hydrogen shift from position 4 of the pyran system 36 to the triple bond (the acetylene–allene rearrangement). Treatment of 37 with Ac$_2$O/HClO$_4$ yields 2,6-diphenyl-4-phenylethynylpyrylium perchlorate (38) rather than the 4-styryl derivative that might be expected. Oxidation of 37 is thus an easier process than its protonation which was never observed.

The organometallic synthesis was also successfully applied to introduce aromatic[38] and heterocyclic (2-thienyl, 2-benzothiazolyl)[49,72] substituents or residues of sterically hindered phenols[73] and carborane systems[74,75] into position 4 of the pyrylium ring. The same approach was used to prepare 2,4,6-tri-t-butylpyrylium salts.[73]

The usual dehydrogenation reagents, acetyl perchlorate and triphenylmethyl perchlorate, fail to dehydrogenate 4-carboranylpyrans to the corresponding pyrylium salts[74,75] although reactions of this type together with many other similar dehydrogenation reactions are traditionally treated as hydride ion transfer or elimination. However, in fact, nucleophilic hydrogen substitution reactions are rather rare because of exceedingly high heterolytic cleavage energy of the C—H bond. At present, the reasonable suggestion that hydride ion elimination is a stepwise process including a number of one-electron transfer events is being increasingly accepted.[76]

2,6-Disubstituted 4-carboranylpyrans may only be dehydrogenated to pyrylium salts by such a typical one-electron oxidizing agent as hexachloroantimonate of tris(p-bromophenyl)aminium which cannot bind hydrogen, whereas 2,4,6-tri-t-butyl-4H-pyran gives the corresponding pyrylium salts under the action of another one-electron acceptor, 4-bromo-2,4,6-tri-t-butylcyclohexadienone (39; quinobromide). On the basis of these two observations a stepwise mechanism for the dehydrogenation of 4H-pyrans was suggested.[73–75]

(39)   (40)

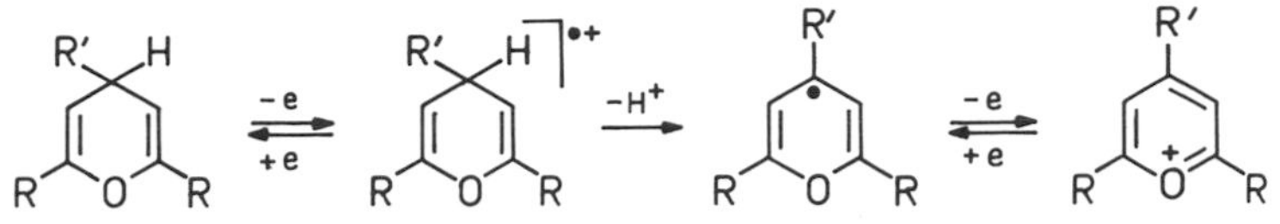

(41)   (42)

In the first step, the pyran is oxidized by **39** to a cation-radical **41** which then eliminates H^+. The resulting radical **42** in turn undergoes a one-electron oxidation to give a pyrylium salt. If the substrate contains an acceptor of elemental hydrogen, the conversion of **41** to a pyrylium salt may be a one-step process.

The occurrence of the corresponding phenoxyl radical **40** in dehydrogenation of 4*H*-pyrans with quinobromide **39** (proved by coloration and ESR spectra[73]) confirms the suggested mechanism. In the next step, the radical acts as hydrogen acceptor and gives 2,4,6-tri-*t*-butylphenol which may be isolated from the reaction mixture. The cation-radical **41** was also detected by ESR[77] which is a strong argument in favor of the reaction scheme cited above.

Dehydrogenation of γ,γ'-bipyranyl **43** with hydride acceptors such as triphenylmethyl perchlorate or acetyl perchlorate leads to γ,γ'-bipyrylium salts **44**.[66,78] However, when applied to bisisochromenes the interring C—C bond is cleaved and 2 mol of benzo[*c*]pyrylium salt are obtained instead of 1 mol of bisbenzo[*c*]pyrylium salt.[79]

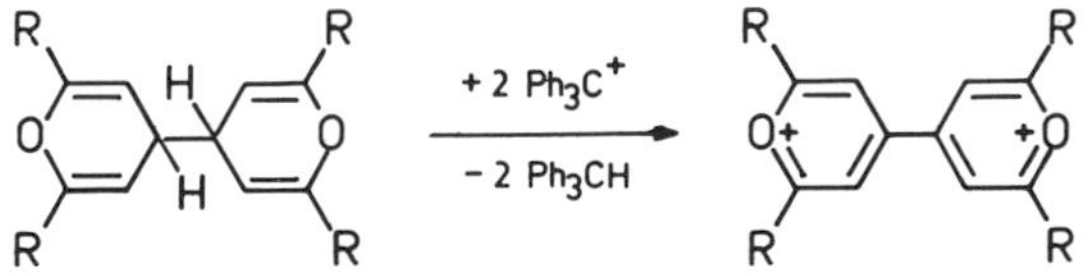

(43)   (44)

2. From n-*Pentane Open-Chain Derivatives*

a. *From 2-Pentene-1,5-diones and Their Derivatives.* The unsubstituted pyrylium cation was first synthesized by Klages and Träger[80] in 1959 starting from pyridine: reaction with chlorosulfonic acid and alkaline hydrolysis affords sodium glutacondialdehyde enolate which cyclizes to pyrylium perchlorate on treatment with perchloric acid in methanol. Recently, Gordzeevich and Skrovachevska[81] suggested a modification of this technique and obtained pyrylium perchlorate in 36% yield by ring opening of N-(2,4-dinitrophenyl)pyridinium chloride (45) to Zincke's aldehyde 46 followed by treatment with perchloric acid at 100°C. Likewise 3-methylpyrylium perchlorate was prepared from 3-methylpyridine in 5% yield.[81]

Mechanistic studies by Williams[82] on the formation of the 2,4,6-triphenylpyrylium cation from 1,3,5-triphenyl-2-pentene-1,5-dione will be discussed in more detail in Section III,B,2,a (cf. also Bunting's review[83]).

The cyclization of 5-chloropentadien-1-ones 47 to 2,6-disubstituted pyrylium salts 48 (R and R' may be different)[84] and the synthesis of 2-aminopyrylium salts 50 from unsaturated nitriles 49 (R = CO_2Et, $CONH_2$; X = OH, NHEt, NHPh, $NAlk_2$)[85,85a] were described by Hartmann and co-workers. According to Schroth and Spitzner[85b] 2-aminopyrylium salts are also accessible by cyclization of nitriles of type $RC(OEt)$=$CHC(NR_2)$=$CXCN$ (X = CO_2Et, CN) which are obtained from acyl-ketene-S,N-acetals and malonic acid derivatives in a multi-step synthesis.

b. *From Pentane-1,5-diones.* Triphenylmethyl hexachloroantimonate[86] generated directly in the reaction mixture from triphenylmethyl chloride and $SbCl_5$, or iodine, bromine and quinones in hydrochloric acid[87] were successfully applied to prepare pyrylium salts from saturated 1,5-diketones. Balaban and co-workers[88] synthesized pyrylium hexachloroantimonates and perchlorates using this technique. Kharchenko et al. studied oxidative dehydrogenations of saturated 1,5-diketones and various 4H-pyrans under the action of $BF_3 \cdot Et_2O$,[89,90] hydrogen halides,[89,91] and other reagents. The reactions were carried out without precautions and possibly involved air oxygen as oxidant. Nevertheless the authors claim that pyrans disproportionate in the presence of an acidic reagent to give pyrylium salts and hydrogenated pyrans.[89-91]

Dorofeenko et al.[92,93] suggested a new route to 2,6-diarylpyrylium salts **52** from phenol ethers and glutaric acid in the presence of polyphosphoric acid (PPA). Probably this reaction also involves oxidative dehydrogenation of the 4H-pyran intermediate **51**.

$$2 \text{ Ar-H } + \underset{\text{COOH}}{\overset{\text{COOH}}{(\text{CH}_2)_3}} \xrightarrow[-3\text{H}_2\text{O}]{\text{PPA}} \text{(51)} \xrightarrow{-\text{H}^-} \text{(52)}$$

B. Two-Component Syntheses

1. C_1 + C_4 Syntheses

a. *Acylation of Unsaturated Ketones.* Relatively few papers on the synthesis of pyrylium salts by monoacylation of α,β- and β,γ-unsaturated ketones were published during recent years. Earlier attempts to prepare pyrylium salts by acylation of ethylideneacetone and crotonaldehyde with AcCl + AlCl₃ failed.[94] It was, however, shown that ethylideneacetone underwent acylation with Ac₂O + SbCl₅ to give 2,6-dimethylpyrylium hexachloroantimonate in 35% yield,[95] while butylideneacetophenone (**53**) proved easy to acylate with (RCO)₂O + HClO₄,[95] or to formylate with HC(OEt)₃ + HClO₄[96] affording pyrylium salts **54** (R = H, Me, Et, Pr) in yields of 31–62%.

$$\text{(53)} \rightleftharpoons \text{ } \xrightarrow[-\text{H}_2\text{O}]{\text{RCO}^+ \text{ ClO}_4^-} \text{(54)}$$

Formylation of dypnone with dichloromethyl butyl ether in the presence of AlCl₃ according to Rieche, Gross, and Höft[97] leads to low yields of 2,4-diphenylpyrylium salt.[98] Starting from cyclohexenylacetophenones **55** under the same conditions 3-aryl-5,6,7,8-tetrahydroisochromylium salts **56** are formed.[99] The latter may be converted almost quantitatively into the corresponding tetrahydroisoquinolines (see Section III,C,3,c). These products, accessible only with difficulty otherwise, may be applied to synthesize analogs of morphine.[100,101]

Acylation of cyclopentylidenecyclopentanone (**57**) with aroyl chlorides in the presence of AlCl$_3$ followed by treatment with 70% perchloric acid leads to the tricyclic cations **58**.[102] Acylation of cycloalkenylacetic acids to pyrylium salts of type **59** (n = 4, 5) occurs likewise.[103]

As shown recently, heating of hydrocinnamic acid (**60**) in PPA followed by addition of aliphatic or aromatic acids results in good yields of diindeno[b,d]pyrylium salts **61**.[104]

b. From Unsaturated Ketones and Aldehydes by Dehydrogenation. Like aldehydes,[105] their acetals react with α,β-unsaturated ketones in the presence of triphenylmethyl perchlorate to give pyrylium salts.[95,106] It seems likely that dehydrogenation occurs under the action of triphenylmethyl perchlorate. Thus, dypnone yields 2,4-diphenylpyrylium salts containing various 6-substituents (H, alkyl, Ar). Mesityl oxide reacts with cinnamaldehyde acetal to give an α-styryl-substituted pyrylium salt. γ-Unsubstituted pyrylium salts 63 containing the trityl group in the β-position of the ring may be prepared from triphenylmethanol and ethylideneacetone or ethylideneacetophenone 62 (R = Me, Ph). Possibly, the reaction proceeeds as shown in Scheme 1.

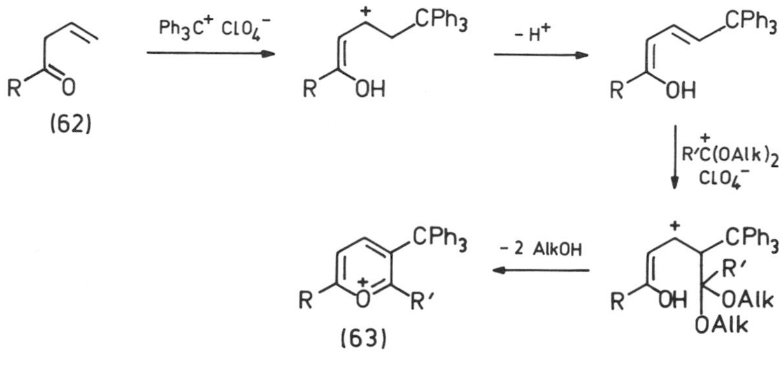

SCHEME 1

2. $C_2 + C_3$ Syntheses

a. *From 1,3-Dicarbonyl Compounds and Methyl(ene) Ketones.* Further developments of the synthesis of pyrylium salts by acidic condensation of 1,3-dicarbonyl compounds with methyl(ene) ketones[107-109] included the application of more reactive 1,3-ketoaldehydes and their derivatives. The yields proved rather sensitive to the amount of the aldehyde form of the 1,3-ketoaldehyde present in the reaction mixture, e.g., as determined by ^{1}H-NMR technique. The yields of the desired products in these reactions range from 0% (hydroxymethylenecamphor, 100% enol) to 60% (2-formylcyclohexanone). The use of ketosteroid hydroxymethylene derivatives opened the way to the condensation of pyrylium and, hence, pyridine rings to rings A and D of steroid molecules, and made possible the synthesis of certain bissteroidopyrylium salts.[110-113]

A very promising reaction is the synthesis of γ-unsubstituted pyrylium salts 65 from benzoylacetaldehyde acetal (64). The latter is prepared from acetophenone, ethyl orthoformate and perchloric acid added in catalytic quantities.[114] The reaction gives high yields.

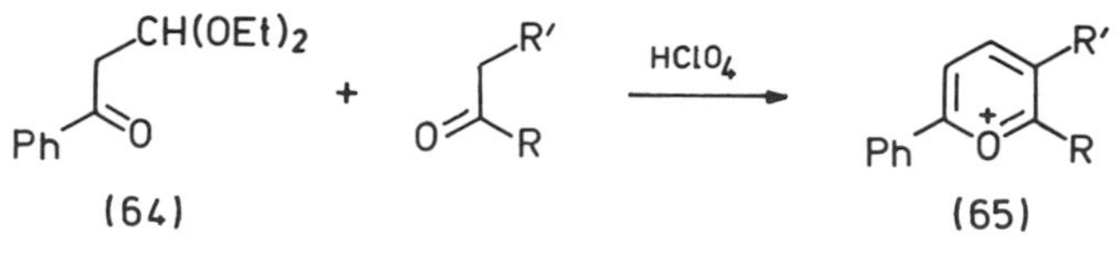

Later, the technique was extended to acetals of other aroylacetalde-
hydes[96,115] and applied to fuse a pyrylium ring to α-tetralones.[116] The
technique was also successfully used in the synthesis of pyrylium salts
containing fragments of two different ketones.[117] In the latter reaction,
the more reactive methylene group of one of the ketones (e.g., cyclohexanone or cycloheptanone) reacts first with orthoformate to give a β-
ketoacetal which then undergoes condensation with a methyl or methyl(ene)
group of the other fragment to give 30–40% yields of pyrylium salts. The
reaction of various β-ethoxymethylene ketones **66** with ketones in the
presence of acidic catalysts leading to 2,3,5,6-tetrasubstituted pyrylium
salts **67** follows the same scheme.[118–120]

Practically the only route to pyrylium salts **69** containing one α-substituent is the condensation of the tetraacetal **68** of malonic dialdehyde
with methyl(ene) ketones.[121] Although the reaction yields are low, the
technique is feasible for these salts which cannot be prepared otherwise.
In recent years, nonfunctionalized 1,3-dicarbonyl compounds have
been used mainly for the synthesis of pyrylium salts with condensed
rings. In continuation of previous work,[108] Schroth and Fischer[122] reported

in detail their acid-catalyzed reaction of 1,3-diketones with aryl methyl(ene) ketones, 1- and 2-indanones, and 2-tetralone, affording pyrylium salts with structures **70–73**.

According to the same reaction principle, oxindole,[123] thioindoxyl[124] as well as barbituric and thiobarbituric acid[125] can be converted into the corresponding pyrylium salts **74–76** (X = O, S). Analogously, starting from oxindole and 2-hydroxymethylene 3-ketosteroids, compounds of type **77** were prepared.[126] Vlasov[127] applied the condensation of 1,3-diketones with pentafluorophenyl methyl ketone to synthesize pyrylium salts containing an α-pentafluorophenyl substituent.

(74)   (75)   (76)

(77)

b. *From β-Chlorovinyl Carbonyl Compounds and Methyl(ene) Ketones or Enamines.* For the synthesis of pyrylium salts after Schroth and Fischer[128] (which had already been mentioned in Part I[1]) from enamines and β-chlorovinyl ketones via ketovinylenamines **78**, detailed experimental data have been published in the meantime.[129,129a] This procedure makes it possible to obtain under mild conditions on the one hand 2,5-disubstituted pyrylium salts **79** which are otherwise difficultly accessible, and on the other hand bi-, tri-, and polycyclic systems of types **80** (*n* = 3–5), **81**, and **82**.[129a] Comparable with this approach is the synthesis of 2-aminopyrylium salts after Gompper and Elser[129b] from β-chlorovinyl ketones and ketene *S,N*-acetals via intermediate ketovinylketene *S,N*-acetals $R^1COCH=CHCR^2=C(SR^3)NR^4_2$; here the acid-catalyzed cyclization proceeds with fission of the alkyl mercaptan. Ketovinylketene dichlorides $R^1COCR^2=CHCH=CCl_2$, easily obtainable from methylene ketones and β,β-dichloroacrolein, react according to Schroth and Burkhardt[129c] smoothly with secondary amines to give the corresponding aminals $R^1COCR^2=CHCH=C(NR^3_2)_2$ which likewise cyclize in the presence of acids to 2-aminopyrylium salts.

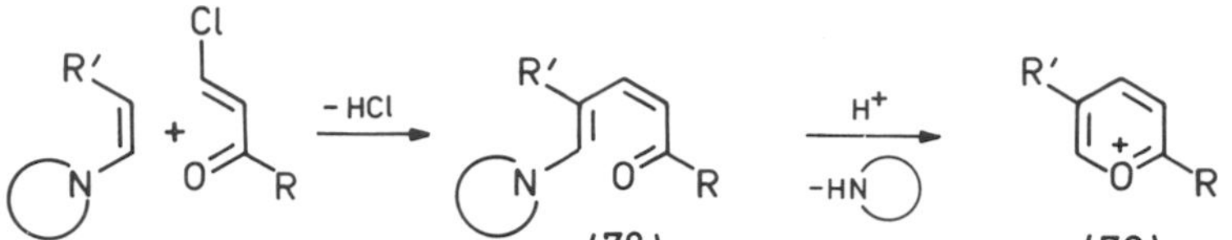

(78)   (79)

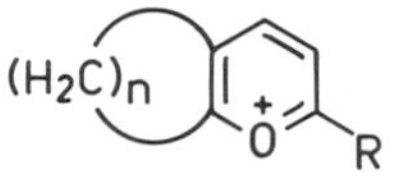

(80) 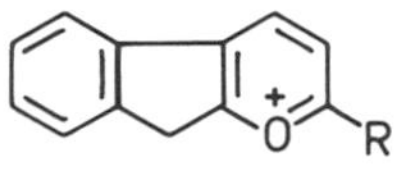   (81) 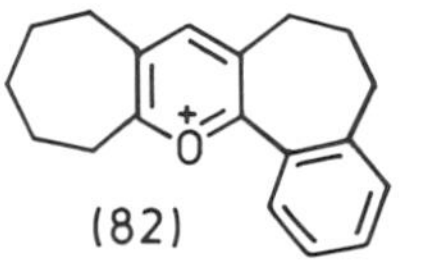   (82)

Under more drastic conditions (heating in acetic acid with perchloric acid), methyl(ene) ketones may also react directly with β-chlorovinyl ketones affording pyrylium salts. Good results have so far been obtained with 2-tetralone (**83**) as the active methylene component, which may be converted into tri-, tetra-, and pentacyclic pyrylium salts **84–87**.[130]

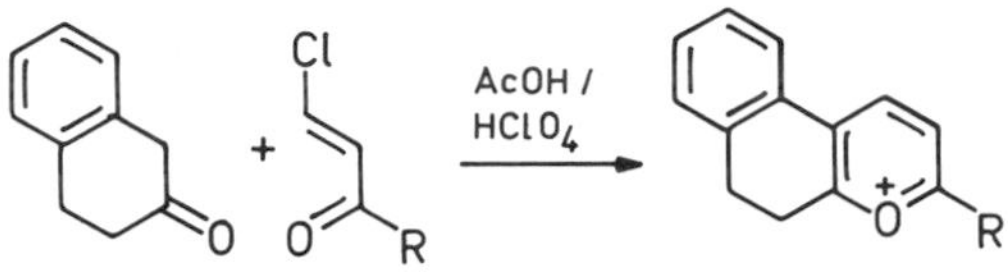

(83)   (84) 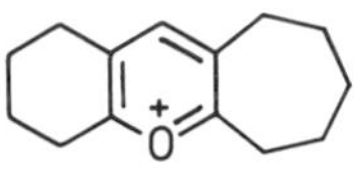   (85)

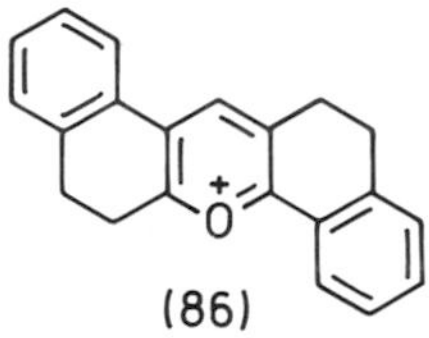

(86) 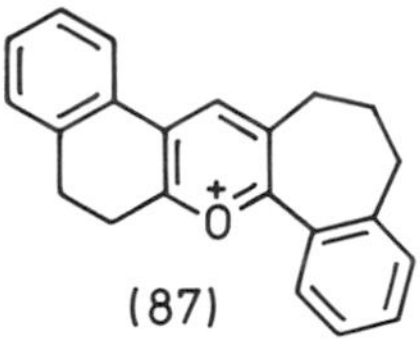   (87)

During the last years this principle was extended successfully to β-chlorovinyl aldehydes, which are readily available by Vilsmeier–Haack formylation of methyl(ene) ketones.[131–133] Thus, Dorofeenko and Pyshchev[134,135] and Andrieux et al.[136] have, simultaneously and independently, worked out a general technique for the synthesis of γ-unsubstituted pyrylium salts **89** based on the condensation of β-chlorovinyl aldehydes **88** with methyl(ene) ketones in the presence of acidic catalysts (HClO₄) or Lewis acids (SnCl₄, SbCl₅, etc.). The reaction proceeds under mild conditions exclusively at the aldehyde group of β-chlorovinyl aldehydes to give high yields of various pyrylium salts, including those containing functional groups in the β- (COR, CH₂COOH, Cl) and α- (COOR, COOH) positions of the pyrylium ring.[124,137] According to the same scheme polycyclic pyrylium salts have also been obtained.[136,138]

Ketone formylation by the Vilsmeier–Haack reagent leads not only to the addition of one more carbon atom to the chain but also to the replacement of the carbonyl oxygen with chlorine. This was utilized to prepare 4-chloroisochromylium salts **90**, which readily undergo hydrolysis to isoflavones[139] (cf. Section III,A,5,b).

It was shown that β-chloro-β-ferrocenylacrylic aldehyde can be used to prepare pyrylium and pyridine derivatives of ferrocene.[140] Salts containing cyclopentadienylmanganesetricarbonyl (cymantrenyl) residues were obtained likewise.[141]

c. *From β-Chlorovinyl Aldimonium Salts and Methyl Ketones.* β-Chlorovinyl aldimonium salts **91** are easily isolable intermediates in the synthesis of the corresponding β-chlorovinyl aldehydes by Vilsmeier–Haack formylation of methyl(ene) ketones and may likewise be used as reactive derivatives of 1,3-dicarbonyl compounds.[142] Hartmann and Förster[84,143] treated **91** with acetophenone in the presence of sodium methoxide to obtain 5-chloropentadienones **92**, which lead, after treatment with perchloric acid, to 2,6-disubstituted pyrylium salts **93** (cf. Section II,A,2,a).

d. *Dehydrogenating Condensation of Olefinic Ketones with Methyl(ene) Ketones.* Only few papers on the addition of methyl(ene) ketones to β,γ-unsaturated ketones leading to pyrylium salts appeared during recent years. Barker and Riley[144] studied the reaction of β-methylchalcones with fatty aromatic ketones in the presence of Lewis acids (SnCl$_4$, ZnCl$_2$,

$SbCl_5$) and showed that in this case pyrylium salts are also formed, but in very low yields.

Strzelecka and Simalty[145] synthesized the bipyrylium perchlorate **95** from 2,6-diphenyl-4-phenacylpyrylium perchlorate (**94**) and chalcone.

Benzylideneacetophenone was shown to add pyruvic acid under mild conditions to give 2-carboxy-4,6-diphenylpyrylium tetrafluoborate.[146] The same ketone reacts with aliphatic compounds, ethyl methyl ketone, acetylacetone, and dibenzoylmethane, at reduced temperatures (0–10°C) to give the pyrylium cations **96–98**.[147] Acetaldehyde reacts under the same conditions to give 2,6-diphenylpyrylium perchlorate (yield 8%), acetone gives 2-methyl-4,6-diphenylpyrylium perchlorate (yield 25%),[147] and acetoacetate reacts with chalcone in the presence of $BF_3 \cdot Et_2O$ to give the fluoborate of **99**.[148] The synthesis of 2,4,6-triarylpyrylium salts containing perfluorophenyl substituents was also described.[149]

As found by Dorofeenko et al.,[150] ethyl orthoformate and perchloric acid can split chalcones as in the reverse Michael reaction, to give the initial aromatic aldehyde and fatty aromatic ketone. The latter reacts with excess unsymmetric chalcone to produce 2,6-diphenyl-4-aryl-substituted pyrylium salts. The reaction may follow the other mechanism involving diethoxycarbocations formed from the aromatic aldehyde and ethyl formate (cf. Section II,C,1,b).

C. THREE-COMPONENT SYNTHESES

1. $C_2 + C_1 + C_2$ Syntheses

a. *Dehydrogenating Condensation of Aromatic Aldehydes with Two Moles of Methyl(ene) Ketones.* The method of acidic condensation of aromatic aldehydes with two moles of methyl(ene) ketones followed by dehydrogenation has found wide applications in the synthesis of 2,4,6-triarylpyrylium salts.[1,7] Boron trifluoride in glacial acetic acid was found to be an effective catalyst of the reaction.[151] The reaction was also applied to synthesize 2,6-diphenylpyrylium salts containing sterically hindered phenol residues in the 4-position,[73] and pyrylium salts condensed with the thionaphthene nucleus.[124] Dimroth and Mach[152] succeeded in synthesizing the difficultly accessible 2,4,6-tri-*t*-butylpyrylium cation by condensing pinacoline with pivalic aldehyde.

b. *From Orthoesters and Two Moles of Methyl(ene) Ketones.* More than 100 years ago Claisen described the formation of acetals from carbonyl compounds and orthoformate. In 1967, Mezheristkii and Dorofeenko[114] studied this reaction and found that the formation of ketals is accompanied by formylation of ketones at the methyl(ene) group if the reaction is carried out in the presence of catalytic amounts of 70% perchloric acid. The latter process leads to diethyl acetals of β-ketoaldehydes.

The reaction of orthoformate with two moles of an aryl methyl(ene) ketone in the presence of one mole of $HClO_4$ or $BF_3 \cdot Et_2O$ directly gives high yields of 2,6-diarylpyrylium salts **100** unsubstituted at the γ-position.[96,114,153]

(100)

The reaction proved to be a general route to 2,6-di-, 2,3,6-tri-, and 2,3,5,6-tetrasubstituted pyrylium salts and therefore to the corresponding pyridines (cf. Section III,C,3,c).[96,114] One mole of orthoformate reacts with two moles of methyl(ene) ketones in the presence of mineral acids (HCl, HBr, HI, H_2SO_4) and Lewis catalysts ($FeCl_3$, $AlCl_3$, $SnCl_4$, etc.) to give excellent yields of 2,6-diarylpyrylium salts containing various anions.[154] The method also provides a convenient route to 2,6-di-*t*-butyl-,[114] 2,6-diferrocenyl-,[50,51] and other pyrylium derivatives.[73,124,155]

Strzelecka and Simalty[156] condensed 2,6-diphenyl-4-phenacylpyrylium perchlorate (**94**) with orthoformate to obtain the trispyrylium salt **101**.

(94)   (101)

According to several authors,[14,157] the first step of the reaction between orthoesters and mineral or Lewis acids involves the formation of dialkoxycarbocations which may further react with carbonyl compounds.[118] Dorofeenko and Luk'yanov showed that dialkoxycarbocations may be generated also in reactions of various acetals with triphenylmethyl perchlorate[158,159] or other dehydrogenating agents[160] in acetic acid, nitromethane or similar inactive solvents. These cations then react with two moles of an aliphatic or fatty aromatic ketone to give pyrylium salts 102 in moderate yields (20–45%).

$$R^3\text{--}CH(OEt)_2 \ + \ Ph_3C^+ \ ClO_4^- \ \longrightarrow \ R\text{--}\overset{+}{C}(OEt)_2 \ ClO_4^- \ + \ Ph_3CH$$

(102)

The advantage of the technique just described over the usual three-component condensations (see Section II,C,1,a) is the possibility of synthesizing γ-unsubstituted pyrylium salts and compounds containing alkyl substituents in the pyrylium ring. Acylals of aromatic aldehydes show a similar behavior in this reaction.[161,162] They even give somewhat higher yields of pyrylium salts than acetals do in a number of cases.

Alkoxycarbocations 103 generated by dehydrogenation of of ethers[163,164] and acyloxycarbocations 104 formed from esters[165,166] also proved quite

$$\overset{+}{RCHOR'} \qquad \overset{+}{RCHOCOR'}$$

(103)   (104)

suitable as starting materials for the synthesis of pyrylium salts. Some of these reactions may involve the formation of α,β-unsaturated ketones from the initial cations and one of the cations, 103 or 104, followed by the addition of methyl(ene) ketones. The use of inexpensive and accessible ethers and esters may extend the range of pyrylium salts considerably and facilitate their production.

c. *From Haloalkanes and Two Moles of Methyl(ene) Ketones.* Luk'yanov and Dorofeenko[167] studied the reaction of *gem*-benzylidene dichloride and benzoyl chloride with two moles of acetophenone in the presence of triphenylmethyl perchlorate as extension to their work on the condensation of benzotrichloride with acetophenone.[168] Both reactions gave low yields (13–20%) of 2,4,6-triphenylpyrylium perchlorate. Dichloro- and chlorobenzoyl cations are believed to appear in these reactions as intermediates, and these may react like dialkoxy- and alkoxycarbocations, respectively. Rather unexpectedly, acetophenone reacts with equimolar quantities of triphenylmethyl perchlorate in CHCl₃ to give 4-methyl-2,6-diphenylpyrylium perchlorate; this finding requires further investigation.[167]

2-Methyl-4,6-diphenylpyrylium perchlorate is formed in low yield (12.6%) in the reaction of acetophenone with α,α-dichloromethyl ether in the presence of triphenylmethyl perchlorate.[169] Chlorodimethyl ether reacts with acetophenone and triphenylmethyl perchlorate in acetic acid–nitromethane to give 15–46% yields of the difficultly accessible 2,4-diarylpyrylium salts.[170] Dypnone was detected in the reaction mixtures as intermediate in some of these reactions.

2. $C_1 + C_3 + C_1$ Syntheses

A rather large number of papers on the bisacylation of olefins or of their precursors (discovered almost simultaneously by Balaban and Nenitzescu[94,171] in Roumania, and Praill [172,173] in England) had been published when Part I[1] came to light. In the meantime some original synthetic extensions were reported.

Simple preparative routes to 2,4,6-trimethylpyrylium perchlorate,[174,175] tetrafluoborate,[176] and trifluoromethanesulfonate[176] by bisacylation of t-butanol with acetic anhydride in the presence of the corresponding acids were developed. 2,4,6-Trimethylpyrylium sulfoacetate is easily obtained in large amounts and possesses several advantages over the preceding cations (higher solubility, lower cost, no danger of explosion).[177] New syntheses of deuterated pyrylium salts were reported[178] (for more detail see Section III,A,7,b). 2,3,6-Triphenyl-4-methyl- and 2,4,6-trimethyl-3-phenylpyrylium perchlorates were prepared by bisbenzoylation and bisacetylation, respectively, of 2-methylpropenylbenzene with acyl chlorides in the presence of AlCl₃.[179] As is well known,[94,171] 2,3-dimethyl-2-butene gives 2,6-dimethyl-4-isopropylpyrylium salts under the action of acetyl chloride and AlCl₃. However, Balaban, Bota, and Stanoiu[180] failed to isolate the expected 2,4,6-triisopropylpyrylium salt by bisacylation of the same olefin with isobutyryl chloride and AlCl₃; instead, trisacylation of the olefin occurred, yielding the pyrylium salt 105 (R = i-Pr). Isomeric ethyltrimethylpyridines and 2,6-dimethyl-3-n-propylpyridines were prepared through the corresponding pyrylium salts from methylpentenes and hexenes, respectively.[181]

(105)

The selectivity of the alkene diacylation was investigated by Arnaud, Roussel, and Metzger[182] using 2-methylpentene, AcCl and $AlCl_3$ in chloroform. The $AcCl:AlCl_3$ ratio plays an important role in the selectivity: with a ratio of 1.5–5, the less substituted pyrylium salt (2,6-dimethyl-4-n-propylpyrylium) predominates but with a ratio equal to 1.0, the more substituted pyrylium salt (3-ethyl-2,4,6-trimethylpyrylium) predominates. Therefore not only the nature of the catalyst, but also the ratio of catalyst to acid derivative governs the selectivity. In agreement with this observation it was found that pure 2,3,4,6-tetramethylpyrylium perchlorate can be conveniently prepared from excess acetic anhydride, one mol of t-amyl alcohol, and 0.5 mol of perchloric acid.[182a]

Dulenko et al.[183] carried out bisacylation of methallyl chloride (2-chloromethylpropene) and 4-methyltetrahydrophthalic acid with Ac_2O + $HClO_4$ to obtain salt 106 and the hydrogenated benzo[c]pyrylium salt 107, respectively, containing functional groups as substituents.

(106)

(107)

Balaban and Badilescu[184] reported the synthesis of the new bridged system 108, which was obtained by bisacylation of cyclodecene with Ac_2O in the presence of sulfoacetic acid. Bisacylation of cyclododecene leading to 3,5-bridged pyrylium salts containing the nonamethylene bridge was described in earlier papers.[185,186]

(108)

Monoacylation of mesityl oxide with butyryl chloride in the presence of $AlCl_3$ gave 2-isopropyl-4,6-dimethylpyrylium perchlorate in low yield

(12%)[187]; further reaction with o-toluidine affords atropisomeric pyridi-nium salts, as shown by chemical shift nonequivalence of isopropyl methyls (cf. Sections III,C,3,c and IV,A,2,a).

Recently, simple routes to 2,6-di-t-butylpyrylium trifluoromethanesul-fonate, chlorostannate, and perchlorate involving bisacylation of t-butyl chloride with pivalic acid derivatives were described.[188,189] Dorofeenko et al.[190] obtained pyrylium salts by bisacylation of cyclic tertiary alcohols obtained from cyclopentanone, cyclohexanone, cycloheptanone, 1-tetra-lone, and methyl- or ethylmagnesium iodides. Quite a number of pyrylium salts and corresponding pyridine bases were obtained using this technique in satisfactory yields. Bisacylation of 6-methoxy- and 6,7-dimethoxy-1-methyl-2-tetralone proceeds similarly.[116] Tertiary carbinols of the het-erocyclic series, dimethyl-2-thienyl- and dimethylbenzothiazolylcarbinols undergo bisacylation to give 2,6-dimethyl-4-(2-thienyl)- and 2,6-dimethyl-4-(2-benzothiazolyl)pyrylium salts in yields of 55 and 74%, respectively.[191]

According to Earnest and Brown[192] compounds containing the strained cyclopropane ring may undergo conversions to pyrylium salts 109 and 110 under the action of acylium cations. The preliminary communication cited contains no yield data. The synthetic potential of the reaction is to be determined in further studies.

(109)   (110)

Arnaud, Pedra, Roussel, and Metzger[193] have recently described the synthesis of pyrylium salts by bisacylation of isoparaffins (isopentane, 2- and 3-methylpentane as well as 2,3-dimethylbutane) with AcCl and AlCl$_3$. Hydride transfer reactions lead first to the formation of alkenes which are then converted to pyrylium salts, along with mono- and tri-acetylation products. Tabushi et al.[194] and other authors[195] had studied such reactions earlier but they had only investigated the "organic" layer that resulted after decomposition of the reaction mixture with ice, containing the monoacylation products and not the aqueous layer containing the di- and triacetylation products. The French authors did not isolate the pyrylium salts but converted them to pyridines which were analyzed by gas chromatography. The selectivity of the reaction depends strongly on the nature of the initial carbocation, on the $AcCl:AlCl_3$ ratio (the higher this ratio, the higher the relative amount of triacylation products), and on the solvent (AcCl or chloroform): isopentane (1 mol) with 8 mol AcCl and 2 mol $AlCl_3$ yielded 25% 2,3,4,6-tetramethylpyridine, 13% 2,6-dimethyl-4-ethylpyridine, and 60% triacetylation products while with 0.6 mole AcCl and 0.4 mole $AlCl_3$, it yielded 18% 2,3,4,6-tetramethylpyridine, 68% 2,6-dimethyl-4-ethylpyridine, and 13% triacetylation products.

III. Reactions of Pyrylium Salts

A. REACTIONS WHICH CONSERVE THE PYRAN RING

1. Anion Exchange Reactions

As stated in Part I,[1] the exchange of the anion of pyrylium salts by another one may be regarded as a one-component synthesis. Such exchange reactions are usually carried out for the purpose of characterization or identification, and also for modifying the solubility, stability, or physical properties of the first formed species. Due to the low solubility of most of the pyrylium perchlorates, in many cases simple treatment of a pyrylium salt solution with 70% perchloric acid leads to the precipitation of the corresponding pyrylium perchlorate in crystalline form. But also the replacement of Cl^-, Br^-, I^-, I_3^-, $SbCl_6^-$, $SnCl_6^{2-}$, BF_4^-, and ClO_4^- with each other was reported.[2,88,90,137,160,190,196–208] Most frequently, this is done by converting the pyrylium salts into the ring-opened pseudobase (cf. Section III,B,2,a) followed by recyclization by means of the desired mineral or Lewis acid.

An interesting anion exchange takes place in the reaction between pyrylium or thiopyrylium iodides 111 (X = O or S) and tetracyanoquinodimethane in acetonitrile.[137,209] The iodide ion undergoes oxidation and pyrylium salts 112 with an anion-radical as counterion are formed. Being charge-transfer complexes (cf. Section IV,A,1,c), salts of this type are deeply colored crystalline compounds having low solubility and high conductivity.

2. Reactions of Alkyl Substituents

a. *Introduction.* As will be shown in Section III,A,7,a, α- and γ-oriented side chains CH_3, CH_2R, or CHR_2 of pyrylium salts easily undergo deprotonation affording α- or γ-alkylidenepyrans (anhydrobases of pyrylium salts) (e.g., **113**).

When both α- and γ-deprotonation may occur, unlike pyridines or pyridinium salts where α-deprotonation prevails, with pyrylium salts γ-deprotonation prevails. This effect has been explained by quantum-chemical calculations yielding a lower energy for the γ-methylenepyran (cf. Section IV,E).

As indicated by the resonance structure **113B**, alkylidenepyrans possess an electron-rich exocyclic carbon atom which is able to react by nucleophilic attack with aldehyde, keto, amide, nitroso, and other groups, affording condensation products. In all these reactions, described in the following subsections, the pyrylium ring system itself remains unchanged (for a review, see Ref. 9).

b. *Reactions with Aldehydes.* The condensation of monocyclic pyrylium salts with aldehydes was first performed by Dilthey and Fischer[107,210] who treated 2-methyl-4,6-diphenyl- (**114**) and 4-methyl-2,6-diphenylpyrylium salts (**116**) with aromatic aldehydes to obtain styrylpyrylium salts of type **115** and **117**, respectively.

More recently, it has been shown for a large number of aromatic aldehydes that electron donor substituents in the aromatic nucleus of the aldehyde facilitate the condensation.[211–215] This can be explained by a mechanism assuming that the aldehyde reacts in protonated form (which is stabilized by donor substituents) with the methylenepyran (e.g., **113**).

Kelemen and Wizinger[212] observed that 2,6-diethyl-4-methylpyrylium perchlorate reacts with aromatic aldehydes exclusively at the γ-methyl group, and that the latter is more reactive in the 2,6-diisopropyl-4-methylpyrylium cation than in the 2,6-diethyl-4-methylpyrylium cation.

Other examples exist showing that in pyrylium salts γ-methyl groups are generally more reactive than α-methyl groups. Thus, the reaction of 2,4,6-trimethylpyrylium and 2,4-dimethyl-6-phenylpyrylium perchlorates with p-dimethylaminobenzaldehyde only involves the γ-methyl groups.[216] These observations are in agreement with the data on isotopic exchange in pyrylium salts containing alkyl substituents[217,218] (cf. Section III,A,7,b).

In the presence of both α-methyl and α-methylene groups, the condensation involves the latter.[219] As shown for the reaction between 2-methyl-4,6-diarylpyrylium salts and aldehydes of the azulene series, aryl substituents in the pyrylium ring enhance the reactivity of the α-oriented methyl group in the order phenyl < 2-thienyl < 3,4-dimethoxyphenyl.[211] In the presence of two or three active methyl groups in the pyrylium salt, di- and tristyrylpyrylium derivatives may be synthesized, but under more forcing conditions than for monostyryl derivatives.[220] Terephthalic aldehyde reacts with two moles of 4-methyl-2,6-diphenylpyrylium to give the corresponding bis-product.[221] Condensations with a variety of aliphatic, unsaturated, and heterocyclic aldehydes were also reported.[213–215,222]

c. *Reactions with 4-Pyrones.* 4-Pyrones react with pyrylium salts containing α- or γ-methyl or methylene groups to yield cyanine dyes. Usually, the process occurs in refluxing acetic anhydride. With a 1:1 ratio of the reactants, pyrylium salts having two or three activated methyl(ene) groups (e.g., **118**, $R^1 = R^2 = Me$) react through the γ-methyl(ene) group affording compounds of type **119**.[216,223,224] Analogous reactions were reported for 1-thio-4-pyrones.[225–227]

The participation of two or three methyl groups can be achieved by changing the reactant ratio.[220] In this way, for example, compounds **120** and **121** were obtained. On treating 4-methoxy-2,6-dimethylpyrylium perchlorate (**122**) with inorganic or organic bases (e.g., pyridine) an α-methyl

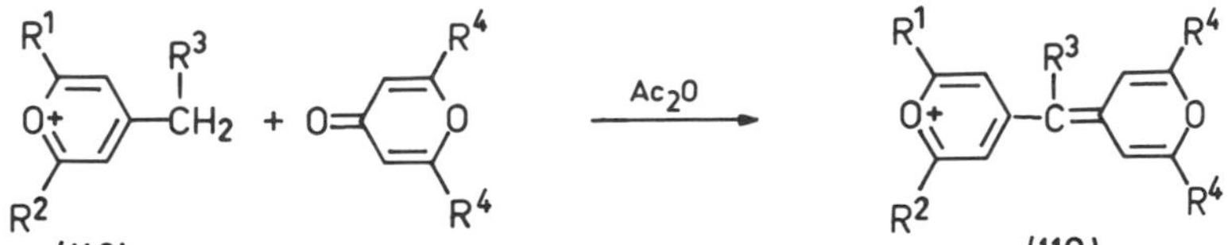

(118) (119)

group reacts with a second mole of **122** (or with 2,6-dimethyl-4-pyrone formed from **122** by dealkylation) to yield the cyanine dye **123**.[28,228] In phosphorus oxychloride the condensation of γ-methylpyrylium salts with 4-pyrones leads to trinuclear dyes like **124**.[38,229]

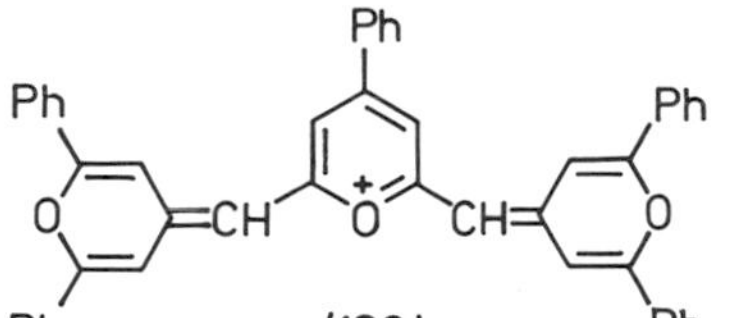

(120)

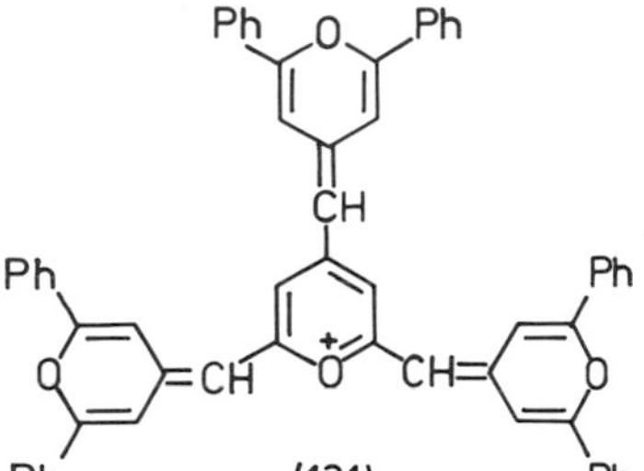

(121)

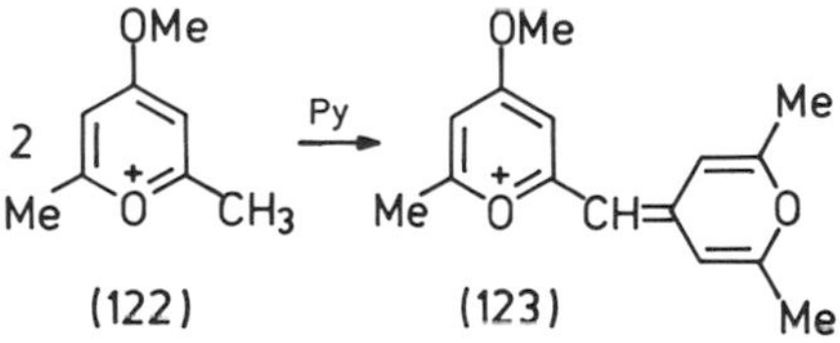

(122) (123)

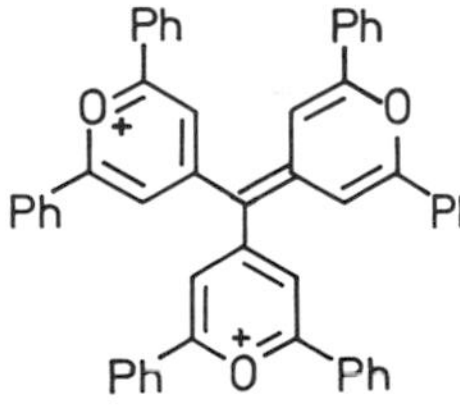

(124)

d. *Reactions with Carboxylic Acid Amides.* γ-Methylpyrylium salts **125** react readily with dimethylformamide in hot acetic anhydride to give 4-(*N,N*-dimethylaminovinyl)pyrylium salts **126**.[230,231] In certain cases the presence of alkali metal cations (e.g., Li$^+$) is advantageous in improving the yields. It was supposed that alkali ions shift the equilibrium Me$_2$NCHO + Ac$_2$O $\rightleftharpoons$ Me$_2$Ṅ=CHOAc AcO$^-$ by binding the acetate ion, and thus increase the concentration of the electrophilic species.[232]

In the presence of POCl$_3$, dimethylformamide leads (obviously via **126**) to the biscondensation products **127**, whereas *N,N*-dialkylamides of aryl or alkyl carboxylic acids under the same conditions form the monoam-inovinyl derivatives **128**.[230,231,233] The latter are also obtained from corresponding thioamides, e.g., *N,N*-dimethylthioacetamide.[230] Vinylogs of dimethylformamide[230,234,235] react to produce highly conjugated systems like **129**. *N*,*N*-Diphenylformamidine was reported to yield anilinovinyl derivatives **130**.[236] The dimethylformamide products of type **126** are hydrolyzed by aqueous alkali to afford 4-pyranylideneacetic aldehydes, whereas acidic hydrolysis leads to the initial 4-methylpyrylium salts **125**.[230,232,237]

Pyrylium salts with an α-oriented methyl(ene) group can also react with dimethylformamide yielding the corresponding 2-(*N*,*N*-dimethylaminovinyl)pyrylium cations.[230,231] If these condensations are performed under standard conditions (15 min refluxing in acetic anhydride), the reaction yields reflect the relative methyl(ene) group reactivities. Thus, e.g., in the series **131**, **116**, **132**, **114** the yields of the corresponding *N*,*N*-dimethylaminovinyl derivatives are 91, 90, 73, and 38%, respectively, indicating once more the relatively low reactivity of the α-oriented methyl(ene) group unless it is part of a condensed saturated six-membered ring. As will be shown in Sections III,C,3,a and c, 2-(*N*,*N*-dimethylaminovinyl)pyrylium salts are interesting starting materials for various ring-transformation reactions.

e. *Reactions with Nitroso Compounds.* Suitable aromatic nitroso compounds can condense like aromatic aldehydes with α- or γ-methyl(ene) groups of pyrylium salts (cf. Section III,A,2,b). Starting from 2-methyl-4,6-diphenyl- and 4-methyl(ene)-2,6-diphenylpyrylium perchlorates, Simalty *et al.*[238] obtained by reaction with *p*-nitrosodimethylaniline azomethines of structure 133 and 134 (R = H, Ph, COPh), respectively. As reaction medium acetic anhydride is necessary; in acetic acid or alcohol no reaction was reported to occur.[239]

Acids hydrolyze the ketimines 134 (R = Ph, COPh) to the corresponding acylpyrylium salts 135, whereas the aldimine 134 with R = H under the same conditions proved to be stable.[238]

f. *Reactions with Dimethyl sulfoxide.* 4-Methyl(ene)pyrylium salts 136 condense with dimethyl sulfoxide under reflux in an acetic anhydride/methylene chloride mixture to give pyrylium salts 137 with S(IV) in the side chain.[240] As in the reactions with dimethylformamide or aromatic nitroso compounds, acetic anhydride plays the part of an activating agent toward both components. In the case R' = H with excess dimethyl sulfoxide biscondensation products 138 are formed whose yields increase in the presence of alkali metal ions.[240,241] The analogous reaction of 2-methyl-4,6-diphenylpyrylium perchlorate with dimethyl sulfoxide could not be stopped at the stage of monocation formation.

g. *Reactions with Orthoesters.* Reactions between γ-methyl(ene)pyrylium salts 136 and orthoesters (R"O)$_3$CH may lead to trimethinecyanine dyes 140 (like the well-known trimethinecyanine synthesis from methyl-substituted benzopyrylium or quinolinium salts) or terminate at the formation of alkoxyvinyl derivatives **139**. The latter are formed when the starting pyrylium salts are heated with a large excess of orthoester in acetic acid or anhydride for a short period of time.[242–246] Conversely, excess pyrylium salt is required to synthesize **140**. In this way trimethinecyanines from ethyl orthoformate and 4-methyl-2,6-di-phenyl-,[236,247,248] 4-benzyl-2,6-diphenyl-,[224,249] and 4-phenacyl-2,6-diphen-ylpyrylium salts[156] were obtained. In the latter case, the trimethinecyanine **140** (R = Ph, R' = PhCO) can be converted by perchloric acid into the trispyrylium perchlorate **101**,[156] mentioned in Section II,C,1,b.

α-Methylpyrylium salts **141** react analogously with orthoesters afford-ing alkoxyvinylpyrylium salts **142**[242–244] or trimethinecyanines **143**.[236,247,248] The former are hydrolyzed in aqueous acids to yield 2-pyranylideneacetic aldehydes **144**,[243,245] whereas the reaction with arylamines leads to 2-(aminovinyl)pyrylium salts **145**,[243,250] mentioned in Section III,A,2,d. Thus, the reaction of equimolar amounts of a 2-methylpyrylium salt, ethyl orthoformate, and an aromatic amine in acetic acid can be used as a one-step route to compounds of type **145**.[250] The reaction of com-pounds possessing active methylene groups CH_2XY (X,Y = COR, COOR, CN, etc.) with 2-alkoxyvinylpyrylium salts **142** leads to 2-pyr-anylidene derivatives **146**.

h. *Other Reactions.* On heating 4-methylpyrylium salts **125** with benzo[*b*]pyrylium salts in pyridine/acetic acid mixtures monomethinecyanines **148** are formed.[251] The reaction involves dehydrogenation of the intermediate **147** by hydride transfer (cf. Section III,A,6,f).

The dehydrogenating condensation of 4-phenacyl-2,6-diphenylpyrylium perchlorate (**94**) with benzylideneacetophenone affording the bipyrylium salt **95**[145] was mentioned in Section II,B,2,d.

Van Allan et al.[252] reported the reaction between 2,6-di-t-butyl-4-methylpyrylium perchlorate (**149**) and 2,4-dinitrochlorobenzene yielding the benzylidenepyran **150**. As found by Balaban,[253] a similar reaction takes place on treating **149** with aroyl chlorides in pyridine; however, in this case the primarily formed pyranylidene derivatives **151** react spontaneously with a second mole of **149** to give the trimethinecyanines **152** (Ar = Ph, p-MeC$_6$H$_4$) as isolable end products.

2-Methylpyrylium salts **141**, amines, and formaldehyde undergo the Mannich reaction in refluxing methanol or acetic acid yielding aminoethylpyrylium salts of structure **153**.[254] With sodium nitrite or isoamyl nitrite 4- or 2-methyl(ene) pyrylium salts form nitrosoalkylpyrylium salts **154** or **155**, respectively.[255-258] However, because the nitroso structure CHN=O was not confirmed the reaction products may also exist in the isomeric oxime form C=NOH. Under the action of fuming nitric acid in glacial acetic acid or of tetranitromethane in pyridine 2-nitromethylpyrylium salts **156** were obtained.[256] Furthermore, α-oriented methyl groups can enter azo coupling leading (via intermediate azo compounds **157**) to arylhydrazones **158**.[239,259]

A U.S.S.R. patent describes the bromination of 2,6-diphenyl-4-methyl- and 2-methyl-4,6-diphenylpyrylium in acetic acid in the presence of Hg(AcO)$_2$ yielding dibromomethyl derivatives.[260] In the absence of any catalyst 2,6-di-*t*-butyl-4-methylpyrylium perchlorate affords with bromine in acetic acid the corresponding 4-tribromomethyl derivative.[261]

3. Reactions of Aryl Substituents

Because of the deactivating effect of the positive charge, no electrophilic substitution is known for the pyrylium ring; however, aryl substituents may be substituted electrophilically.

Le Fèvre *et al.*[262,263] studied the nitration of 2,4,6-triphenylpyrylium perchlorate (**159**) and found by oxidation (cf. Section III,A,8,a) that the α-oriented phenyl rings are nitrated meta, while the γ-oriented phenyl is nitrated para. This is readily explained by the higher positive charge at the α-positions (cf. Sections IV,A,2,a, IV,A,2,b, and IV,E).

Electrophilic alkylation with *t*-BuCl + AlCl$_3$ is possible only for aryl rings which are not deactivated. Balaban, Katritzky, and Semple[264] studied the structure of the mono- and di-*t*-butylation products obtained from 2,4-diphenyl-5,6,7,8-tetrahydrobenzo[*b*]pyrylium, *t*-butyl chloride and aluminum chloride employing both ¹H-NMR and chemical oxidation. As shown by structures **287** and **288** (presented in Section III,A,8,a) only the γ-phenyl group is *t*-butylated. Since the α-phenyl is strongly deactivated both in the diketone **160** and in the pyrylium salt **161**, but the γ-phenyl is not deactivated in the diketone **160** and mildly deactivated in the pyrylium salt **161**, control experiments were effected in order to see whether **287** or **288** may be obtained from **161** with *t*-BuCl + AlCl$_3$. The negative result indicates that the diketone **160** is *t*-butylated before the formation of the pyrylium salt **161**.[264]

4. Reactions of Carboxyl Substituents

Betaine (zwitterion) salts result when pyrylium salts possessing a carboxyl substituent are treated with weak bases. Both α-[69,70,124,146] and γ-carboxyl-substituted pyrylium salts[66,146,202,265,266] are known.

The decarboxylation of γ-carboxyl derivative **162** in the presence of Vaska's compound (Ph$_3$P)$_2$IrCl(CO) leads to the cation-radical **163** (cf. Section IV,C,3).[267] It is assumed that a complex iridium carboxylate is first formed, which loses CO$_2$ to form a metallocarbene; the fission of the carbon–metal bond then yields a carbene (which may either dimerize to a 4,4'-bipyranylidene which is oxidized to **163** by the medium) or the monocyclic cation-radical (which dimerizes and then undergoes reduction to **163**).[267–273]

On the other hand, decarboxylation of **164** yields a free carbenoid species **165** which can be trapped either by ferrocene,[269] by carbonyl derivatives (benzaldehyde, acetophenone, alkyl benzoate and ring-substituted derivatives thereof)[271,274] yielding oxoniabenzyl alcohols **166** (R

= Me, Ph; X = H, Ph, OMe), or by acetylenes yielding α-styrylpyrylium salts **167**.[270]

5. Reactions Involving Substituent Exchange

a. *Reactions of Alkoxypyrylium Salts.*   As described in Part I,[1] 4-alkoxy groups of pyrylium salts are very easily replaced by other alkoxy groups (e.g., on recrystallization from a corresponding alcohol), by alkylmercapto groups or by dialkylamino groups, whereas the reaction with water, sulfide, selenide, and primary amines leads to 4-pyrones, 4-pyranthiones, 4-selenopyrones, and 4-pyroneimines, respectively. The latter may be formed also among ring transformation products (cf. Section III,C,3,c).

At elevated temperatures, 4-methoxy-2,6-diphenylpyrylium perchlorate (**168**) reacts with aromatic compounds ArH activated by electron donor substituents (e.g., dialkylanilines, *N*-alkylindolines, *N*-alkyltetrahydroquinolines, etc.) to give 2,4,6-triarylpyrylium salts **170**.[275] The yields are higher than those from "pyrylation" of aromatic compounds with 2,6-diphenylpyrylium perchlorate (cf. Section II,A,1,c). The explanation lies in the easier rearomatization of the 4-alkoxy-4*H*-pyran intermediates **169** by elimination of the alkoxy group compared with the hydride abstraction from 4*H*-pyrans of type **29**.

In 2,4-dialkoxypyrylium salts like **171** which may act as effective O-

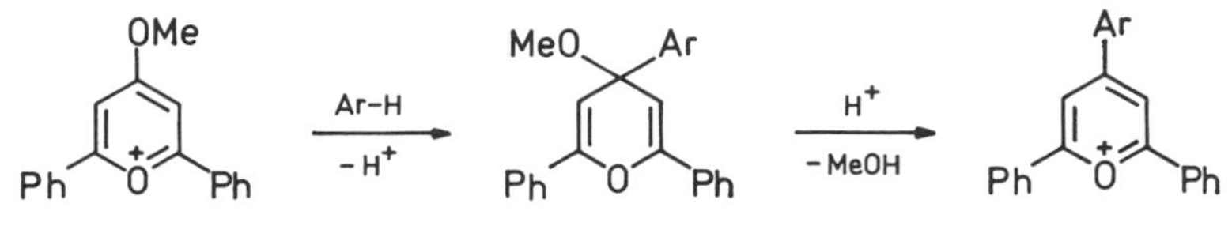

(168)   (169)   (170)

and N-alkylating agents the α-oriented alkoxy group shows the higher
reactivity.[276–278] This is most clearly seen from the reaction of **171** with
2,6-dimethyl-4-pyrone (**11**) resulting in 4-methoxy-6-methyl-2-pyrone (**172**)
and 4-methoxy-2,6-dimethylpyrylium salt (**122**). With secondary amines,
first the α-methoxy group, then the γ-methoxy group is replaced leading
to pyrylium salts **173** and **174**. Tertiary amines are quaternized; similarly,
nitriles undergo alkylation to give nitrilium salts.

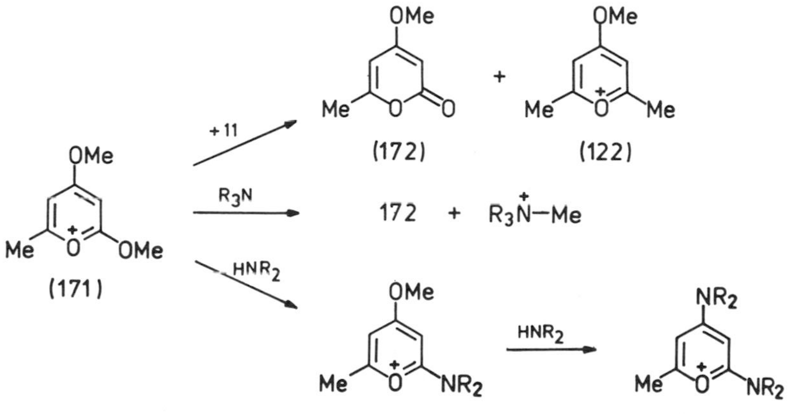

(172)   (122)

(171)

172 + $R_3\overset{+}{N}$–Me

(173)   (174)

An interesting elimination of ethylene was observed in the reaction
of the 2-ethoxypyrylium salt **175** with pyrones **176** (X = O; R = Mc,
Ph) and their *N*-phenylimine derivatives **176** (X = NPh; R = Me, Ph).[276]
Other examples for characteristic reactions of alkoxy groups of pyrylium
salts are mentioned in Sections III,A,6,a, III,A,6,e, and III,B,2,a.

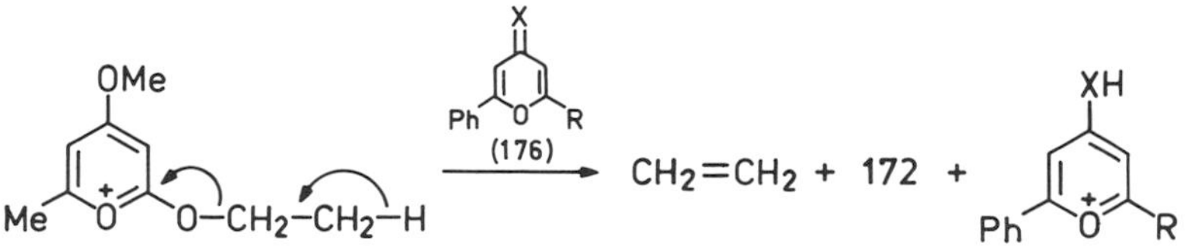

(176)

CH_2=CH_2 + 172 +

(175)

b. *Reactions of Chloropyrylium Salts.* Since 4-chloropyrylium salts
10 are more reactive than the corresponding 4-alkoxypyrylium salts (cf.

Section III,A,5,a) they allow the introduction of alkoxy and aryloxy groups as well as acyloxy groups leading to pyrylium salts of type **6** (R = alkyl, aryl) and **177**, respectively.[41,279] Ethylene glycol reacts with two moles of 4-chloro-2,6-diphenylpyrylium (**178**) to give the bispyrylium salt **179**.[279] The nucleophilic replacement of the chloro substituent by dialkylamines leads to 4-dialkylaminopyrylium salts.[280] Analogously, from **178** and the heterocyclic amines indoline, hexahydrocarbazole, and tetrahydroquinoline the pyrylium salts **180**, **181**, and **182**, respectively, were obtained.[281]

With 3,4-dihalo-substituted pyrylium salts, e.g., 3-bromo-4-chloro-2,6-diphenylpyrylium, only the 4-halo atom is replaced by dialkylamines.[282]

Like 4-alkoxypyrylium salts, 4-chloropyrylium salts react with suitable aromatic, heteroaromatic, and azulene compounds to give the corresponding "pyrylated" systems **170** (cf. Section III,A,5,a).[283] The same reaction occurs with **178** and sterically hindered phenols, e.g., 2,6-di-*t*-butylphenol (**183**), which does not lead to a 4-phenoxypyrylium salt of type **6**, but to the quinoid system **185**. Protonation of the latter with perchloric acid yields the pyrylium perchlorate **186**.[34,279]

Reactions of γ-chloro- and γ-bromopyrylium salts with arylidenepyrans **187** (X = O, S) result in pyrylocyanines **188**.[61,284] Azolidines **189** (R = H, Ph; X = O, S) react with **178** to give pyranylidene derivatives of type **190**.[285] A similar reaction between intermediately formed 4-chloropyrylium salts and 2-phenyloxazol-5-one was already mentioned in Section II,A,1,b.

6. Additional Reactions Leading to Stable Pyran Systems

a. *Reactions with Oxygen Nucleophiles. (i.) Hydroxyl.* As will be discussed in more detail in Section III,B,2,a α- or γ-hydroxypyrans usually undergo a ring opening to 1,5-enediones or their tautomers (pseudobases). However, Griot, Royer, and Dreux[286] claimed that at pH > 11 2,4,6-triphenylpyrylium (**159**) hydrolyzes to a mixture of 1,5-enedione **192** and of 2-hydroxy-2H-pyran **191** which can be separated by extraction (first the former in hexane, then the latter in chloroform). Physical data (IR, UV, ¹H-NMR and melting points) differ significantly and agree with the proposed structure. In the class of benzopyrylium salts such cyclic pseudobases have been known for a long time.[287–289] The cyclic pseudobase **191** is obtained pure either from **159** at pH ≥ 14, or by isomerization of **192** with 0.5 N HCl in dioxane; **191** gives **159** on treatment

with perchloric acid, reacts with Grignard reagents to a mixture of α-
and γ-pyrans, and with potassium borohydride to a 2*H*-pyran.

(159)   (191)   (192)
                        m.p. 148 °C   m.p. 119 °C

On the other hand, 2,6-diphenylpyrylium was shown by Stetter and
Reischl[290] by cryoscopic molecular weight determination to afford in basic
medium a product with twice as many carbons as the expected pseudobase. It seems that Krivun and Dul'skaya's[291] postulated γ-pyranol
formation followed by elimination of water to afford an ether is not
confirmed by X-ray crystal structure determination (cf. Section III,B,2,a).
    Lithium salts **22** of 4-hydroxy-4*H*-pyrans were isolated from the re-
action of 2,6-dimethyl-4-pyrone with organolithium reagents (cf. Section
II,A,1,b).
    In the case of 2- or 4-alkoxypyrylium salts the addition of a hydroxide
ion leads to the corresponding 2- or 4-pyrones (cf. Section III,A,5,a).
Kinetic studies of hydrolysis of 4-ethoxypyrylium and 2,6-dimethyl-4-
ethoxypyrylium perchlorate have shown that at low pH the hydroxide
ion adds exclusively at the γ-position, whereas at higher pH it attacks
the α-position leading via ring-opened intermediates likewise to the corresponding 4-pyrones,[292,293] as will be discussed in Section III,B,2,a. The
kinetics of methoxide addition to 2,6-diphenyl- and 4-methoxyl-2,6-di-
phenylpyrylium cations was also reported.[294]

    (*ii.*) *Alkoxy and Aryloxy Groups.*   Pedersen, Buchardt and co-work-
ers[295,296] investigated the structure of the products formed from alkali
nitrites and 2,4,6-triarylpyrylium in methanol or ethanol. They established by X-ray molecular structure determination that, when the reaction
is performed in the presence of air, one can isolate products which have
an uncommon Δ³-tetrahydropyran structure **193** (R = Me, Et), and which
under more drastic conditions are converted to isoxazole derivatives (cf.
Section III,C,2,a).

(193)   (194)

Katritzky and co-workers[297] recently investigated by means of ^{13}C-NMR spectra in DMSO solution, without isolation, the reaction products **194** (R = Me) of sodium methoxide with 2,4,6-triarylpyrylium perchlorates, where the aryl group is phenyl, *p*-tolyl, or *p*-fluorophenyl. Since there is no carbonyl signal above 152 ppm they favor a 2*H*-pyran structure over a ring-opened dienonic structure. Earlier investigations of Balaban and Silhan[298] of IR and ^{1}H-NMR spectra did not allow a definitive choice (cf. Section III,B,2,b); however, from 2,4,6-triphenylpyrylium perchlorate and sodium isopropoxide a colorless crystalline adduct was obtained, whose structure is assumed to be that of an α-pyran.

On acid-catalyzed cyclodeamination of the vinylogous amide **195** in methanolic solution, Jutz and co-workers[299] isolated the 2-methoxy-3,6-diphenyl-2*H*-pyran **197** which results obviously from the intermediately formed pyrylium cation **196** by addition of methanol to the unsubstituted α-position. In the light of this finding it may be assumed that the nitrogen-free solvolysis products of unknown structure, obtained previously by Fischer and Schroth[129a] on treating acylvinylenamines **198** (R = alkyl, aryl, *n* = 4, 5) with acetic acid in ethanol, are likewise alcohol adducts of the corresponding pyrylium salts **80** (cf. Section II,B,2,b).

4-Methoxy-2,6-diphenylpyrylium (**168**) reacts with alkali methoxide, according to Bersani *et al.*,[300] forming a mixture of the 4*H*-pyran **199** and the 2*H*-pyran **200**, whose instability precluded their separation.

Recently, Fischer, Zimmermann, and Weissenfels[301] found that the alkoxide addition to 2,3,4,6-tetrasubstituted pyrylium salts **201** occurs regioselectively, leading to colorless crystalline 2*H*-pyrans **202** (R = Me, CH₂Ph, Ph; R' = Me, Et). The latter are also formed simply on refluxing **201** in the corresponding alcohol with a trialkylamine (e.g., triethylamine) as proton acceptor. The regioselective attack of the nucleophile in the 2-position of the asymmetrically-substituted cation **201** is due to the stronger positive character of this position; in turn, this positive nature may be plausibly explained by the sterically-conditioned stronger tilting of the 2-aryl group than of the 6-aryl group. Then, in agreement with ^{13}C-NMR data[302] (cf. Section IV,A,2,b) a lower electron density results

at the 2-position. 3,5-Dialkyl-2,4,6-triarylpyrylylium salts react analogously.[301] Subsequent reactions of the adducts **202** are discussed in Section III,C,3,e.

(168)   (199)   +   (200)

(201)   (202)

cis-α-(*o*-Hydroxystyryl)pyrylium salts **203** form spiropyrans **204** on
reaction with bases.[303–306] These compounds attract at present much attention because of their thermo- and photochromic properties due to the
reversible valence isomerization **204** ⇌ **205**. For references to reviews
on spiropyrans and on thermochromism see Section III,D,2.

(203)   (204)   (205)

b. *Reactions with Sulfur Nucleophiles.* By analogy to the reaction
discussed above (Section III,A,6,a) between 2,6-diphenylpyrylium and
aqueous bases, treatment of the same pyrylium salt **35** with aqueous
sodium sulfide was reported to yield first a γ-pyran mercaptide **206**, then
the γ-pyranthiol **207**, and finally the γ-pyran thioether **208**.[66] However,
in view of the criticism based on X-ray structure determination of the
analogous ether quoted above, it seems reasonable to ask that structure
208 should be rechecked, especially since, in addition to structures analogous to those having oxygen in place of sulfur, in the present case
disulfide S—S bonds may also be involved.

Pyran structures involved in the conversion of 4-methoxy-2,6-di-
methylpyrylium perchlorate with sodium sulfide or potassium hydrogen
sulfide to 2,6-dimethyl-4-pyranthione and subsequent reactions are discussed in a wider context in Section III,C,3,b.

c. *Reactions with Nitrogen Nucleophiles.* Usually, the 2H-pyrans which are the primary addition products of secondary amines to 2,4,6-trisubstituted pyrylium salts cannot be isolated because they isomerize spontaneously under electrocyclic ring opening to vinylogous amides; according to the nature of their substituents, these ring-opened products may be stable (as will be described in Section III,B,3,b) or under the reaction conditions they may undergo a ring transformation leading to benzene derivatives (cf. Section III,C,3,c). Isolation of 2H-pyran derivatives succeeded so far only with pyrylium salts possessing certain substituent patterns. Thus, Van Allan, Reynolds, and Petropoulos[228] could isolate from the addition of secondary amines (e.g., piperidine, pyrrolidine) to 4-dialkylaminopyrylium salts **209** 2H-pyrans as hydroperchlorates **210** whose further conversion to *m*-phenylenediamine derivatives will be described in Section III,C,3,c.

As found by Fischer, Zimmermann and Weissenfels,[307] dialkylamines (e.g., dimethylamine, piperidine, morpholine) add to 3-methyl-2,4,6-triarylpyrylium salts **201** similarly to alkoxides (cf. Section III,A,4,a), i.e., regioselectively yielding colorless crystalline 2H-pyrans **211** (R = Me). 3,5-Dialkyl-substituted 2,4,6-triarylpyrylium salts give analogous products.[307] Further reactions of these adducts are described in Section III,C,3,e.

It has been repeatedly pointed out that whereas 2,4,6-trisubstituted pyrylium salts usually favor α-attack by nucleophiles, 2,6-disubstituted pyrylium cations frequently undergo γ-attack. The reaction of 2,6-diphenylpyrylium perchlorate (**35**) with potassium phthalimide accordingly was reported to yield the γ-imidopyran **212**.[69]

(35)   (212)

d. *Reactions with Phosphorus Nucleophiles.* Märkl and co-workers[308] studied the reaction of 2,4,6-trisubstituted pyrylium salts **213** with phosphorus nucleophiles PR_3 such as tris(hydroxymethyl)phosphine or tris(trimethylsilyl)phosphine (cf. Section III,C,3,d) and observed that the volume of the substituents determines the site of the attack and the course of the reaction. If $R^1 > R^2$, the attack occurs at the 4-position leading to 4*H*-pyrans **214** which dimerize to **215**. If $R^1 < R^2$ but $R^3 > R^1$, the attack occurs at the 2-position but the resulting 2*H*-pyran **216** undergoes ring opening to an acyclic 2,3-*trans*-dienone **217**. If one is interested in increasing the yield of phosphabenzenes **218** (which will be discussed in Section III,C,3,d) it is necessary to have large substituents R^1 and R^2 on the pyrylium ring, such as aryl or *t*-butyl, and/or small substituents R^3 on the phosphine. It will be seen in Section III,C,3,d that indeed with PH_3 under acid catalysis, phosphabenzenes result also with smaller substituents on the pyrylium ring.

(216)   (213)   (214)

(217)   (218)   (215)

Krivun and co-workers[65,309] investigated the reaction of 2,6-diphenylpyrylium (**35**) with triphenylphosphine as well as with di- and triethyl phosphite. The results are presented in Scheme 2. The γ-pyranyltriphenylphosphonium salt **219** formed in high yields can eliminate a proton on treatment with phenyllithium or potassium t-butoxide yielding a γ-pyranylidenephosphorane **220**. This compound results in a bipyranylidene **221** by elimination of triphenylphosphine[310,311] or γ-alkylidenepyrans **223** with aldehydes by a Wittig reaction.[310,312,313] Such compounds were also prepared by Hünig et al.[311] and by Krivun et al.[310] using other reactions (cf. Section IV,C,3).

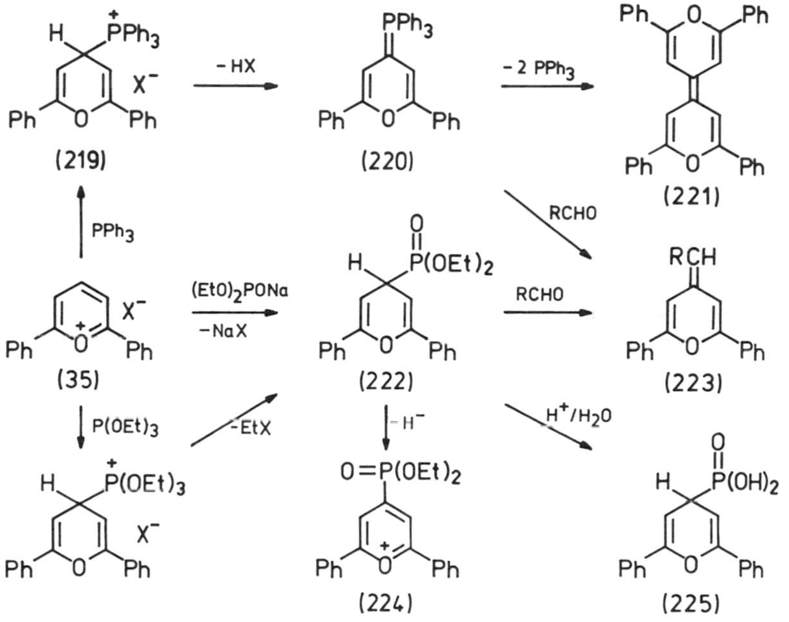

SCHEME 2

The reaction of 2,6-diarylpyrylium salts with triethyl phosphite at 100°C provides an interesting example of the Michaelis–Arbusov rearrangement leading to the diethyl γ-pyranylphosphonate **222**. The same product **222** may be obtained from **35** with sodium diethyl phosphite through a Michaelis–Becker reaction. Under the action of triphenylmethyl perchlorate, **222** eliminates a hydride ion yielding the interesting 4-phosphonyl-pyrylium cation **224**. Hydrolysis of **222** leads to the γ-pyranyl-4-phosphonic acid **225**. By a Horner reaction, deprotonation of **222** followed by treatment with aldehydes provides an alternative route to γ-alkylidenepyrans **223**. Analogous reactions were performed starting from **35** and tributyl phosphite and sodium dibutyl phosphite, respectively.[313a]

e. *Reactions with Carbon Nucleophiles.* (i.) *Grignard Reagents (Mixed Organomagnesium Compounds).* Dimroth and co-workers[314–316] isolated from 2,4,6-triphenylpyrylium (**159**) and benzylmagnesium halides or benzyllithium a γ-pyran adduct **226** which isomerized[315,316] on heating with calcium oxide to a crystalline colorless α-pyran **227**, whose solutions or melt are yellow, possibly due to valence isomerization to **228**. The further cyclization of **228** to 1,2,3,5-tetraphenylbenzene both on heating with calcium oxide or with sodium diethyleneglycolate and on treatment with ethanolic hydrogen chloride, is described in Section III,C,3,e. This reaction occurs even on standing, or in solution in the presence of phenylhydrazine (no phenylhydrazone of **228** could be isolated). The isomerization **226** → **227** occurs also photochemically.[315] The formation of 1,3-diphenylnaphthalene and acetophenone from **226** and acids is likewise described in Section III,C,3,e.

It was observed[315] that pentaphenylpyrylium perchlorate adds benzylmagnesium bromide at the α-position affording a hexaphenylpentadienone which in the presence of sodium diethylenglycolate can cyclize to hexaphenylbenzene.

Dreux, Royer, and co-workers investigated in detail the reaction of Grignard reagents with pyrylium salts in a sequence of three papers: in the first,[317] 2,4,6-trimethylpyrylium perchlorate was treated with a variety of Grignard reagents RMgX, with various R and X groups; in the second,[318] several 2,4,6-trialkyl- or arylpyrylium salts as well as di- and tetraalkyl-substituted salts were treated with methylmagnesium iodide; the third paper[319] was a general theoretical analysis of the results. The nature of the halide X has a minor influence, but the electronic and steric effects of the alkyl groups have a decisive influence on the outcome of the reaction, by favoring usually the α- but in some cases also the γ-addition to 2H- and 4H-pyrans, respectively. When the α-oriented groups in 2,4,6-trialkylpyrylium cations are n-Pr, n-Bu or t-Bu and the γ group is methyl, then there is some (6–12%) addition of methylmagnesium iodide at the γ-position; but when the γ-position is unsubstituted as in 2,6-dimethyl- or 2,3,5,6-tetramethylpyrylium, then with MeMgI a substantial amount of γ-addition (40–60%) occurs. Asymmetrically substituted pyrylium salts like **229**, **230**, and **231** react regioselectively with MeMgI by α-addition. The nature of the substituents and the substitution pattern direct the addition more to one of the two α-positions than to the other, as shown in Scheme 3.

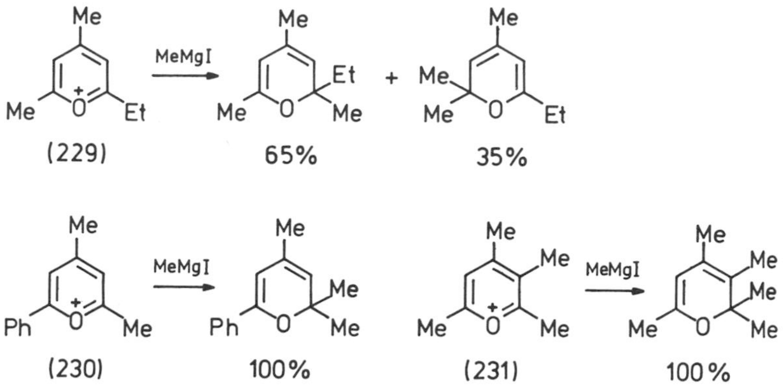

SCHEME 3

The structures of the compounds were assigned on the basis of ¹H-NMR spectra, of diene reaction between maleic anhydride and α-pyrans, and of hydrogenation of the latter to dihydropyrans which were compared to authentic synthetic products.

The variation of the R and X groups in the addition to 2,4,6-trimethylpyrylium perchlorate leads to the results presented in Table I.

Secondary Grignard reagents RMgX (R = *i*-Pr, *sec*-Bu) react exclusively to give γ-pyrans, methyl exclusively to give α-pyrans, while the remaining groups yield a mixture of products. This leads to the following scale of increasing "softness" of the Grignard reagent: CH₃MgI < RCH₂MgI ≅ R₃CMgI < R₂CHMgI.[319] In Pearson's and Klopman's theories of hard and soft acids and bases, the softer the reagent, the more favored should be the γ-attack. The fact that electronic rather than steric factors control the course of the reaction is demonstrated by the regiospecificity of the reaction between 2,3,4,6-tetramethylpyrylium and

TABLE I

RATIO α/γ ADDITION PRODUCT OF RMgX TO 2,4,6-TRIMETHYLPYRYLIUM PERCHLORATE[317,318]

X	Substituent R of the Grignard reagent							
	Me	Et	n–Pr	n–Bu	i–Bu	i–Pr	s–Bu	t–Bu
I	∞	1.08	1.08	0.85	1.04	0	0	1.08
Br	∞	1.50	1.33	1.00	1.50	0	0	1.33
Cl	∞		0.75	0.82	0.85	0	0	1.00

methylmagnesium iodide where the addition occurs at the more hindered α-position.

As a side product in the reaction of 2,4,6-trimethylpyrylium with *t*-BuMgX, hexamethyl-4,4'-bi-4*H*-pyran 232 was obtained[317] (i.e., *t*-BuMgX exerts a reducing effect on pyrylium, favoring a homolytic mechanism). The reaction of MeMgI with 2,6-dimethylpyrylium affords exclusively the γ-addition product, 2,4,6-trimethyl-4*H*-pyran (233)[318] (which can also be obtained from sodium borohydride and 2,4,6-trimethylpyrylium perchlorate, cf. Section III,A,6,f).

The zwitterionic pseudoazulenes 234 (R = alkyl or aryl) (cf. Section III,A,7,a) are attacked by organolithiums R'Li (R' = alkyl or aryl) at the unsubstituted γ-position of the pyrylium system affording 4,9-dihydroindeno[2,1-*b*]pyrans 235[67,68] whose dehydrogenation to the corresponding pyrylium salts 34 was mentioned in Section II,A,1,c.

(*ii.*) *Reactions with Compounds Possessing Active Methyl(ene) Groups.* As will be described in Section III,C,3,e, CH acids such as nitromethane, acetylacetone, ethyl acetoacetate, ethyl cyanoacetate, etc. react, according to Dimroth and co-workers, in the presence of two moles of potassium *t*-butoxide (or triethylamine) with 2,4,6-tri- as well as with higher substituted pyrylium salts to give benzene derivatives. On using, however, only one mole of the base, in some cases intermediates could be isolated. Thus, for the adduct of the ethyl acetoacetate anion to 2,4,6-triphenylpyrylium (159) the 2*H*-pyran structure 236 was suggested,[320] whereas for the reaction product of 159 with acetylacetone an open structure is assumed[321] (cf. Section III,B,4,c). On the other hand, the addition of acetylacetone to 3-alkyl-2,4,6-triarylpyrylium salts 201 (with one equivalent of triethylamine as base) leads regioselectively to crystalline 2H-pyrans 237 (R = Me, Et), as shown recently by Fischer *et al.*[322] The various possibilities for the conversion of 237 to benzene derivatives will be discussed also in Section III,C,3,e.

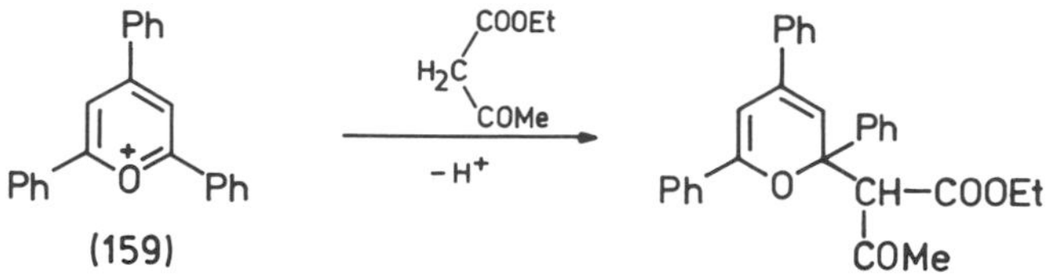

Ph

Ph $\overset{+}{O}$ Ph

(159)

$H_2C\overset{COOEt}{\underset{COMe}{}}$

$\xrightarrow{-H^+}$

Ph

Ph O Ph CH—COOEt COMe

(236)

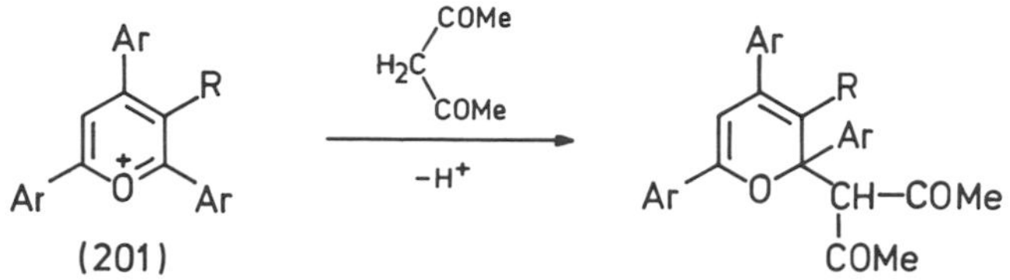

Ar

Ar $\overset{+}{O}$ Ar R

(201)

$H_2C\overset{COMe}{\underset{COMe}{}}$

$\xrightarrow{-H^+}$

Ar

Ar O Ar R Ar CH—COMe COMe

(237)

Unlike the above reactions, the attack of CH acids on 2,6-disubstituted pyrylium salts 238 occurs at the γ-position leading to 4H-pyrans 239. These may be dehydrogenated by 2,4,6-triphenylphenoxyl-catalyzed oxidation with cyanoferrate(III)[323] or by excess 238 [which undergoes conversion to a 2,6-disubstituted 4H-pyran, (cf. Section III,A,6,f)] yielding a γ-alkylidenepyran 242. In reactions 238 → 239 (R = Ar) the CH acids CH₂XY may be, e.g., nitroalkanes,[56,323,324] ethyl cyanoacetate,[56,323] malonitrile, 1,3-dicarbonyl compounds,[56,323] 2-phenyl-2-oxazolin-5-one,[325] and hippuric acid.[326,327]

If the 2,6-disubstituted pyrylium salt 238 has electronegative substituents such as acyl or carbalkoxy in the α-position then the γ-nucleophilic attack succeeds even with monocarbonyl compounds like acetone, acetophenone, or ethyl acetate, leading to corresponding methylenepyrans 242 with X = H, Y = COMe, COPh, COOEt[70,328,329] (cf. Section III,A,7,a). Pyrylium salts of this type react also with compounds possessing C=C double bonds such as styrene or ethyl cinnamate, initiating cationic polymerizations[70] (cf. Section V,D).

Related reactions of 2,6-disubstituted pyrylium salts with activated aromatics such as N,N-dialkylanilines, polyfunctional phenols and their alkoxy derivatives, indoles, pyrroles, azulene, pseudoazulenes, as well as with methylenepyrans (the latter affording pyrylocyanines) have been mentioned in Section II,A,1,c.

4-Methoxy- or 4-methylmercaptopyrylium salts 240 (Z = O or S) add CH acids such as methyl cyanoacetate, malonitrile,[38,329a,329b] 2,4-dinitrotoluene derivatives,[329c,d] benzothiazole derivatives,[330] hippuric acid or acetylglycine[331] to form intermediates of type 241, which easily eliminate methanol or methyl sulfide, respectively, affording also γ-alkylidenepyrans 242.

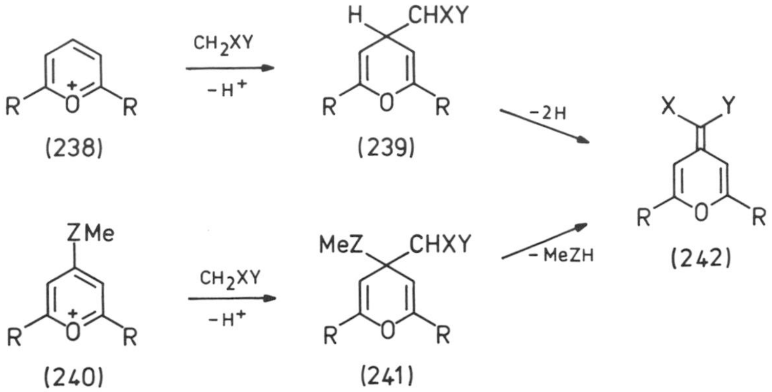

(238) → (239) → (242)

(240) → (241)

If the 4-alkoxy- or 4-alkylmercaptopyrylium salts **240** (Z = O, S) possess α-methyl(ene) groups, these groups may act as CH acids, yielding pyrylocyanines like **121**[28,228,330] (cf. Section III,A,2,c).

(iii.) Reactions with Other Carbon Nucleophiles.   2,6-Diphenylpyrylium salts (**35**) react with ethyl diazoacetate affording ethyl-4-pyranylidene acetate **243**.[332] However, diazomethane reacts with **35** otherwise yielding a reduction product, 2,6-diphenyl-4*H*-pyran, in low yield.[332]

The reaction between 2,6-diphenylpyrylium (**35**) and aqueous sodium cyanide yields the 4-cyano-4*H*-pyran **244** which may be hydrolyzed by concentrated hydrochloric acid to the acid **245**. The reaction of cyanide with 2,4,6-trisubstituted pyrylium salts takes place by α-addition followed by ring opening (cf. Section III,B,4,a).

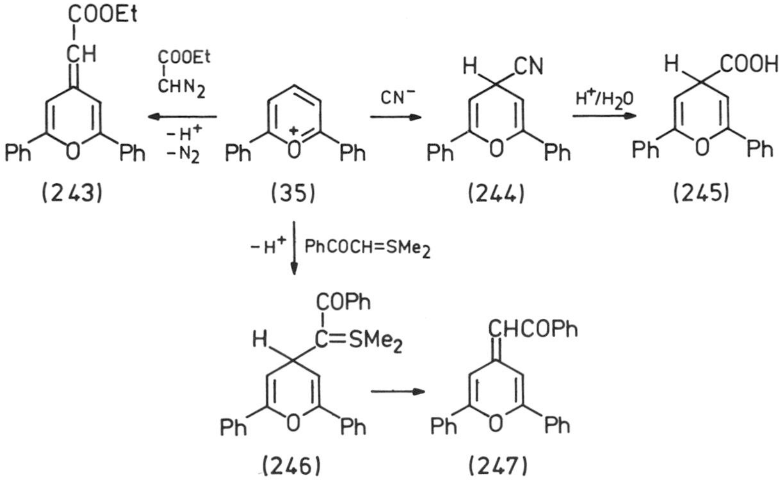

(243)   (35)   (244)   (245)

(246) → (247)

From the various ylids which react with pyrylium salts, the sulfonium benzoylylid $Me_2\overset{+}{S}$—$\overset{-}{C}HCOPh$ deserves to be mentioned here because it affords with 2,6-diphenylpyrylium salts 35 a γ-addition product 246 which in alkaline medium eliminates dimethyl sulfide, leading to 4-phenacyli-dene-2,6-diphenylpyran (247).[333] The reaction with trisubstituted pyrylium salts takes another course (cf. Section III,C,3,e).

f. *Reactions with Hydride Donors.* In Part I[1] (Section II,B,1,e) it was seen that one preparative method for obtaining pyrylium salts was by abstracting a hydride ion from the γ-position of a 4H-pyran (cf. also Section II,A,1,c of the present part). In a reverse reaction, most pyrylium cations react readily with hydride ion donors forming α- and/or γ-pyrans. 2,4,6-Trisubstituted 4H-pyrans may be formed by two alternative γ-ad-ditions: of Grignard reagents to a 2,6-disubstituted pyrylium salt (cf. Section III,A,6,e) or of a hydride ion to a 2,4,6-trisubstituted pyrylium cation. The latter reaction will be discussed here in more detail.

The reduction of pyrylium salts with sodium borohydride was shown[334] to yield two products resulting from γ- and α-addition; with alkyl-sub-stituted pyrylium salts 248 (R = R′ = alkyl) the γ-pyran 251 is more volatile and may be easily separated by fractionation from the α-addition product which is the more polar dienone 250 resulting from valence isomerization of the α-pyran 249 (for the latter products see Sections III,B,4 and III,D,2). The 250/251 ratio depends on the structure of the pyrylium salt 248[334,335] (cf. Section III,B,4).

(248)   (249)   (250)   (251)

(252)   (253)

On the other hand, exclusive γ-attack on hydride reduction is observed when the γ-position is unsubstituted. Thus octahydroxanthylium perchlorate (252) affords octahydroxanthene (253).[196,336,337] 2,6-Diphenyl-4-methylpyrylium perchlorate does not react with borohydride under the above conditions.[335]

Hydride ions may be provided not only by inorganic reducing agents like metal hydrides or complex hydrides as described above, but also by organic molecules through hydride transfer reactions. Sections II,A,1,c, III,A,2,h, and IV,C,3 describe hydride abstraction by pyrylium cations from various organic substrates. Among these reactions, two have a particular significance: (i) those involving pyran, thiopyran, selenopyran, or tropylidene, and the corresponding cations in pairwise combinations allowing relative stabilities to be determined (e.g., pyrylium < selenopyrylium < thiopyrylium[338] and pyrylium < tropylium[328]); (ii) hydride transfers from a 2,4,6-trisubstituted 4H-pyran to a 2,6-disubstituted pyrylium cation leading to a more stable 2,4,6-trisubstituted pyrylium cation and to a 2,6-disubstituted 4H-pyran. The driving force here is the hyperconjugative delocalization of the partial positive character on the substituent bonded to position 4 in the resulting 2,4,6-triarylpyrylium cation. Reactions of this type were mentioned in Sections II,A,1,c, III,A,2,h, and III,A,6,e. Other examples include the hydride abstraction by 2,6-diphenylpyrylium (35) from the 1,5-dione 254 leading to the pyrylium salt 255 and 2,6-diphenyl-4H-pyran (256),[57] or by 2,6-dicarbonyl-substituted pyrylium salts 257 (R = MeO, Ph) from 4H-pyrans 258 (R = MeO, Ph; R' = AcCH₂, p-anisyl) leading to 259 and the more stable 2,4,6-trisubstituted pyrylium salts 248.[70,328] Pyrylium cations like 260 are so reactive that they are able to extract hydride ions even from alcohols (which are converted to carbonyl derivatives)[70] through a Claisen-type rearrangement of the initially formed allyl ether 261.[339]

The hydride abstraction from 1,5-diones leading to pyrylium salts may be effected electrochemically on the rotating platinum electrode, and allows rationalizations of the chemical reactivity on a quantitative basis.[340]

7. Deprotonation and Related Reactions

a. Anhydrobases. Syntheses of pyrylium salts starting from alkylidenepyrans were described in Part I,[1] pp. 259, 262–268. Removal of protons attached to α- and γ-benzylic positions of side chains in pyrylium salts results in neutral α- and γ-pyran systems 262 and 263, respectively. All these deprotonation reactions are reversible; if the products 262 or 263 are acidified, they regenerate pyrylium cations. Anhydrobases may

(254)   +35   (255)   +   (256)

(257)   +   (258)   →   (259)   +   (248)

(260)   Ph₂CHOH / −H⁺   (261)   −Ph₂CO   [structure]

also be formed by an alternative pathway, dehydrogenation of 2-alkyl-2*H*-pyrans and 4-alkyl-4*H*-pyrans (cf. Section III,A,6,f), where the alkyl group is CH_3, CH_2R, or $CHRR'$. Still other routes to anhydrobases are via phosphorus derivatives as described in Section III,A,4,d, or via reactions with CH acids (Section III,A,6,e).

(262)   (263)

If X is an electronegative heteroatom such as oxygen, sulfur, or nitrogen, the products **262** and **263** are isolable stable compounds, pyrones, pyranthiones, and pyroneimines, respectively. The chemistry of such compounds is too vast to be discussed here. If the R groups are simple alkyls and X is also an alkyl CR'_2 group, the product is an unstable alkylidenepyran (anhydrobase); condensation or deuteration reactions (cf. Sections III,A,2 and III,A,7,b) involve the intermediate formation of such alkylidenepyrans. On the other hand, if the electronegativity of the exocyclic carbon is increased by electron-accepting groups, e.g., if X is $CHCOR'$, $CHCN$, $CHNO_2$, $CHCOOR'$, $C(COOR')_2$, or $C(CN)_2$ or vinylogs thereof, the resulting alkylidenepyrans are isolable crystalline compounds, as indicated in Tables XII and XVI (Appendix, Section VII).

Oestensen and Undheim[328] as well as Balaban and Gheorghiu[341] investigated the stereochemistry of vinylogous γ-pyrones 263, X = CHCOMe (anhydrobases of γ-acetonylpyrylium salts). From the two possible rotamers it was shown both for R = COOMe[328] and for R = Me[341] that the only rotamer existing in these cases is the s-cis isomer with the smaller charge separation (see Section IV,A,2,a).

2,6-Diphenyl-4-benzylidenepyran (263, X = CHPh, R = Ph) is formed by deprotonation of 4-benzyl-2,6-diphenylpyrylium (phenylmagnesium bromide acts as a base,[315] not as a nucleophile) or by dehydrogenation of 4-benzyl-2,6-diphenyl-4H-pyran under the action of the triphenylphenoxyl radical.[315]

Other instances of such stable γ-anhydrobases 263 may be observed in Table XVI (Appendix, Section VII). X is an alkyl group (CH$_2$, CMe$_2$, CHPh) but the R groups are carbalkoxy or aryl groups. The X group is a heterocyclic ring (α- or γ-pyrone or xanthone, oxazolone, thiazolone, etc.) or a carbocyclic ring (2,6-cyclohexanedione, cyclopentadiene). In these cases the negative charge of the exocyclic carbon is stabilized through the electronegative substituents (acyl, cyano, nitro) or through the hetero- or carbocyclic ring.

The 2,6-dimethyl-4-tetraphenylcyclopentadienylpyrylium cation 264 reacts with very weak bases (ammonia, diethylamine, isopropylamine, p-aminobenzoic acid, p-nitroaniline) or very strong bases (sodium hydroxide), i.e., with bases having pK_a < 4 or pK_a > 9 by deprotonation affording 265; however, with primary amines of intermediate strength (4 < pK_a < 9) like benzylamine, aniline, and p-toluidine it forms an N-

(264)   (265)

(266)

substituted pyridinium salt (cf. Section III,C,3,c) which may also be deprotonated.[196]

α-Anhydrobases can be stabilized not only by structures discussed above (cf. Table XII) but also by conjugated condensed rings, e.g., **266** (X = O, CHR$_2$ with electron-accepting R groups like CN, carbonyl, or heterocyclic rings, and vinylogs thereof), as shown in Table XIII (Appendix, Section VII).

A very interesting stabilization by such delocalization was described by Boyd[204,342,343] for the case of a condensed conjugated five-membered ring. The deprotonation of cyclopenta[*b*]pyrylium salts **267** results in colored anhydrobases **268** (R = *t*-Bu, Ph, *p*-anisyl; R′ = H, Ph; Ar = Ph, *p*-anisyl; cf. Table XIV, Appendix, Section VII) which are isoelectronic with azulene, involving a lone pair of the oxygen heteroatom.* Such pseudoazulenes ("oxalenes") have also been obtained by Schroth and Fischer[128,345–347] from indeno[2,1-*b*]pyrylium salts **34** (cf. Section II,A,1,c), **72** (cf. Section II,B,2,a), and **81** (cf. Section II,B,2,b) with various substitution patterns (cf. Table XV, Appendix, Section VII). The deprotonation of an indeno[1,2-*b*]thiopyrylium salt to the corresponding pseudoazulene is mentioned in Section III,C,3,b. Syntheses, physical properties, and chemical reactions of pseudoazulenes are subjects of a review[348]; for newer data on pseudoazulenes of the indeno[2,1-*b*]pyran type cf. also Refs. 67 and 349.

Pyrylocyanines benefit likewise from conjugative stabilization and are easily formed on treating the corresponding pyrylium dications, e.g., **269**[350] and **270**,[253] with water which acts as a base.

Pyrylocyanines were intensely studied by Wizinger (cf. Part I,[1] Section II,B,1 and Part II, Section IV,A,1,a), and more recently by Van Allan, Reynolds *et al.*,[351–354] Tolmachev *et al.*,[355–358] and other authors.[359] Besides mono- and trimethinepyrylocyanines like **188** and **152**, azapyrylocyanines[353] as well as penta- and heptamethinepyrylocyanines[354] are known. The latter have electronic transitions in the infrared region. All may serve as photosensitizers in photography (cf. Section V,B). The pyrylium

---

* The question whether compounds of type **268** are protonated at position 7 yielding **267** or at position 5 of the cyclopenta[*b*]pyran system was discussed in terms of molecular orbital theory by Boyd and Ellis.[344]

dications obtained on protonation of pyrylocyanines were studied by electronic and ^{1}H-NMR spectrometry to establish the site of protonation.[357] Both α- and γ-pyran pyrylocyanines are known.

Merocyanines 272 possessing one pyrylium ring are also easily obtained (from 8-ethoxymethylene-2,4-diphenyl-5,6,7,8-tetrahydrobenzo[b]pyrylium salts 271, $n = 0$, or their vinylogs and compounds possessing active methylene groups H₂CXY such as dibenzoylmethane, coumarone, rodaninic acid, malonitrile, p-nitrobenzyl cyanide, phenylmethylpyrazolone, barbituric acid, or 1,3-indanedione, followed by deprotonation).[350] Merocyanines of type 272 are intensely colored compounds with potential application in photography (cf. Section V,B).

When several factors combine, the pyrylium salt deprotonates as soon as it is formed. Such spontaneous deprotonation was, e.g., observed with the intermediate cations 273 and 274.[328] In the latter case one mole of unreacted 2,6-di(carbomethoxy)pyrylium salt (260) acts as a dehydrogenating agent (cf. Section III,A,6,f).

Indeno[2,1-b]pyrylium salts show an increasing deprotonation tendency with increasing phenyl substitution; thus, for example, the acid-catalyzed reaction of 2-indanone with dibenzoylmethane (cf. Section II,B,2,a) leads, by spontaneous deprotonation of the intermediately formed pyrylium salt 275, directly to the pseudoazulene 276.[122]

(273)

(260)   (274)

The deprotonation occurring in mass spectra with formation of an-hydrobases is discussed in Section IV,A,5. In Section B,2,a an example of a γ-methylenepyran formed in preference to a pseudobase from 4-methyl-2,3,5,6-tetraphenylpyrylium and alkali is presented.

(275)   (276)

(277)   (278)

Phenylogs of hydroxypyrylium salts, i.e., 2- or 4-(o- or p-hydroxy-phenyl)pyrylium cations, eliminate the phenolic proton under the action of bases yielding quinopyrans, e.g., 277 and 278.[34,360-362]

2,6-Diphenyl-4-hydroxymethylpyrylium chloride (**279**) readily eliminates hydrogen chloride, affording the corresponding 4-hydroxymethy-
lenepyran **280** which is tautomeric with 4-formyl-4*H*-pyran (**281**).[331]

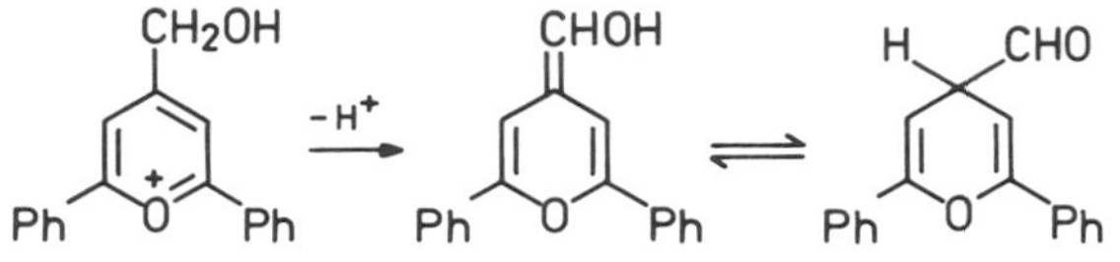

CH₂OH   CHOH   H   CHO

Ph   O   Ph   -H⁺   Ph   O   Ph   ⇌   Ph   O   Ph

(279)   (280)   (281)

R   R   R   R
OH   O⁻   O   O

R   O   R -H⁺ R   O   R   hν   R   O   R   R   O   R

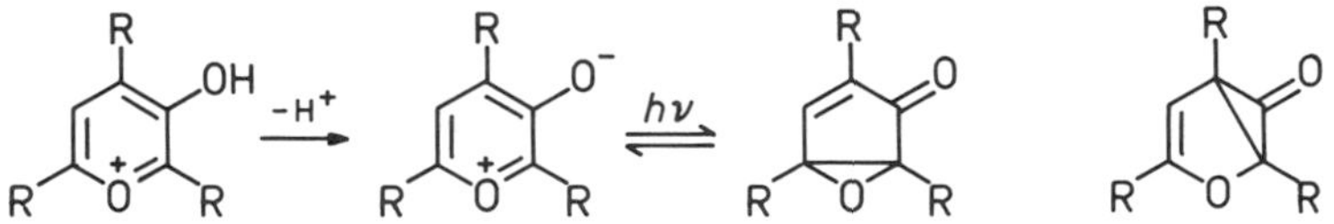

(282)   (283)   (284)   (285)

Deprotonation of 3-hydroxypyrylium salts **282** affords zwitterionic pyrylium 3-oxides **283**. From the two possible valence isomeric structures,
284 and **285**, the former, which is a cyclopentadienonemonoepoxide, is
less strained than the latter which is a dihydrofuran condensed with a
cyclopropanone. The valence isomerization involving **283** and **284** is
described in Sections III,D,1,c and III,D,2.

b. *Hydrogen Isotopic Exchange of Pyrylium Salts.*   Following initial
observations[363,364] of the deuteration of α- and γ-benzylic positions of
alkyl groups attached to pyrylium rings on heating in deuterium oxide,
a ¹H-NMR study by Balaban and co-workers[217] demonstrated that 2,4,6-
trimethylpyrylium perchlorate (**286**) underwent γ-deuteration about ten
times faster than α-deuteration, allowing the preparation of selectively
deuterated 2,4,6-trialkylpyrylium salts according to Scheme 4.

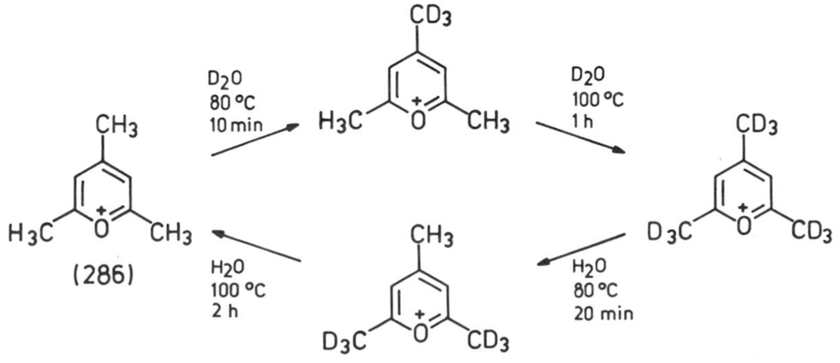

SCHEME 4

Kinetic studies by ^{1}H-NMR methods (cf. Section IV,A,2,a) using buffered media were performed for deuterations and dedeuterations of 2,4,6-trimethyl-,[178,217] 2,6-diethyl-4-methyl-,[365] 2,6-dimethyl-4-ethyl-,[365] 2-ethyl-4,6-dimethyl-,[365] 2,6-diisopropyl-4-methyl-,[366] 2,6-dimethyl-4-isopropyl-,[366,367] 2-isopropyl-4,6-dimethyl-,[368] and 2,6-diaryl-4-methylpyrylium salts (the aryl being phenyl, p-tolyl, and p-anisyl).[369] In all cases the γ-deuteration proceeds faster than α-deuteration. Intramolecular kinetic comparisons with 2-ethyl-4,6-dimethyl- or 2-isopropyl-4,6-dimethylpyrylium salts showed that taking statistical factors into account, an α-methyl hydrogen undergoes isotopic exchange 2.6 times more slowly than the α-isopropyl benzylic hydrogen,[368] or than an α-ethyl benzylic hydrogen.[365] On increasing the pH of the buffer in the range 0–4, the rate of the isotopic exchange increases markedly in all cases. With 2,6-diaryl-4-methylpyrylium, the higher the electron-donating capacity of the aryl group the lower the rate of the deuteration. The isotopic exchange rate of 2,6-diphenyl-4-methylpyrylium is much higher than the exchange rate of the γ-methyl in 2,4,6-trimethylpyrylium under comparable conditions of solvent, buffer, and temperature. An isotopic effect $k_H/k_D = 2.2$ was found if the solvents are H$_2$O and D$_2$O, but if limited amounts of H$_2$O or D$_2$O are used in acetonitrile as the main solvent, the isotopic effect is $k_H/k_D = 1.6$. All these observations agree with the mechanism involving reversible deprotonation to anhydrobases (see Scheme 5).

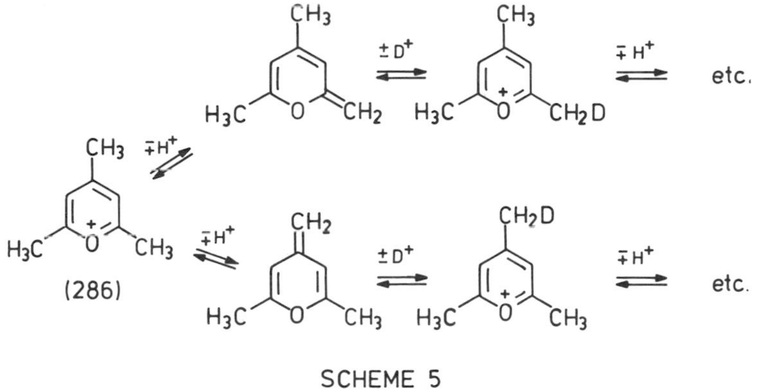

SCHEME 5

Theoretical calculations (cf. Section IV,E) indeed indicate lower energies (i.e., higher stabilities and formation rates) for the symmetrical γ-methylenepyrans than for the nonsymmetrical α-methylenepyrans. 1,2,4,6-Tetramethylpyridinium salts present, however, faster α-deuteration than γ-deuteration. An interesting observation[369a] is that in the deuteration of 2,3,4,6-tetramethylpyrylium perchlorate the relative rates of isotope exchange of the 4-, 2-, and 6-methyls are 35:5:1. The large difference between the deuteration rates of the two α-methyls are accounted for by the stabilities of the corresponding anhydrobases, as indicated by PPP calculations.[369b]

Not only pyrylium-bonded α- or γ-methyl groups undergo this exchange but also tropylium-bonded ones.[370] It was established with 2,3,5,6-tetramethylpyrylium and with 2,3,6-trimethyl-4-phenylpyrylium salts that β-oriented methyl groups are not deuterated.[371] Nor are γ-oriented hydrogens deuterated,[371] but β-oriented hydrogens do undergo deuteration very slowly, e.g., in the latter salt or in 2,4,6-triphenylpyrylium; this β-hydrogen exchange apparently proceeds through the pseudobase.[371]

Once formed, selectively deuterated pyrylium salts are easily converted to the corresponding deuterated pyridines, phenols, furans, azulenes,[372] naphthalenes, mesitylene or their derivatives[373] (cf. Section III,C). Such deuterated compounds are difficultly accessible, or inaccessible, by alternative procedures, and their ready formation indicates some interesting secondary isotopic effects. The lanthanide-induced shifts[374] and careful kinetic studies of the Menshutkin reaction (quaternization of pyridines with CH_3I or CD_3I)[375] revealed that there exists a steric component in the isotopic effect observed when there exist two α-CD_3 substituents in the pyridine (2,4,6-tri-, 2,3,5,6-tetra-, or 2,3,4,5,6-pentamethylpyridine). Due to the smaller volume of the CD_3 group relative to the CH_3 group, accelerations (negative isotope effects) are observed on deuterating either the α-methyls, or the methyl iodide (in the latter case, however, the dissection of the isotope effect into a steric and an electronic effect is no longer possible).

The preparation of side-chain selectively deuterated pyrylium salts and the corresponding pyridines has been reviewed.[375a]

8. Oxidation and Reduction Reactions

a. *Oxidations.*   Few oxidation reactions are known for pyrylium salts, probably because the pyrylium is usually destroyed. In some instances this destruction was useful for structural determinations. Thus, the oxidation in acid medium with permanganate of the trinitration product obtained from 2,4,6-triphenylpyrylium perchlorate afforded a mixture of *m*- and *p*-benzoic acids in a ratio consistent with *p*-nitration of the 4-phenyl group and *m*-nitration of 2- and 6-phenyl groups[262,263] (cf. Section III,A,3) in agreement with newer data on the electron density in the α- and γ-positions (cf. Sections IV,A,2,a, IV,A,2,b, and IV,E).

Analogously the structure of pyrylium salts obtained[264] from phenyl-substituted 1,5-diketones, e.g., **160**, and *t*-butyl chloride in the presence of Lewis acids (the Me_3C^+ cation acts as dehydrogenating agent and as electrophilic reagent for substitution of the phenyl group which is not deactivated) was proved by oxidation to *p-t*-butylbenzoic acid from **287** and to 3,5-di-*t*-butylbenzoic acid from **288** (cf. Section III,A,3).

Wasserman and Pavia[376,377] showed that simple pyrylium salts do not undergo autoxidations. However, 2,4,6-triphenylpyrylium 3-oxide is readily oxidized by air to a dihydrofuran-2-one derivative.[377] This and other reactions involving an oxidative ring transformation (by hydrogen peroxide, iodine, perbromide, air) will be discussed in more detail in Section III,C,2,a.

The 2,4,6-triphenylpyrylium cation (**159**) can react with oxygen if its alcoholic solution is irradiated with UV light, yielding as main products benzaldehyde and benzoic acid.[378] The intermediate **289** was proposed as a more reasonable alternative to attack of pyrylium by the electrophilic singlet oxygen.

b. *Reductions.* One-electron reduction products of pyrylium salts were first isolated by Balaban *et al.*[379] using zinc dust with two-phase (aqueous–ethereal) solutions of alkyl- or aryl-substituted pyrylium salts. The products are 4,4'-bi-4*H*-pyrans, identical to those obtained by electrochemical reduction (cf. Sections IV,A,3 and IV,C,3, which explains why the intermediate pyranyl free radicals afford only 4,4'-dimers, excluding 4,2'- or 2,2'-dimers). Similar reactions take place with other reducing agents such as Mg, Cu, Ag,[66,78,380] VCl$_2$,[381] CrCl$_2$,[382] organometallic compounds,[317,383] 2,6-di-t-butyl phenoxide,[78] or tetramethyl-p-phenylenediamine.[78] The 4,4'-bi-4H-pyran may be reoxidized to the initial pyrylium salt either electrochemically (cf. Section IV,C,3) or chemically, e.g., with chromic anhydride and perchloric acid.[379] When the initial pyrylium salt has no γ-substituent, the bi-4H-pyran may be dehydrogenated to a 4,4'-bipyranylidene[209,311] or in the presence of hydride acceptors like triphenylmethyl perchlorate it may afford a 4,4'-bipyrylium dication.[78]

The yield of the reduction depends markedly on the nature of the reducing agent and on the structure of the pyrylium salt. 2,4,6-Trimethyl- and 2,4,6-triphenylpyrylium perchlorates do not react with VCl$_2$ or CrCl$_2$ but react readily with other reducing agents; 2,4,6-triphenylpyrylium perchlorate does not react with organometallic compounds such as disodium cyclooctatetraene, sodium anthracene, or t-butylmagnesium chloride. Reducing agents like CuCl, Na + NH$_3$, or K in THF give poor or no results. By contrast, reductions with zinc are quantitative.

2,4,6-Triphenylpyrylium fluoborate, a good photosensitizer (cf. Section V,B), is photoreduced to the bi-4H-pyran on irradiation with its x-band absorption wavelength (436 nm, cf. Section IV,A,1,a) in the presence of indene which dimerizes by [2 + 2] cycloaddition.

Two-electron reductions are discussed in Section III,A,6,f.

B. Reactions Involving Ring Opening to Stable End Products

1. Introduction

It was pointed out in the Introduction (Section I) that the pyrylium ring **290** is able to add nucleophiles, according to electronic and/or steric effects of substituents and to the selectivity of the nucleophile, either in α- or in γ-positions affording a 2H- **291** or a 4H-pyran **293**, respectively. Unless the γ-position is unsubstituted or unless the nucleophile is small or unselective (e.g., hydrides, Grignard reagents), α-attack is the preferred pathway because the electron deficiency at the α- is more pronounced than at the γ-position (cf. Section IV,E).

Both the α- and the γ-pyrans may then undergo subsequent reactions converting them to acyclic end products, **292** and **294**, respectively, which then in many cases can recyclize to other ring systems. The present

Section discusses only the former type of reactions leading to isolable acyclic products.

Actually, the term *isolable* needs a brief comment. In many recyclization mechanisms of the pyrylium ring to other products which will be discussed in Section III,C, plausible intermediates are involved. They will not be discussed in the present Section, unless they are stable enough to be isolated in substance or demonstrated in solution by a reliable physical method. Possibly, with the advent of more sophisticated techniques, such evidence will increase in the future for shorter lived intermediates which so far escaped direct detection, refining thereby our understanding of the reaction mechanism.

In general, α-pyrans produced by addition of nucleophiles can undergo electrocyclic rearrangements to substituted pentadienones **292** with cis configuration at the 2,3-C=C double bond. This process is thermally allowed by the Woodward–Hoffmann rules[384–387] because it has a six-membered conjugated transition state, i.e., a concerted process involving $4n + 2$ ($n = 1$) π-electrons (in the hexatriene–cyclohexadiene case such a thermal process is disrotatory). The *cis*-pentadienones **292** may cyclize involving the nucleophile, a side-chain atom, or by intramolecular Michael reaction, yielding 5-, 6-, or 7-membered conjugated ring systems, or may undergo geometric isomerization to a *trans*-pentadienone which is no longer able to cyclize.

On the other hand, γ-pyrans **293** may react, as vinyl ethers, by hydrolysis to yield 1,5-pentanediones **294**.

It should be stressed that the facile valence isomerization of primarily formed α-pyrans to open-chain dienones renders uncertain the structural assignments on the basis of chemical reactions. Even simple physical methods are sometimes unreliable, since ¹H-NMR spectra do not easily distinguish between these two isomers: carbonyl stretching bands in the IR spectra of adducts formed with strongly donor nucleophiles like R_2N are strongly shifted (below 1620 cm^{-1}). With pure crystalline compounds the electronic absorption spectra and the ^{13}C-NMR spectra are a more reliable structure proof (e.g., α-pyrans are colorless, dienones with donor substituents are colored; the presence or absence of a ^{13}CO peak in ^{13}C-NMR spectra demonstrates one of the alternative structures). However, with compounds which because of their instability or low melting point have only been investigated in solution, the situation is much more complex because often the two isomers coexist.

2. Reactions with Oxygen Nucleophiles

a. *Hydroxyl.*   In principle, addition of a hydroxide ion (or of water followed by deprotonation) to a pyrylium salt **290** can take place at α or γ-positions leading to *true* α- **296** and γ-pseudobases **295**, respectively, which are pyranols. The α-pyranol **296** is a hemiacetal and can undergo ring opening to form a 1,5-enedione (acyclic pseudobase) **298** by a thermally allowed electrocyclic process leading to the enolic form **297** of the 1,5-enedione. It was proposed[83] that the term pseudobase should be reserved for the pyranols **295** and **296**, but since this is contrary to established custom, we shall employ the term pseudobase, as is done in the literature, indiscriminately for cyclic or acyclic tautomers.

The reaction of 2,4,6-tri-, 2,3,4,6-tetra-, 2,3,5,6-tetra-, and 2,3,4,5,6-pentaarylpyrylium salts with a hydroxide ion converts them to stable, crystalline 1,5-enediones, e.g., **192**. Infrared spectral studies by Berson[388] confirmed the 1,5-enedione structure of **192**. The pseudobase **299** obtained from 2,6-di-*t*-butyl-4-methylpyrylium is crystalline at room temperature (mp 60°C, from ether) but undergoes self-condensation to a green oil on standing in air in a few minutes.[389] Baeyer and Piccard[44,390] had obtained from 2,4,6-trimethylpyrylium under careful conditions the pseudobase **300**, 4-methyl-4-heptene-2,6-dione, which self-condenses easily (intermolecularly to polymers on standing, and intramolecularly to 3,5-xylenol on heating in alkali hydroxide solution, cf. Section III,C,3,a). Physical methods (IR, ^{1}H-NMR) indicate that the liquid nonpurifiable **300** is an equilibrium mixture of cis–trans stereoisomers **300a** $\rightleftharpoons$ **300b**.[82]

Williams[82] investigated by UV absorption spectra the hydrolysis of 2,4,6-trimethylpyrylium perchlorate, 2,4,6-triphenylpyrylium fluoborate, and 2-methyl-4,6-diphenylpyrylium chloride, over the range of pH values between 3 and 10 with various buffer concentrations in water or deuterium oxide as solvents. He found that the first step is a general base-catalyzed reaction yielding an intermediate α-hydroxypyran (cyclic hemiacetal) which then decomposes via a pH independent pathway, and that the rate k_f of the forward reaction obeys the empirical equation

$$k_f = k_{H_2O} (1 + a_H/K_a) + k_{OH^-} [OH^-] + k_B[B]$$

indicating that the reaction involves an equilibrium $k_f/k_r = K$. The values k_{H_2O}, k_{D_2O}, k_{OH^-}, k_{OD^-}, and k_B for the buffer bases were determined, and an exponent α = 0.45 was obtained for the Brönsted relationship for 2,4,6-trimethylpyrylium. The trimethylpyrylium cation yields at equilibrium (pH > 6) 100% hydrolysis (K_{eq} > 500), while 2,4,6-triphenyl- and 2-methyl-4,6-diphenylpyrylium have lower K_{eq} values for the hydroxypyran $\rightleftharpoons$ diketone equilibrium. Interestingly, from the similar hydrolysis rates of 2,4,6-trimethyl- and 2-methyl-4,6-diphenylpyrylium (about ten times faster than for 2,4,6-triphenylpyrylium) it was concluded that the latter undergoes nucleophilic attack at the position adjacent to the α-methyl, not to the α-phenyl group.

Salvadori and Williams[292,293] similarly studied the kinetics of hydrolysis of 4-ethoxypyrylium salts **301** (R = H, Me) in H_2O, D_2O, and $H_2^{18}O$ leading to the corresponding 4-pyrones **304**. They demonstrated two parallel mechanisms, one at low pH, via nucleophilic attack of water at the γ-position through a 4-ethoxy-4-hydroxypyran **303** which is then converted to **304** without ring opening, and the second at higher pH (involving

a detectable acyclic intermediate through UV spectra) via nucleophilic
α-attack. The intermediate is the pseudobase **302**.

(301)   (302)

(304)   (303)

The oxygen exchange between water and 2,4,6-trisubstituted pyrylium
salts which had been enriched with ^{18}O was studied at various pH values
in buffered solutions.[391] The exchange reaction rate at 100°C increases
with increasing pH value in the pH range 0.6 to 4.0. The results were
interpreted as involving reversible ring opening to the pseudobase. Deuterium exchange at the β-ring carbon also involves reversible ring opening
to pseudobases.[217,371]

Basselier[392] obtained from 2,3,5,6-tetraphenylpyrylium chloroferrate
(**305**) and aqueous sodium hydrogen carbonate under carefully controlled
conditions (no heating above room temperature) a ketoenolic form **306**
of the corresponding pseudobase **307** (UV and IR evidence). This form
regenerates the pyrylium cation easily on treatment with acids (strength
at least equal to that of oxalic acid). With certain acids (HBr, HCl,
$ArSO_3H$, H_2SO_4, oxalic acid, trichloroacetic acid) one may also obtain
a double salt (with two moles of acid per mole of pseudobase) which on
heating eliminates one mole of acid, leaving the pyrylium salt. With
alkali, the ketoenol **306** gives a deep-red solution which on standing
undergoes C—C bond fission to benzoate and α,β-dibenzoylstyrene and
with oxygen affords an unstable hydroperoxide. On heating, ketoenol
306 isomerizes to the crystalline 1,5-dione **307** which is slowly converted
by acids to simple (pyrylium) or double salts.

(305)   (306)   (307)

Unlike the previous cation **305**, 4-methyl-2,3,5,6-tetraphenylpyrylium **308** affords with alkali the anhydrobase **309**; the diketonic pseudobase **310** yields the pyrylium cation **308** on treatment with acids, and a methylenic isomer **311** on treatment with alkali.[392]

Rio and Fellion[393] studied the two crystalline pseudobase isomers obtained earlier from 2,3,4,6-tetraphenylpyrylium (**312**) by Dilthey and Böttler,[394] and showed that they differ by their cis–trans configuration. *cis*-Tetraphenylpentene-1,5-dione (**314**) is the product of mild hydrolysis of a pyrylium ring or oxidative ring fission of a cyclopentenediol system **313**; it isomerizes to the trans product **315** on UV irradiation or on treatment with alkali. A similar cis–trans isomerization of pseudobases was observed by the same authors starting from 3-methyl-2,4,6-triphenylpyrylium salts.[393]

On mild treatment of 2,6-diphenylpyrylium (**35**) with aqueous sodium hydrogen carbonate, a solid red product **317** is obtained: it is formed by condensation of two molecules of pseudobase **316**, as indicated by its molecular weight (determined cryoscopically).[290] The structure **318** given

by Stetter and Reischl[290] and the dipyran ether structure **319** proposed
by Krivun and Dul'skaya[291] are disproved by an X-ray crystal structure
analysis which agrees with formula **317**.[395,396]

(35)   (316)   (317)

(318)   (319)

Strzelecka and Simalty[145] have observed that the pyrylocyanine monocation **320** which is a pseudobase of the trispyrylium cation **101** (cf.
Sections II,C,1,b and III,A,2,g) forms this latter cation only with anhydrous acid; traces of water induce ring opening of **101**, even in
concentrated acid medium. The reason for this sensitivity toward water
was ascribed to the nonplanarity of **101**, whereas **320** is planar and
stabilized by the extended resonance.

(320)   (101)

(321)   (322)

In a recent report, Ukhin *et al.*[397] showed that the 2,2′,6,6′-tetra-*t*-butylbipyrylium dication **321** affords a crystalline pseudobase **322** on treatment with aqueous ammonia, instead of a pyridine. On standing, **322** undergoes condensation and becomes an oil. Its IR and ¹H-NMR spectra indicate the presence of carbonyl and enolic groups. On heating with bases (e.g., sodium acetate in aqueous acetone), however, **321** undergoes a ring transformation yielding a spiran system (cf. Section III,C,2,a).

In Section VII (Appendix, Table XVII) a list of stable acyclic pseudobases of pyrylium salts is given.

b. *Alkoxides.*   Dilthey[398,399] proposed an ether structure for the methylation product of 1,3,5-triphenylpentene-1,5-dione (**192**, 2,4,6-triphenylpyrylium pseudobase) with methyl iodide under alkaline conditions. Rio and Fellion[393] showed, however, that the reaction is not an O-methylation but a C-methylation, since the product gives with acids 3-methyl-2,4,6-triphenylpyrylium.

Nevertheless, ethers may be obtained from pyrylium salts and alkoxides: 2,4,6-triphenylpyrylium salts yield on treatment with anhydrous sodium alkoxides in the respective alcohols deep red solutions whose IR and ¹H-NMR spectra seemed to indicate a keto dienic structure **324**.[400] However, the crystalline isopropoxy derivative obtained from the red solution of 2,4,6-triphenylpyrylium with sodium isopropoxide is colorless[298] and Katritzky's ¹³C-NMR study in DMSO of the reaction product of the same cation with methoxide agrees with an α-pyran structure **323**,[297] as described in Section III,A,6,a. It appears that the red alcoholic solutions may contain the acyclic valence isomer.

In a recent report

(159)   (323)   (324)   R = Me, Et, i-Pr, t-Bu

3. *Reactions with Nitrogen Nucleophiles*

a. *Ammonia.*   Balaban and Toma[401] isolated in crystalline form the intermediate in the conversion of 2,4,6-triphenylpyrylium to 2,4,6-triphenylpyridine, a reaction which had been discovered by Baeyer[29,44,390] and performed many times since then (cf. Section III,C,3,c). On shaking 2,4,6-triphenylpyrylium perchlorate (**159**) with a two-phase mixture of ether and aqueous ammonia and concentrating the ether layer, a solid product is deposited, which is much less soluble in ether than 2,4,6-triphenylpyridine. It melts with dehydration and resolidification to 2,4,6-

triphenylpyridine. The dehydration takes place easily in solution in the presence of acids or bases. The solid dehydrates spontaneously at room temperature in a few days. On the basis of UV, IR, and ^{1}H-NMR data, the most probable formula for this compound seems to be that of an iminoenol, **325**, but ^{13}C-NMR spectra should provide more reliable evidence.

As shown in more detail in Section III,C,3,c, the conversion of 4-acetonyl-2,6-diphenylpyrylium salts, under the action of ammonia, to 2-methyl-4-phenacyl-6-phenylpyridine is another proof that an acyclic intermediate is involved in this reaction.[402,403]

b. *Primary and Secondary Amines.* Primary and secondary amines in equimolar amount, or tertiary amines under most conditions behave as bases in aqueous or ethanolic solution toward pyrylium salts leading to pseudobase formation.[404–406] An excess of primary amine usually converts pyrylium salts to pyridinium salts (discussed in Section III,C,3,c). Lombard and Kress,[406] Toma and Balaban,[407] and later Susan and Balaban[408] identified the intermediate acyclic ketodienamine. 2,4,6-Triphenylpyrylium **290** (R = Ph) reacts with methylamine yielding a very unstable tautomeric ketodienamine identified only by IR,[408] but the intermediates **326** formed in the reaction of **290** (R = Ph) with cyclohexylamine[406] or of **290** (R = Me) with *n*-octadecylamine[407] are more stable, but dehydrate slowly to a pyridinium salt **327**. When R' = *t*-Bu, even with R = Me, cyclization to a pyridinium salt is not possible.[408] Infrared data[407] (v_{OH} at 3620 cm^{-1}) indicate that the structure of the products is iminoenolic; ^{1}H-NMR spectra confirm this structure.[407]

Katritzky and co-workers[13,409] investigated by means of ^{13}C-NMR the structures of the reaction products between primary or secondary amines and 2,4,6-triarylpyrylium salts (^{13}C-NMR assignments were facilitated by *p*-fluorophenyl groups). In agreement with earlier studies, they found that secondary amines afford the open-chain divinylogous amide. Interestingly, significant chemical shift differences suggest that the piperidine compound assumes predominantly structure **328**, while the pyrrolidine analog exists mainly as **329**.

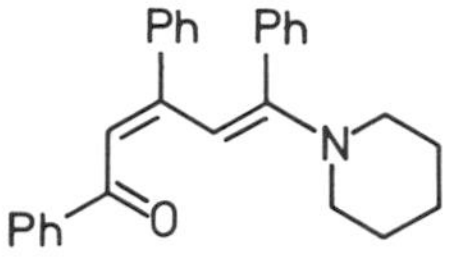

(328) 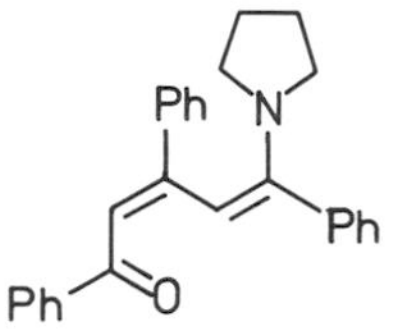   (329)

The reaction between 2,4,6-triarylpyrylium salts, e.g., **159**, and primary amines is more complex and the reaction sequence, as indicated by detailed kinetic studies using ^{13}C-NMR[13,409] and UV spectroscopy,[410] is shown in Scheme 6. The first step of the reaction involves α-addition of the amine, affording **330**, followed by deprotonation and thermally allowed ring opening of the α-pyran derivative to the divinylogous amide **331**. This step is base-catalyzed and no pyran intermediate lives long enough to be observable. For aliphatic amines, one mole of the amine (if for 1 mol of pyrylium one takes more than 2 mol of amine) acts as the base for deprotonation of **330**; for amines of pK_a lower than 8, such as aniline or *p*-nitroaniline, this step becomes fast and preparatively useful in the presence of triethylamine. With equimolar amounts of pyrylium salt and amine, part of the pyrylium salt (~50%) is converted to the pseudobase **192** by the water formed in the reaction. Compound **192** reacts with amines much more slowly than the pyrylium cation.

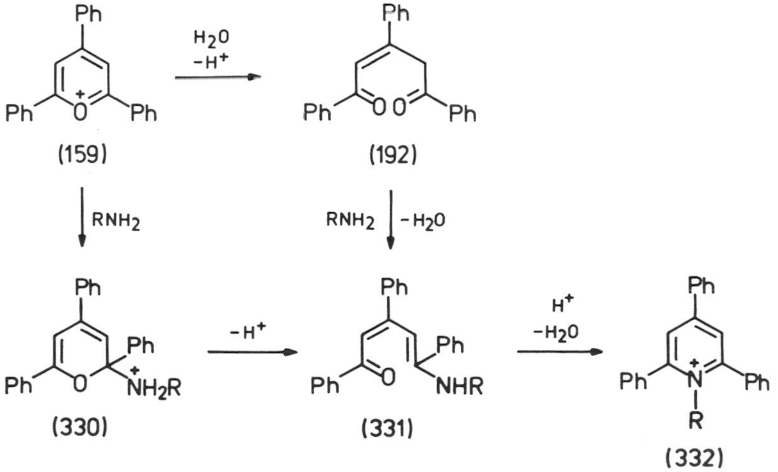

(159)   (192)

(330)   (331)   (332)

SCHEME 6

In a second step, the acyclic divinylogous amide **331** cyclizes to a pyridinium salt **332**, (described in Section III,C,3,c).

Diels and Alder[411] found that secondary amines (dimethylamine, piperidine) convert α-methylpyrylium salts to aniline derivatives (discussed in Section III,C,3,c). Lombard and Kress[406] were the first to isolate acyclic products when they treated 2,4,6-triarylpyrylium salts with secondary aliphatic amines, showing that these red products **333** have a large contribution of the dipolar structure in agreement with electronic and vibrational absorption spectra (no IR absorption band in the usual carbonyl stretching range, 1620–1800 cm^{-1}). In the cases Ar = Ph, R = Me or Et it could be shown also that the ^{1}H-NMR spectra agree with the acyclic structure.[298]

(333)

(127)   (334)   (335)

The structure of the reaction product between the dication **127** (R = Ph; cf. Section III,A,2,d) and three moles of piperidine is either a 2H-pyran **334** or a ketodienamine **335** (only electronic spectra are given, without IR or NMR data).[231]

The fact that pyrylium salts possessing 2-(2-dialkylaminovinyl) groups with secondary amines cyclize to acylbenzene derivatives, which include the vinylene carbons in the benzene ring, also indicates the intermediate ring opening of the pyrylium system (cf. Section III,C,3,c).[230]

A stable ring-opened product **337** is obtained from N-methylaniline and the 4-alkoxypyrylium salt **336** with a free α-position.[412,413] 4-Methoxypyrylium perchlorate (**338**) reacts with N-methylaniline to give the pentamethinecyanine **339** which may be used as starting material for an azulene synthesis[413] (cf. Section III,C,4,b).

(336)   (337)   (338)   (339)

c. *Hydroxylamine, Hydrazine, Substituted Hydrazines.*   On treating
2,4,6-triphenylpyrylium perchlorate (159) with hydroxylamine, Balaban[414]
obtained a colorless crystalline compound whose IR, UV, and ^{1}H-NMR
spectra indicated that it is the monoxime of the corresponding pseudobase. Although as a solid it is stable, in solution it isomerizes readily
to the isoxazoline 341 as discussed in more detail in Section III,C,2,c.
From the two possible isomeric monoximes formula 340 with a conjugated carbonyl and a nonconjugated oxime function agrees with the
experimental IR and UV data.

(159)   (340)   (341)

(342)   (343)

Treatment of 159 at room temperature with hydrazine in an
aqueous–ethereal two-phase system followed by vacuum evaporation of
the ether layer yields the crystalline monohydrazone 343 of the pseudobase. As indicated by IR spectra, a thermally allowed six-membered
transition state 342 would favor the formation of the isomer with the
carbonyl group adjacent to the methylene group, not to the double bond.
This isomer dehydrates readily in solution to a 1,2-diazepine (discussed
in Section III,C,4,a).

The other isomer of the monohydrazone was assumed to be the non-isolable intermediate **344** in a different cyclization (when the starting materials are the pseudobase **192** and hydrazine) leading to the pyrazoline **345** as described in Section III,C,2,c.

(192)   (344)   (345)

Schneider and co-workers[415–418] first investigated the reaction of pyrylium salts with phenylhydrazine; 2,4,6-triphenylpyrylium (**159**) affords a crystalline "α-pyranolhydrazide" which on refluxing in acetone is converted to an isomeric "β-pyranolhydrazide". Each isomer affords with an excess of phenylhydrazine what is now known to be the phenylhydrazone of each initial isomer. Only the "α-pyranolhydrazide" can cyclize in acetic acid to a pyridinium salt **348** (cf. Section III,C,3,c). The difference between the two isomeric "pyranolhydrazides" was first believed to involve valence-isomeric pyran (cyclic) and diene (acyclic) structures.[415–418] Then since both isomers presented carbonyl stretching bands, it was thought that it involves azo–hydrazo isomerism,[406] then cis–trans isomerism.[419] Only on the basis of ^{1}H-NMR spectra was it finally possible to solve this problem, when Balaban[414,420] showed that the "α-pyranolhydrazide" has the acyclic form **346** (cis-monophenylhydrazone of the pseudobase, resulting from a hydrogen transfer involving a six-membered transition state as in the preceding reaction with hydrazine),[298] whereas the "β-pyranolhydrazide" is a pyrazoline **347** (cf. Section III,C,2,c). Treatment of the pseudobase with phenylhydrazine affords directly **347**, possibly through the isomeric monophenylhydrazone having a conjugated COCH=C system.

4. Reactions with Carbon Nucleophiles

a. *Cyanide.* Balaban and Nenitzescu[421] showed that 2,4,6-trisubstituted pyrylium salts react rapidly with aqueous alkali cyanides under ring opening, without any noticeable thermal effects (this indicates how readily the α-pyran intermediate is formed and opened). The products from trialkylpyrylium salts **248** (R = R' = alkyl) are liquid 5-cyano-2,4-pentadienones **349**. Their stereochemistry is cis as depicted, because hypobromite oxidation converts them to cis-cyanosorbic acid as indicated by ^{1}H-NMR spectra.[422]

(159) → (346) "α-pyranolhydrazide" → (347) "β-pyranolhydrazide"

(348)

Two interesting reactions were observed with these cyanodienones **349**: (i) they dissolve in hydrochloric acid, and due to the free rotation in the conjugated acid **350**, on dilution with water trans isomers **351** are obtained (some of which are crystalline at room temperature) and (ii) the functional derivatives **352** (oximes, 2,4-di-, 2,6-di-, or 2,4,6-trinitro-phenylhydrazones, but not the phenylhydrazone or the *p*-nitrophenylhy-drazone) of the *cis*-cyanodienones **349** (but not those of the trans isomer **351**) cyclize on heating with cleavage of hydrogen cyanide and afford

(248) → (349) → (350)

(353) ← (352) ← (351)

pyridinium derivatives **353**, i.e., pyridine N-oxides and pyridinium N^+,N^--betaines. This interesting cyclization is in agreement with the cis stereochemistry of the 2,3-double bond in **349** and their functional derivatives **352**.

2-Methyl-4,6-diphenylpyrylium sulfoacetate yields a cyanodienone whose 2,4-dinitrophenylhydrazone eliminates hydrogen cyanide (on heating at 200°C for 30 min) yielding an N^+,N^--pyridinium betaine identical to that obtained from the same pyrylium salt and 2,4-dinitrophenylhydrazine.[419] A similar reaction was performed with 2,4,6-triphenylpyrylium[419]; the crystalline cis-cyanodienone in this case (**349**, R = R' = Ph; the ^{1}H-NMR spectrum is described in Ref. 298) is not isomerized into a trans isomer by acids, but regenerates the pyrylium salt and eliminates hydrogen cyanide. 2,6-Diphenyl-4-methylpyrylium, which yields a crystalline cis-cyanodienone, and 2,6-diisopropyl-4-methylpyrylium, which gives a liquid cyanodienone, also do not undergo isomerization into trans products.

Refluxing with aqueous ammonia converts the cis-cyanodienone **349** (R = R' = Me) to 2,4,6-trimethylpyridine (35% yield) with elimination of hydrogen cyanide.[421]

Summing up, the reaction of pyrylium salts with alkali cyanides constitutes a convenient method for obtaining 1,5-cyanodienones with definite stereochemistry. Hydrolysis of the nitrile group leads to carboxylic acids. For **349** with R = Me, hypobromite oxidation converts the other end of the molecule to a COOH group;[422] thus alkyl cyanosorbic and alkyl muconic acids with definite stereochemistry become readily available.

  b. *Organometallic Compounds.* Köbrich and Wunder[423,424] obtained from 2,4,6-trimethylpyrylium perchlorate (**286**) and p-dimethyl-aminophenyllithium a mixture of the acyclic α-adduct **354** and the biphenyl derivative **355**. As will be discussed in Section III,C,3,e, the latter is formed from **354** by cyclodehydration. Dimroth and co-workers[315] formulated the reaction product from pentaphenylpyrylium (**356**) and benzylmagnesium chloride as a dienone **357**.

(286)   (354)   (355)

(356)   (357)

Also Dreux and Royer[317–319,425] could show that pyrans obtained in the reactions of pyrylium salts with Grignard reagents (cf. Section III,A,6,e) undergo ring opening to dienones. Thus, 2,6-dimethylpyrylium perchlorate (358) and methylmagnesium iodide afford besides the main product 360 with γ-pyran structure an α-addition product 359 which is unstable and is valence isomerized to a stereoisomeric mixture of 6-methyl-3,5-heptadiene-2-ones 361 (*s-cis, trans, s-trans*) and 362 (*s-cis, cis, s-trans*). With excess Grignard reagents the dienones react further, yielding a tertiary unsaturated alcohol.[426]

(358)   (359)   +   (360)
                          5%   45%

(361)   (362)
25%   25%

c. *Compounds Possessing Active Methylene Groups.* As mentioned in Section III,A,6,e, intermediates in ring transformations of pyrylium salts by anions of CH acids such as nitroalkanes, 1,3-dicarbonyl compounds, ethyl cyanoacetate, malonitrile, etc., could be isolated only rarely. An acyclic structure 363 (R = Ph) is assumed for the primary product of the reaction of 2,4,6-triphenylpyrylium perchlorate with acetylacetone in the presence of one equivalent of potassium *t*-butoxide,[321] whereas with other pyrylium salts and acetylacetone or with 2,4,6-triphenylpyrylium salts and ethyl acetoacetate under similar conditions 2*H*-pyrans were obtained (cf. Section III,A,6,e).

(363)

(364)   (365)   (366)

In the reaction of 2,4,6-triarylpyrylium salts with 2-phenyl-2-oxazolin-5-one leading to benzene derivatives (cf. Section III,C,3,e) acyclic intermediates 365 (R = Ph and p-Cl-C$_6$H$_4$) formed by isomerization of the dienones 364 could be isolated in two cases.[427] Acyclic products 366 (e.g., R = Ph; R' = COOH) were obtained as intermediates of the ring transformation of pyrylium salts into benzene derivatives by alkylidenetriphenylphosphoranes[428] (cf. Section III,C,3,e).

5. Reactions with Metal Hydrides

According to the resonance structures of pyrylium cations, it could be expected that a nucleophilic hydride anion would add to α- or γ-positions of pyrylium salts in ratios depending on the electronic requirements of the substituents bonded to the ring. The initial reaction products could then react by ring opening.

Indeed, Balaban, Mihai and Nenitzescu[334] reported that 2,4,6-trialkyl-substituted pyrylium salts 248 react readily with sodium borohydride in aqueous medium producing in over 90% yield a liquid mixture of three products: two major products which behave like ketones, and a small amount (≤5%) of an alcohol produced by subsequent reduction of a ketone. By working rapidly at 0°C in a two-phase aqueous–ethereal mixture, alcohol formation is suppressed; 2,4,6-trimethylpyrylium perchlorate thus affords only two products which can be separated easily by fractionation: a volatile 4H-pyran 251a (20% yield) which is readily hydrolyzed to a 1,5-pentadienone 367a, and a higher boiling product 249a

$\rightleftharpoons$ **250a** which gives a 2,4-dinitrophenylhydrazone whose λ_{max} indicates a 2,4-pentadienone structure. Since it is not identical with the 3-*trans*-4-methylhepta-3,5-dienone described earlier,[429] the corresponding ketone must possess the 3-cis structure **250a**. Interestingly, the ratio of α/γ addition products depends strongly on the nature of the substituents: the more electron-donating substituents Me < Et < i-Pr increase the rate of addition at the carbon to which they are bonded (Scheme 7).[334,335]

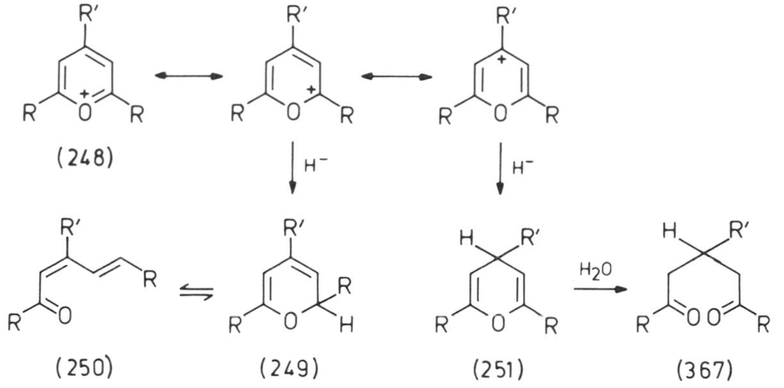

	a	b	c	d	e	f	g	h
R	Me	Me	Et	i-Pr	n-Pr	n-Bu	t-Bu	Me
R'	Me	Et	Me	Me	Me	Me	Me	H
Ratio 250/251								
according to Ref. 334	4.5	1.0	7.0	10.0				
according to Ref. 335	7.3	2.3	11.5		15.7	19.0	∞	0.4

SCHEME 7

This reaction is the counterpart of one-component pyrylium syntheses described in Part I[1]: dehydrogenation of pyrans (Section II,B,1,e), of 2,4-pentadien-1-ones (Section II,B,2,e), and of 1,5-pentanediones (Section II,B,2,f, all in Part I[1]) and is a convenient means of obtaining 2*H*- or 4*H*-pyrans and their acyclic counterparts, pentadienones and 1,5-pentanediones, respectively, by obtaining pyrylium salts from two-component or three-component syntheses, followed by reduction with borohydride.

Subsequent investigations by Marvell, Gosink *et al.*[430–432] brought additional evidence for the correctness of the previous mechanism: the γ-pyran **251a** was isolated in pure state and its ^{1}H-NMR spectrum confirmed the structure. By performing the borohydride reduction at 0°C for 20–30 sec in water/*n*-pentane and recording the UV spectrum, or in water/ carbon tetrachloride and recording the ^{1}H-NMR spectrum, the α-pyran **249a** was detected at $-20°C$ as a species which is converted rapidly at room temperature ($k_{13°C} = 3 \times 10^{-3} sec^{-1}$) to the 2,3-cis-4,5-trans isomer **250a** as shown by ^{1}H-NMR. At room temperature only **250a** is stable.

The equilibrium between an α-pyran and a dienone (see also Sections III,A,6,f, III,B,2,a, and III,D,2) was studied also for compounds **369** ⇌ **370** which like **249** ⇌ **250** possess a hydrogen atom in a position α to the oxgygen heteroatom.[433] However, these compounds were obtained by a different reaction: partial reduction of a C≡C bond in **368**. The unstable α-pyran **370** was identified by UV and ^{1}H-NMR spectra; total hydrogenation afforded **371** and **372**, proving the existence of **369** and its transient valence isomer **370**.

Another confirmation for the structures and stereochemistry of products **249–251**, using GLC separation and IR or ^{1}H-NMR techniques,[335] demonstrated also that 2,6-di-*t*-butyl-4-methylpyrylium affords only the dienone **250g** while 2,6-dimethylpyrylium yields more 4*H*- than 2*H*-pyran. It was argued that the theory of hard and soft acids and bases classifies the hydride anion as a soft reagent (while Grignard reagents are hard reagents) and that therefore the former anion should attack in larger amount the soft γ-site of pyrylium rings; indeed this approach, together with the increasing relative positive charge at the γ-position (calculated[319] by CNDO/2 methods), explains qualitatively the increasing γ-attack in the series **248a, 248b, 248h** (cf. table on p. 85).

The same authors showed that the α-pyran **249b** isomerizes not only to a dienone **250b** but also to *cis*- and *trans*-4-ethylidene-2,6-dimethyl-3,4-dihydro-2*H*-pyrans **373** and **374**, separable by GLC.

Compound	R_α	R'_γ	Positive charges		Ratio of H$^-$ attack
			α	γ	α/γ
248a	Me	Me	3.713	3.791	7.3
248b	Me	Et	3.713	3.793	2.3
248h	Me	H	3.708	3.825	0.4

A reaction which interconverts pyrylium salts and open-chain 1,5-pentanediones involves hydride transfer reactions. Farcasiu, Vasilescu, and Balaban[88] showcd that 2,6-dimethylpyrylium hexachloroantimonate (358) reacts with 1,3,5-triphenylpentane-1,5-dione (375) by hydride transfer leading to the 2,6-hcptanedione (376) and the more stable trisubstituted pyrylium salt 159 (cf. Section III,A,6,f).

C. RING TRANSFORMATION REACTIONS

1. Survey

Unlike the reactions described in Section III,A, which conserve the pyran skeleton, ring transformation reactions lead to modifications of the ring skeleton by breaking old σ-bonds and forming new ones (ANRORC*

* Addition of Nucleophile—Ring Opening—Ring Closure.

mechanism[434]). The primary step of most such reactions consists of the addition of a nucleophile Y to one of the two α-positions of the pyrylium cation (cf. Section III,A,6). As indicated in Section III,B, the 2*H*-pyrans **377** thus formed isomerize reversibly and easily by a thermally-allowed electrocyclic process to their acyclic valence isomers **378**; in certain cases both **377** and **378** may be isolated or demonstrated by spectral methods, especially ¹H-NMR or IR. The latter valence isomers **378**, which are double vinylogs of carboxylic or carbonic acid derivatives are able to undergo a wide variety of synthetically useful inter- or intra-molecular reactions with electron-deficient or electron-rich centers. In ring transformation reactions, valence isomers **378** often cyclize spon-taneously (under mild conditions in acid or base catalysis) forming a new ring system with aromatization as a driving force. The aromatic end products (benzene, pyridine, pyridinium derivatives, etc.) have in most cases a higher delocalization energy and a more even charge distribution than the initial pyrylium salt.

In a few cases, ring transformation reactions proceed through attack of the nucleophile at the γ-position of the pyrylium cation resulting in a 4*H*-pyran **379**; this may either isomerize to a 2*H*-pyran **377** or may undergo ring opening hydrolytically to a pentane-1,5-dione **380**; subse-quent reactions may lead then to new ring systems.

The variety of pyrylium ring transformation reactions is based on the large number of possibilities for recyclization of the acyclic intermediates, especially those of type **378**. Characteristic reaction pathways are pre-sented schematically in Table II. For demonstrating structural relation-ships, in this table and in following formulas, the numbering of carbon atoms C-2 to C-6 from the pyrylium ring will be conserved; and for

TABLE II

POSSIBILITIES FOR THE RECYCLIZATION OF RING-OPENED INTERMEDIATES IN RING
TRANSFORMATION REACTIONS OF PYRYLIUM SALTS

Recyclisation mode of intermediate [a,b]	Incorporated portion of the pyrylium chain	Ring systems so far obtained	Reaction types so far known [c]
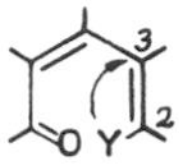 2,3-linkage	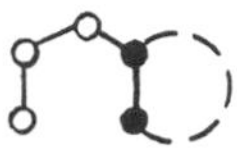 C_2-moiety	 heterocyclic	$2,3-[C_2+N_2C]$
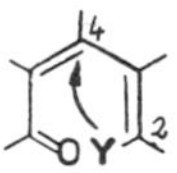 2,4-linkage	 C_3-moiety	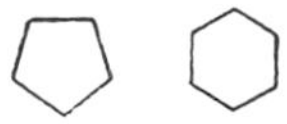 heterocyclic	$2,4-[C_3+NO]$ $2,4-[C_3+N_2]$ $2,4-[C_3+NCN]$ $2,4-[C_3+C_2N]$
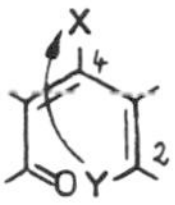 2,4-linkage	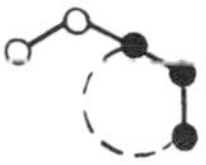 C_3-moiety	 heterocyclic	$2,4-[C_3S+S]$
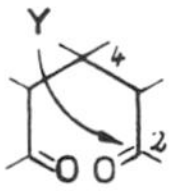 2,4-linkage	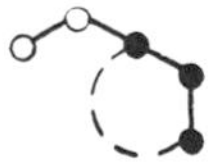 C_3-moiety	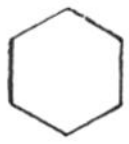 carbocyclic	$2,4-[C_3+C_3]$
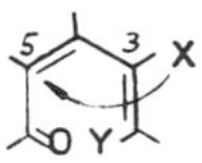 3,5-linkage	 C_3-moiety	 heterocyclic	$3,5-[C_3+NO]$
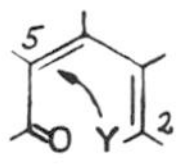 2,5-linkage	 C_4-moiety	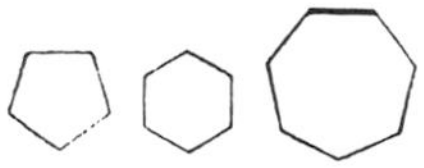 carbocyclic and heterocyclic	$2,5-[C_4+O]$ $2,5-[C_4+S]$ $2,5-[C_4+C]$ $2,5-[C_4+NC]$ $2,5-[C_4+N_2]$ $2,5-[C_4+C_2]$ $2,5-[C_4+C_3]$

(continued)

TABLE II (continued)

Recyclisation mode of intermediate [a,b]	Incorporated portion of the pyrylium chain	Ring systems so far obtained	Reaction types so far known [c]
3,6-linkage	C_4-moiety	heterocyclic	$3,6-[C_4O]$
2,4-linkage	n=1 : C_4-moiety n=2 : C_5-moiety	heterocyclic	$2,4-[C_4+O]$ $2,4-[C_4+S]$ $2,4-[C_5+N]$
4,6-linkage	n=2 : C_5-moiety	heterocyclic	$4,6-[C_5+N]$
2,6-linkage	C_5-moiety	carbocyclic and heterocyclic	$2,6-[C_5+O]$ $2,6-[C_5+S]$ $2,6-[C_5+N]$ $2,6-[C_5+P]$ $2,6-[C_5+C]$ $2,6-[C_5+N_2]$ $2,6-[C_5+C_2]$
2,5-linkage	n=2 : C_6-moiety	carbocyclic	$2,5-[C_6]$
2,6-linkage	C_6-moiety	carbocyclic	$2,6-[C_6]$

TABLE II (continued)

Recyclisation mode of intermediate [a,b]	Incorporated portion of the pyrylium chain	Ring systems so far obtained	Reaction types so far known [c]
2,6-linkage	C_6-moiety	heterocyclic	$2,6-[C_6O+N]$

[a] The arrows indicate only which atoms become linked on recyclisation, and they are not meant to imply donor activity

[b] Letter X indicates a heteroatom side-chain, while C and C_n indicate a carbon side-chain with one or n carbon atoms

[c] The notation of the various reaction types uses the following symbols

$$2,3 - [C_2 + N_2C]$$

Ring-incorporated moiety of the nucleophile with the first indicated atom as the attacking nucleophilic centre

Ring-incorporated moiety of pyrylium salt (including exocyclic substituents)

Positions of the pyrylium salts which become linked (through the nucleophile Y, the substituent X, or the oxygen heteroatom)

simplifying the classification of reaction types it will be assumed that the α-carbon atom being attacked by Y is the C-2 atom.

As seen from Table II, depending on the nature of the nucleophile Y, the substituent X already present on the pyrylium ring, and the α/γ-position of the nucleophilic attack, the pyrylium C_5 carbon chain can be incorporated totally or partly into the new ring so that ring-synthetically pyrylium can act as a C_2, C_3, C_4, or C_5 synthon, and even as a C_6 synthon, when one carbon of an α-substituent is also incorporated. In these recyclization reactions, the nucleophile Y or the substituents X or C (side chains) can participate in the intramolecular ring closure. In some cases the originally present oxygen heteroatom of pyrylium can participate in the recyclization leading to a new ring system, e.g., a furan $(3,6-[C_4O]$

reaction type*; the explanation of this notation is provided by footnote *b* of Table II).

The ring transformation reactions which will be described in more detail in Sections III,C,2 through 4 are arranged first according to the magnitude of the newly formed ring; secondly, for practical purposes they are arranged according to the nature of the *primarily* attacking nucleophile, irrespective of which structural element is incorporated into the newly formed ring. In other words, the nucleophile Y which becomes attached to the C_5 chain of the pyrylium ring by a σ-bond may play two roles: (i) it provides structural elements (atoms) which become incorporated into the skeleton of the new ring; (ii) it only serves to dearomatize and ring-open the pyrylium cation, and then either appears as a side chain of the newly formed ring, or is completely removed on recyclization.

2. *Formation of Five-Membered Rings*

a. *Reactions with Oxygen Nucleophiles.*   Under appropriate conditions, pyrylium salts undergo a ring contraction under attack by oxygen nucleophiles leading to five-membered heterocycles. Thus Balaban and Nenitzescu[435] showed that 2,4,6-trialkylpyrylium salts **381** treated with hydrogen peroxide lead through a 2,5-$[C_4 + O]$ or 3,5-$[C_4O]$ synthesis to 2-acyl-3,5-dialkylfurans **386**. The most plausible mechanism involves the formation of a 2*H*-pyran hydroperoxide **383** which is converted by acids irreversibly to a cation **384**. Recyclization of the resonance-stabilized acyclic valence isomer **385** of this cation leads then to the furan ring whose formation is favored by its aromaticity. In preparative applications of this reaction,[436] e.g., for the preparation of isotopically-labeled compounds[372,437] such as 3,5-di[D_3]methyl-2-acetylfuran (**387**)[438] it is advisable to start from pyrylium salts with $R^1 = R^3$, otherwise there results a mixture of isomers: for $R^1 = R^2 = $ Me, $R^3 = $ Et almost equal amounts of the two acylfurans are formed; they may be separated by preparative gas–liquid chromatography.[439]

As shown by Dimroth and Mach[152] by synthesizing the furyl ketone **388** from 2,4,6-tri-*t*-butylpyrylium fluoborate, pyrylium salts with bulky substituents may also undergo this ring contraction with hydrogen peroxide. Surprisingly, the action of hydroxylamine hydrochloride on 2,4,6-tri-*t*-butylpyrylium fluoborate also affords **388**.[152] Here hydroxylamine does not react as an N-nucleophile (cf. Sections III,C,2,c and III,C,3,c) but as an oxygen nucleophile yielding the intermediate **382** ($R^1 = R^2$

---

* The product of this transformation can be identical to that of the 2,5-$[C_4 + O]$ reaction type, when Y = O.

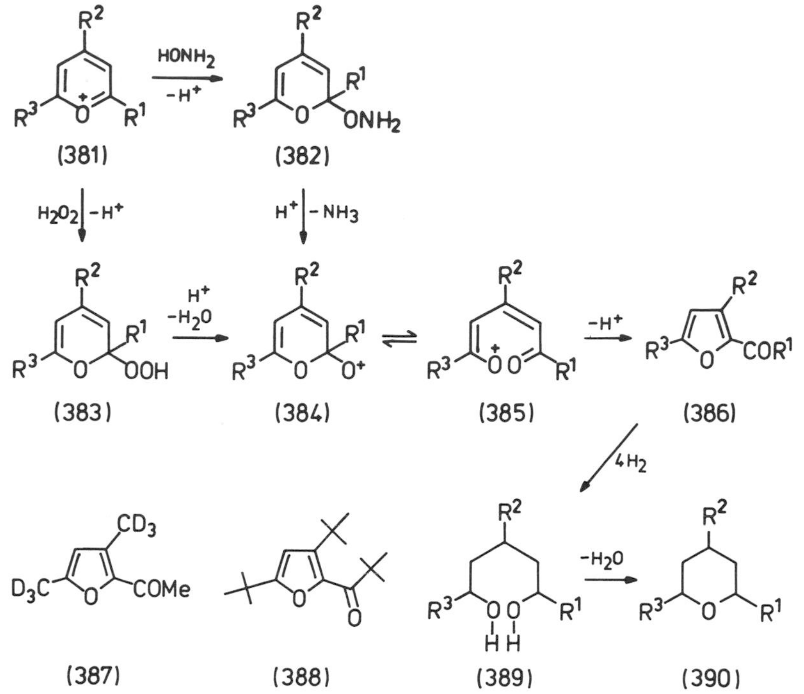

(381) (382)

(383) (384) (385) (386)

(387) (388) (389) (390)

= R^3 = t-Bu) which eliminates ammonia under acid catalysis leading likewise to a cation of type **384**.

The furfuryl ketones possessing a 2-pivaloyl group do not form functional derivatives, whereas those with a 2-isobutyryl group react slowly with 2,4-dinitrophenylhydrazine affording hydrazones whose first electronic absorption maximum has a considerable hypsochromic shift relative to those of 2-propionyl- or 2-acetylfurans.[435]

From 2-acylfurans **386** one can regenerate via **389** the pyran skeleton in the form of tetrahydropyrans **390**.[440] Also the action of ammonia on 2-acylfurans takes place with ring enlargement leading to 3-hydroxypyridines.[440]

The ring transformation of 2,4,6-triphenylpyrylium (**159**) to 2-benzoyl-3,5-diphenylfuran (**392**) was accomplished by Pedersen[441] by oxidizing the anion **391** of the pseudobase **191** with iodine in acetone.

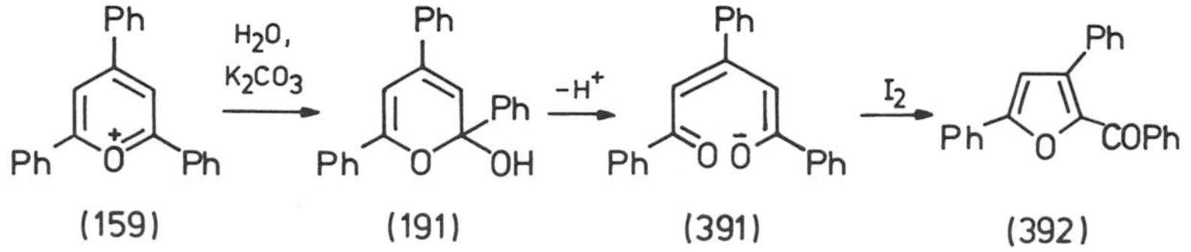

(159) (191) (391) (392)

For the similar formation of 2-benzoyl-3,4,5-triphenylfuran (**394**) during the alkaline hydrolysis of 2,3,4,6-tetraphenylpyrylium perbromide, Quint, Pütter, and Dilthey[442] formulated the reaction course shown in Scheme 8, having succeeded in isolating the intermediate **393**.

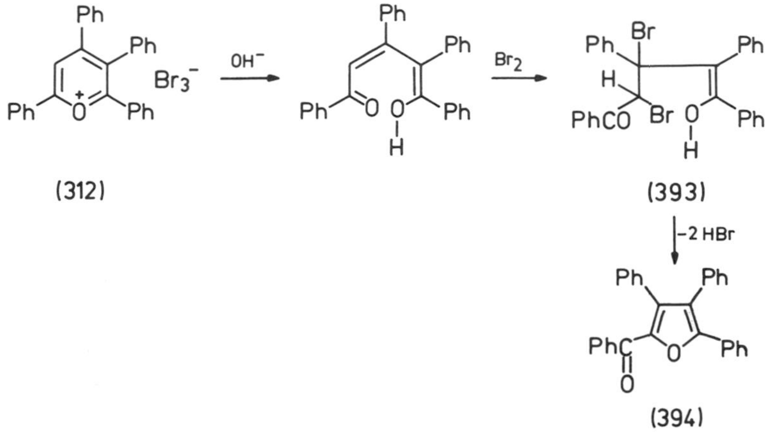

SCHEME 8

On oxidizing 2,4,6-triphenylpyrylium 3-oxide (**395**) with air oxygen, Wasserman and Pavia[377] observed a ring contraction to 2-benzoyl-2,4-diphenylbutenolide (**398**) which resulted by way of a 2,5-linking of the acyclic intermediate **396**, affording a dipolar intermediate **397**; a 1,2-acyl migration stabilizes this intermediate, leading to the final product **398**.

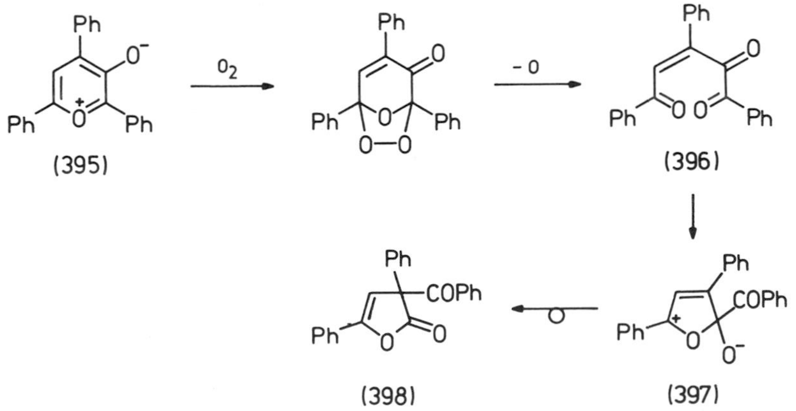

An oxidative ring contraction is also involved in the reaction of 2,4,6-triarylpyrylium salts **399** with alkali nitrites in acetonitrile in the presence of air, studied by Pedersen and Buchardt.[293,295] The recyclization of the oxime intermediate **400** (which can be isolated if air is excluded) leads, presumably through **401** and **402** or **403**, to 3,5-diaroyl-4-arylisoxazoles **404** in yields up to 65%. In alcohols as solvents, the formation of adducts **193** (which can be isolated) predominates (cf. Section III,A,6,a). Their acid hydrolysis leads to **401** which may also be converted to diaroylisoxazoles **404** in low yield. Both reactions represent 3,5-[C_3 + NO] transformations and constitute the first example of a 3,5-linkage.

The formation of 3-(acylmethyl)-5-alkylfurans **406** on refluxing 2,6-dialkyl-4-chloromethylpyrylium salts **405** in dilute solutions of alkali hydroxides in dimethylformamide proceeds without participation of oxidizing agents; this reaction was studied by Dulenko and co-workers[443,444] and represents a 2,4-[C_4 + O] synthesis with participation of an exocyclic carbon atom in the C_4 chain.

Another recyclization accompanied by halide elimination occurs in the alkali-initiated ring contraction of 3-bromocumalic acid (**407**, $R^1 = R^3$ = H; R^2 = COOH)[445] and of other 3-bromo-substituted 2-pyrones[446] to furan derivatives **408**. Unlike the reaction **405** → **406**, here a 2,5-[C_4 + O] transformation is involved. With **407** (R^2 = COOH; $R^1 = R^3$ = Me) the ring contraction is accompanied by decarboxylation of the 2-carboxyl group in **408** leading to 2,4-dimethylfuran-3-carboxylic acid.[447]

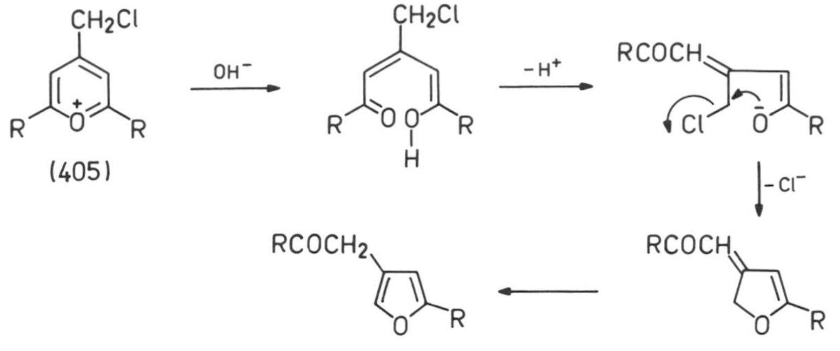

(405)

(406)

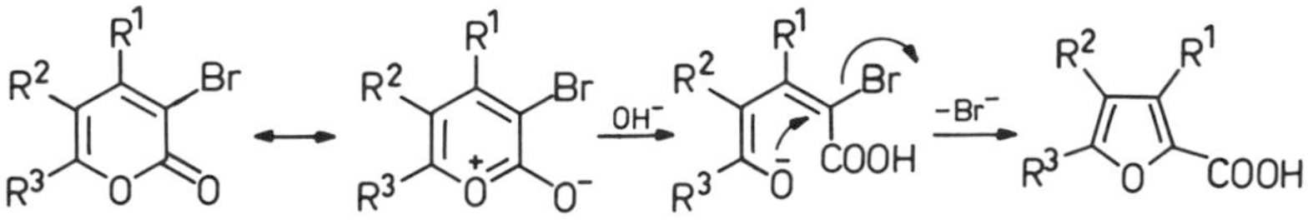

(407)   (408)

If the 2,2',6,6'-tetra-*t*-butylbipyrylium dication **321** is treated with bases (e.g., sodium acetate in aqueous acetone), the intensely colored reaction mixture deposits white crystals which by X-ray diffraction analysis were shown to possess the spiran structure **410**.[397] This reaction represents a 2,4-[C$_4$ + O] transformation by intramolecular γ-attack of the anionic oxygen atom of the pseudobase **409** to its second pyrylium nucleus.

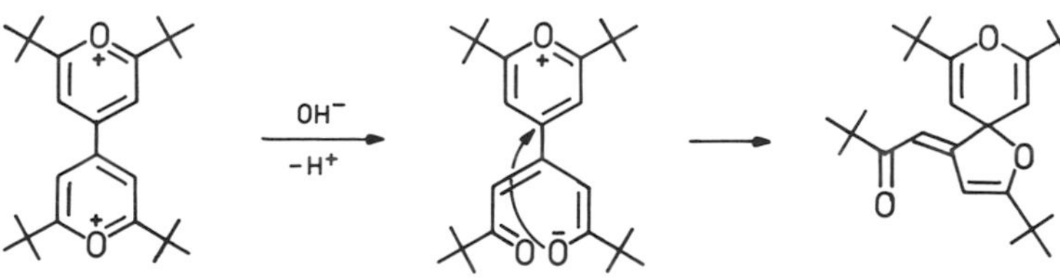

(321)   (409)   (410)

b. *Reactions with Sulfur Nucleophiles.* Analogously to the oxidative ring contraction leading to 2-benzoyl-3,5-diphenylfuran (**392**), described earlier in Section III,C,2,a, Pedersen[441] found that the 2,4,6-triphenyl-pyrylium cation **159** may be converted to 2-benzoyl-3,5-diphenylthiophene (**411**) by reaction with sodium sulfide in acetone followed by oxidation with air or iodine (2,5-[C$_4$ + S] synthesis). The 2,5-linkage by sulfur has a parallel reaction in the oxidation of thiopyrylium (**412**) by manganese dioxide to thiophene-2-aldehyde (**413**) described by Degani,

Ph
    a) Na₂S/Me₂CO
    b) O₂ or I₂

Ph   O⁺   Ph

(159)

Ph

Ph   S   COPh

(411)

S⁺   MnO₂   S   CHO

(412)   (413)

CH₂Cl

R   O⁺   R

Na₂S/DMS

(405)

RCOCH

Cl   S   R

–Cl⁻

RCOCH₂

S   R

(414)

Fochi, and Vincenzi.[448] For a 2,6-linkage leading from 2,4,6-triarylpyrylium salts and sodium sulfide to thiopyrylium salts, see Section III,C,3,b.

The procedure for converting 2,6-dialkyl-4-chloromethylpyrylium salts **405** to 3-(acylmethyl)-5-alkylthiophenes **414** patented by Alekseev, Golyak, and Dulenko[449] consists of refluxing with sodium sulfide in dimethylformamide, and follows the pattern of the furan synthesis described earlier in Section III,C,2,a, representing here a 2,4-[C_4 + S] transformation, where the C_4 chain includes one atom of the γ side chain.

Treatment of 4-pyranthiones **415** with alkali sulfides or hydrogen sulfide, followed by acidification leads to unstable acyclic products for which Traverso[450–453] proposed a 1,5-bismercapto structure **416**, hence a 1,2-dithiepin-5-one structure **417** was assigned to the heterocycle formed therefrom by air oxidation, and a corresponding thione structure **418** to the reaction product of **417** with phosphorus pentasulfide. Arndt et al.[454] had already proposed structure **418** for the reaction product obtained from phosphorus pentasulfide and 2,4,6-heptanetrione. Later IR investigations[455] and X-ray structure analyses[456,457] demonstrated, however, that these heterocyclic compounds had structures **420** and **421**, respectively.* Therefore the acyclic intermediate must be a 1,3-bismercapto derivative **419**. The formation of **420** represents an oxidative ring contraction of a pyran ring which corresponds to a 2,4-[C_4S + S] transformation. The theoretically interesting bonding problems in **421** (no-bond resonance or a central S(IV) atom in a trithiapentalene structure) have been reviewed.[459–461]

---

* Analogously it can be assumed that also in the case of the product obtained[458] from 2,6-dimethyl-4-selenopyrone and sodium selenide, the structure contains a five-membered and not a seven-membered ring.

c. *Reactions with Nitrogen Nucleophiles.* The reaction of pyrylium salts with nitrogen nucleophiles with the general formula H_2NXH (X = O, NH, NR) may afford, according to reaction conditions and substitution pattern of the reactants, five-membered, six-membered, or seven-membered nitrogen heterocycles, or mixtures of such products (cf. Sections III,C,3 and III,C,4). In reactions proceeding with ring contraction or ring enlargement, these reagents function as 1,2-bifunctional nucleophiles, but in reactions affording six-membered rings they function as monofunctional nitrogen nucleophiles, i.e., as primary amines.

On treating 2,4,6-triphenylpyrylium perchlorate (**159**) with hydroxylamine, Balaban[414] demonstrated the intermediate formation of the acyclic pseudobase monoxime **340** (cf. Section III,B,3,c); this monoxime recyclizes rapidly, however, under mild conditions affording the stable crystalline 3,5-diphenyl-5-phenacyl-2-isoxazoline (**341**). On treatment with mineral acids, this isoxazoline eliminates acetophenone and aromatizes to 3,5-diphenylisoxazole (**422**). The same 2,4-[C₃ + NO] transformation was described by Kumler, Pedersen, and Buchardt[462] for pyrylium salts with 2,4,6-triaryl substituents (aryl = substituted phenyl): in acetic acid in the presence of sodium acetate, the formation of isoxazolines related to **341** is replaced by 2,4,6-triarylpyridine *N*-oxide formation, especially when the pyrylium salt bears more than three aryl substituents[463] (cf. Section III,C,3,c). On the other hand, the reaction of 4-(2-benzothiazolyl)-2,6-dimethylpyrylium salt (423) with hydroxylamine leads directly to the isoxazole 424 and acetone.[47]

The reaction of pyrylium salts with monosubstituted hydrazines H_2NNHR may take a course analogous to the preceding reaction with hydroxylamine leading to a 2,4-[$C_3 + N_2$] transformation. The reaction of 2,4,6-triphenylpyrylium salts with phenylhydrazine was first studied by Schneider and co-workers[416–418,464] who obtained two isomeric products ("α- and β-pyranolhydrazide", the crystalline α-compound isomerizes to the β-compound on refluxing in ethanol) to which they ascribed cis and trans acyclic structures which were later criticized by Lombard and Kress[406] who proposed an azo–hydrazo isomerism. The attempt to elucidate their structure by using IR spectra[419] was only partly successful (cf. Section III,B,3,c), but ^{1}H-NMR spectra recorded by Balaban and Silhan[298] showed clearly that the α-isomer was acyclic, whereas the β-isomer was 1,3,5-triphenyl-5-phenacyl-2-pyrazoline (426, R^1–R^4 = Ph). On heating with mineral acids, this compound eliminates acetophenone, yielding 1,3,5-triphenylpyrazole (427, R^1 = R^2 = R^4 = Ph). 1,3,5-Triphenylpentene-1,5-dione (2,4,6-triphenylpyrylium pseudobase 192) also reacts with phenylhydrazine or with hydrazine, yielding phenacylpyrazolines of type 426; this is the only pathway allowing the synthesis of pyrazolines 426 with R^4 = H since 2,4,6-triarylpyrylium salts react with hydrazine affording diazepines (cf. Section III,C,4,a).

Dorofeenko and co-workers[465,466] found that with excess hydrazine pyrylium salts yield directly 3,5-disubstituted pyrazoles **427** (R^4 = H) by spontaneous cleavage of methyl ketones from the 2-pyrazolines **426** (R^4 = H). The pyrylium cation is hereby used ring-synthetically as a potential 1,3-diketone. Similarly, Snieckus and Kan[467] obtained directly 1-methylpyrazoles **427** (R^4 = Me) whose intermediate precursors **426** (R^4 = Me) are stable only under special conditions (reaction at 0°C without a solvent).

Pedersen and Buchardt[468] found that along with the formation of 2-pyrazolines which aromatize more or less easily to pyrazoles, 2,4,6-triarylpyrylium salts **425** (R^1–R^3 = Ar) react with excess phenylhydrazine in hot ethanol affording (apparently in a reaction catalyzed by the excess base), by cyclization of the intermediate 2-pyrazoline **426** (R^4 = Ph), derivatives of pyrazolo[2,3-*a*]quinoline **429**. These aromatize under dehydrogenation (in the presence of iodine) and 1,2-migration of the R^2 group to the fully conjugated system **430**.

Lempert-Sréter and Lempert[469,470] also observed interesting reactions subsequent to the formation of 2-pyrazolines **428** from pyrylium salts **425** and benzenesulfonylhydrazide: on treatment with mineral acids, instead of eliminating a methyl ketone, according to the substitution pattern of the initial pyrylium salt, the 2-pyrazolines either reform the initial pyrylium ring (425, if R^1–R^3 = Ph), or eliminate benzenesulfonic acid yielding 431 (R^1 = R^3 = Ph; R^2 = H), or finally afford pyridinium salts 432 (R^1 = R^2 = Ph; R^3 = Me), a reaction pertaining to Section III,C,3,c.

The reaction of ethyl hydrazidecarboxylate with 2,4,6-triphenylpyrylium perchlorate (159) leads to a mixture of the five-membered 1-carbethoxy-3,5-diphenyl-5-phenacyl-2-pyrazoline (433) and the six-membered 1-ethoxycarbamoyl-2,4,6-triphenylpyridinium perchlorate (434).[471] Also the reaction of hydrazines with 4-(2-benzothiazolyl)-2,6-dimethylpyrylium perchlorate (423) leads to a mixture of pyrazoles 435 (formed by ketone elimination) and of pyridinium salts 436.[47]

For the mechanism of the ring contraction of the 2,4,6-triphenylpyrylium cation (159) to five-membered heterocycles of type 439 under the action of nitrogen nucleophiles H_2NXH (X = O, NH, NR), Balaban[420] proposed a general scheme (Scheme 9). Therein, the 2,4-linkage may be formed through the structural element X when X = O or X = NPh either through the isolable intermediate 437 (cf. Section III,B,3,c), or through its tautomeric form 438. However, when X = NH, the isolable monohydrazone 437 does not lead to a pyrazoline by intramolecular Michael addition, but in a 2,6-[C_5 + N_2] transformation to the seven-membered 1,2-diazepine system (cf. Section III,C,4,a). Therefore the ready formation of **439** from hydrazine and the ring-opened pseudobase **192** excludes **437** and favors **438** as acyclic intermediate.

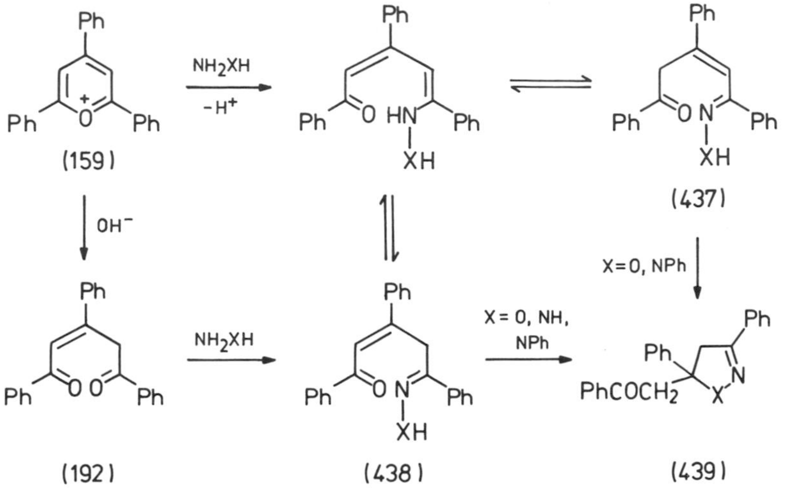

SCHEME 9

Van Allan and co-workers[228] showed for the case of the reaction sequence **440** → **441** involving hydrazine as nucleophile that pyrylium salts possessing substituents like NR_2' in the 4-position (which can exchange with nucleophiles) do not need to eliminate a methyl ketone to form an aromatic five-membered ring, because they can eliminate the NR_2' substituent. A similar reaction course explains the formation of pyrazole-5-acetaldehyde hydrazones **442** (R = H, Ph, 4-$O_2NC_6H_4$) in the reaction of hydrazines with 4-pyrone.[472,473]

An interesting ring contraction was observed by Toda et al.[474] in the reaction of α-phenylpyrylium salts **443** (R = H, Ph; R' = H, Ph) with pyridine-N-imine leading to 3-substituted 2-phenylpyrazolo[1,5-a]pyridines **445**. The reaction represents the only 2,3-[C_2 + N_2C] transformation known so far, i.e., the only reaction in which two carbons of a monocyclic pyrylium C_5 carbon chain appear in the newly formed ring. As reaction intermediate one can assume an acyclic addition product **444**, which is an enamine, and therefore has a manifest nucleophilic character at the C-3 position.

In Section III,C,3,c the reaction of hydrazines with various pyrone derivatives leading to mixtures of five- and six-membered nitrogen heterocycles will be examined in more detail.

d. *Reactions with Carbon Nucleophiles.*   On treating 2,4,6-triphen-ylpyrylium perchlorate (**159**) with sulfonium acylylids, Katritzky and co-workers[475,476] obtained furan derivatives with structure **447**. This represents the first example of a ring contraction initiated by a carbon nucleophile. The alternative structural assignment involving an oxepine ring **448** was ruled out on the basis of X-ray spectra since the NMR, IR, and mass spectral data did not prove sufficient for structure determination. The recyclization of the betaine intermediate **446** to the final product **447** occurs through a 1,3-shift of a hydrogen atom. Since no structural fragment of the ylid nucleophile enters the newly formed ring, this reaction can be classified as a 3,6-[C_4O] transformation. Reactions between pyrylium salts and sulfur ylids leading to benzene derivatives will be described in Section III,C,3,e.

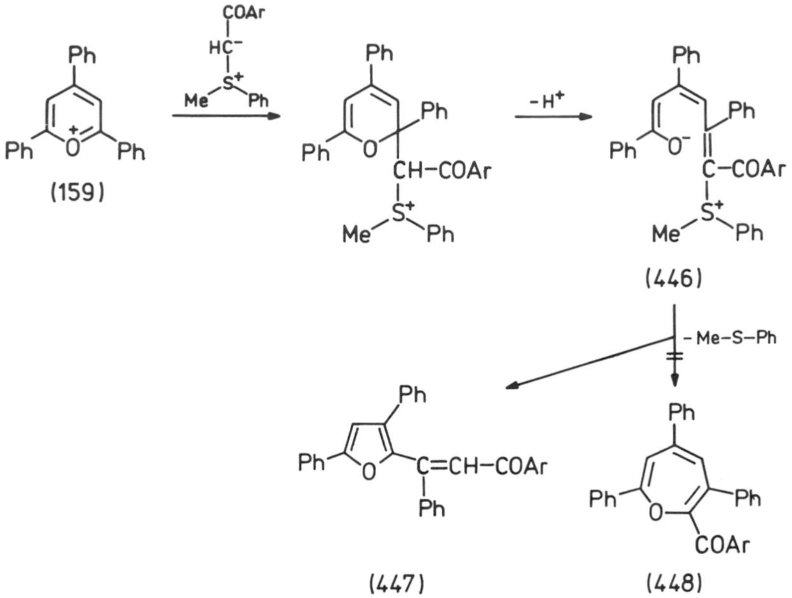

(159)

(446)

(447)

(448)

3. Formation of Six-Membered Rings

a. *Reactions with Oxygen Nucleophiles.*   The isotopic exchange reaction between 2,4,6-trimethylpyrylium perchlorate (286) and $H_2^{18}O$ to
an ^{18}O-labeled pyrylium salt 451 studied by Balaban et al.[391] can be
viewed as a 2,6-[C_5 + O] transformation. As expected on the basis of
mechanistic considerations (cf. Section III,B,2,a), the reaction rate of
the exchange increases with increasing pH in the range 0.65–4.0. The
acyclic intermediate (pseudobase 449 and 450, respectively) is unstable
toward condensation at higher pH when the side chains are alkyl groups.

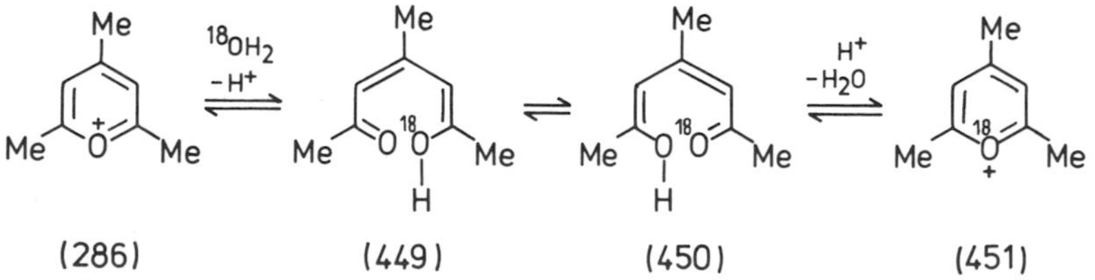

(286)   (449)   (450)   (451)

The conversion of pyrylium salts to benzene derivatives was observed
for the first time by Baeyer and Piccard[390] in the case of 2,4,6-trimethyl-
pyrylium perchlorate (286) which on refluxing with aqueous sodium hydroxide affords 3,5-dimethylphenol (453). Since the 2,6-linkage takes

place incorporating also one carbon atom of an α-oriented side chain (intramolecular condensation of the pseudobase 452), the pyrylium cation functions as a C_6-synthon (2,6-[C_6] transformation).

Balaban and Nenitzescu[94] used this reaction for cations 454 and synthesized the more highly substituted phenols 455. This reaction was also used for obtaining 3,5-di[D_3]methylphenol (456)[372,438] and 3,5-[1,3-$^{14}C_2$]dimethylphenol,[477] for converting the pyrylium salt 423 to 2-(3-hydroxy-5-methylphenyl)benzothiazole (457),[47] as well as for the conversion of the pyrylium salt 458, which can be obtained from piperidine and 4-methoxy-2,6-dimethylpyrylium perchlorate (122) (cf. Section III,A,5,a), to 3-methyl-5-(N-piperidino)phenol (459).[228] Thienopyrylium,[478,479] selenopyrylium,[480,481] benzofuropyrylium,[482] and benzo[c]pyrylium salts [483] containing α-methyl groups were also converted to the corresponding phenols under the action of alkali. Contrary to literature reports[484] which do not consider this reaction as synthetically useful due to its low yield, it was possible to increase[485] the yield to 70% relative to the pyrylium salts. Optimizations were carried out for pyrylium synthesis and selectivity, and for conversion to prehnitenol, i.e., 2,3,4,5-tetramethylphenol (461) arriving thus at a technically useful synthesis of this phenol from simple starting materials (46% overall yield relative to carbinol 460).

The regioselectivity of the reaction between 2,3,4,6-tetramethylpyry-lium perchlorate and sodium hydroxide leading to a mixture of 2,3,5-trimethylphenol (predominantly) and 3,4,5-trimethylphenol[486] was rein-vestigated,[486a] and it was found that the relative amounts of these two phenols are 87 and 13%, respectively. This is in agreement with other studies of the regioselectivity of the nucleophilic attack with the same pyrylium cation and other nucleophiles such as Grignard reagents,[318] cyanide,[486a] borohydride,[486a] and dialkylamines.[486a]

Along with its preparative interest, this reaction has also proved useful for structural assignments of pyrylium salts before the advent of NMR techniques when the conversion to the corresponding pyridine did not solve the assignment.[486] 2-Methyl-6-phenyl-substituted pyrylium salts do not lead to 3-hydroxybiphenyl derivatives, however, because under the drastic reaction conditions the pseudobase is cleaved to benzoic acid (olefin deacylation).[94]

Another formation of benzene derivatives initiated by oxygen nucleo-philes was observed by Reynolds and Van Allan[230] on treating 2-(2-di-methylaminovinyl)-4,6-diphenylpyrylium perchlorate (462) with sodium methoxide or dilute sodium hydroxide. With the former reagent one obtains (through the nonisolable intermediates 463–465) 4-methoxy-2-phenylbenzophenone (466). With the latter reagent one obtains a mixture of the pyran aldehyde 467 and of 4-dimethylamino-2-phenylbenzophe-none (471). The last product results probably via 468–470 through the reaction of 467 with the dimethylamine formed during the hydrolysis, leading to 467; indeed, in a separate reaction it was shown that aldehyde 467 readily forms 471 on treatment with dimethylamine.[230] The action of methanolic potassium hydroxide on 462 yields a mixture of 466 and 471. Both these benzene derivatives result through a 2,5-[C$_6$] transfor-mation, which formally consists of replacing the —CH=O— ring portion by the exocyclic ring fragment —CH=CH—; in both cases an electrocyclic rearrangement is the crucial step. The reaction $463 \rightarrow 466$ has a direct counterpart in the 2,5-$[C_4 + C_2]$ transformation of pyrylium salts under the action of enamines, where the —CH=CH— fragment participating in the cyclization originates from the nucleophilic enamine (cf. Section III,C,3,e).

b. *Reactions with Sulfur Nucleophiles.* By treating 2,4,6-triarylpyrylium salts **399** with sodium sulfide in acetone and then precipitating the product with mineral acids, Wizinger and Ulrich[487] developed a first simple synthesis of thiopyrylium salts **473**. The yellow to blue–red intermediate colors were ascribed to acyclic anions **472**. Their recyclization corresponds to a 2,6-$[C_5 + S]$ transformation.

This reaction was subsequently employed by various authors[228,488–495]; it succeeds with 2-methyl-4,6-diphenylpyrylium salts[225] and also with the indeno[1,2-b]pyrylium salt **474**; in the latter case, the thiopyrylium salt **475** that is produced may be deprotonated by bases to the deeply colored pseudoazulene **476**[205] (cf. Section III,A,7,a).

Earlier claims for the conversion of thiopyrylium cations (e.g., **473**, Ar = Ph) with phenyllithium to 1,2,4,6-tetraphenylthiabenzene[488] were

(399)   (472)   (473)

(474)   (475)   (476)

disproved by Mislow and co-workers.[495a] The chemistry of thiopyrylium salts has been reviewed.[16,17,461]

When instead of containing aryl substituents, the pyrylium salt is 2,4,6-trialkyl-substituted, there are practically no data concerning the O → S exchange. 2,6-Disubstituted pyrylium salts devoid of a 4-substituent react differently, adding the nucleophile in the 4-position, leading to 4-mercapto-4H-pyrans (cf. Section III,A,6,b).

Traverso[450] and later Kato et al.[496] obtained 1-thio-4-pyranthione 478 from 4-methoxy-2,6-dimethylpyrylium perchlorate (122) and sodium sulfide or potassium hydrogen sulfide through the isolable 2,6-dimethyl-4-pyranthione (477, X = S; cf. Section III,A,5,a). The same 1-thio-4-pyranthione 478 is formed also on treating the selenopyrone 477 (X = Se) with sodium hydrogen sulfide.[458] Earlier in the laboratories of Arndt[497] and Traverso[451,452,498] the conversion of 4-pyranthiones to 1-thio-4-pyranthiones had been investigated. 2,6-Diphenyl-4-(N-piperidino)pyrylium perchlorate reacts with sodium sulfide affording the thiopyrylium salt 479 by a normal O → S exchange of the ring heteroatom.[228]

(122)   (477)   (478)   (479)

  c. Reactions with Nitrogen Nucleophiles. (i.) Ammonia.   The earliest ring interconversion reaction of pyrylium salts is represented by

the conversion of 4-methoxy-2,6-dimethyl-, 2,4,6-trimethyl-, and 2,6-di-
methyl-4-phenylpyrylium perchlorate with aqueous ammonium carbonate
to the corresponding pyridines, reported by Bayer and Piccard.[29,44,390]
This pyridine synthesis proved to be the prototype for analogous conversions of pyrylium salts to pyridinium salts on reaction with a host of
primary amines, as will be described in the next subsections.

As acyclic intermediate of this 2,6-[C_5 + N] transformation one must
assume the valence isomer **480** and tautomers thereof (i.e., **481**). Recyclization (i.e., of **481** by an electrocyclic process) leading to the 2,6-
linkage, then affords **482** which dehydrates irreversibly yielding the aromatic pyridine system **483**. This mechanism is also valid for the reactions
to be described in the following subsections between pyrylium salts and
primary amines in the general sense (including hydroxylamine and hydrazines, with formula H_2NR, with R = Alk, Ar, OH, NH_2, OR, NHR',
NR_2', etc.) so that the mechanism will not be repeated separately in each
subsection.

As mentioned in Section III,B,3,a, an experimental proof for the intermediate formation of acyclic adducts was provided by Balaban and
Toma[401] who isolated a crystalline iminoenol of type **481** from the reaction
of 2,4,6-triphenylpyrylium perchlorate and ammonia.

The conversion of pyrylium salts to pyridine derivatives proceeds
usually with good or excellent yields. Sometimes the replacement of the
aqueous ammonium carbonate or ammonia solution by alcoholic ammonia improves the yield; this technique can consist of bubbling gaseous
ammonia through a suspension of the pyrylium salt in an appropriate
alcohol (methanol,[499] t-butanol[2,3]). As reported by Dorofeenko et al.[500]
in certain cases urea or thiourea may function as ammonia donors when
they are heated with pyrylium salts in dimethylformamide.

Unlike the O → S exchange which was described in Section III,C,3,b, the O → N exchange by means of ammonia is not accompanied by an attack at the 4-position (when this is unsubstituted) or by a replacement of a 4-alkoxy group, so that this Baeyer pyridine synthesis is of almost universal applicability. Its preparative use followed closely the development of novel pyrylium salt syntheses so that most of the new pyrylium salts were converted to the corresponding pyridines. Thus were reported not only 2,4,6-triaryl-, 2,4,6-trialkyl-, other trialkyl-, or 2,4,6-trisubstituted alkyl/aryl pyridines, but also tetra-, penta-, as well as mono- and disubstituted monocyclic pyridines, as can be seen from Table XVIII (Appendix, Section VII). Since the unsubstituted pyrylium cation is very sensitive to hydrolysis, Klages and Träger[80] converted it to pyridine in low yield by reaction in molten ammonium carbonate. In many instances, the conversion to known pyridines served for confirming or assigning the structure of pyrylium cations especially in cases where more than one structure was possible.

Of particular importance is the Baeyer pyridine synthesis in those cases where pyridines cannot be obtained by alternative means, or where other approaches are much more difficult. 2,4,6-Tri[D₃]methylpyridine (**484**)[372,438] and its selectively deuterated cogeners **485** and **486**[374,501] as well as 2,4,6-[2,6-¹⁴C₂]collidine (**487**)[437] represent isotopically labeled pyridines. The synthesis of sterically hindered pyridines **488**[152] and **489**[94,188,189] was possible through the intermediacy of the corresponding *t*-butyl-substituted pyrylium salts; these pyridines are of interest because they are nonnucleophilic bases. Also for obtaining pyridines with heterocyclic substituents like **490**,[191] **491**,[58,502] **492**,[58,502] **493**,[47,191] **494** (X = O, S),[502] or **495**[502] or with ferrocenyl groups[50,51,140] (e.g., **496**[140]) or cyclopentadienyl manganesetricarbonyl groups[141] the approach via pyrylium salts is one of the simplest.

(484)   (485)   (486)

(487)   (488)   (489)

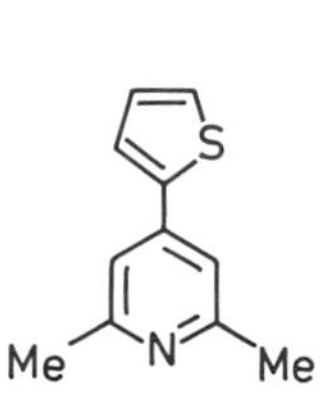

(490) 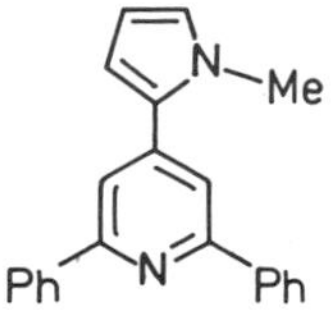   (491) 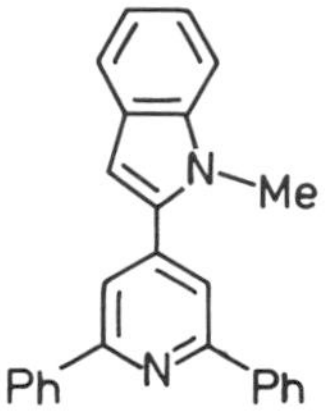   (492) 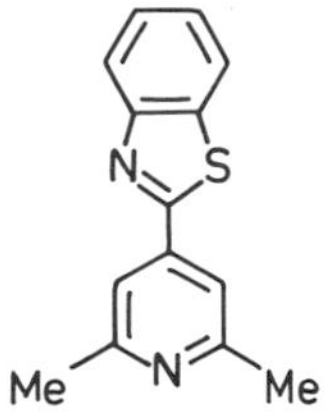   (493)

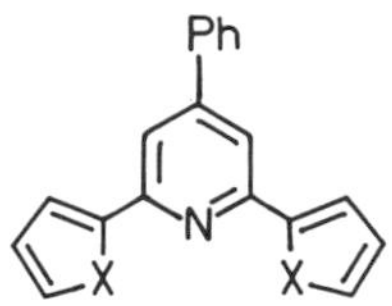

(494) 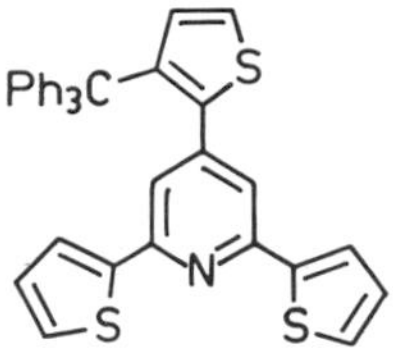   (495) 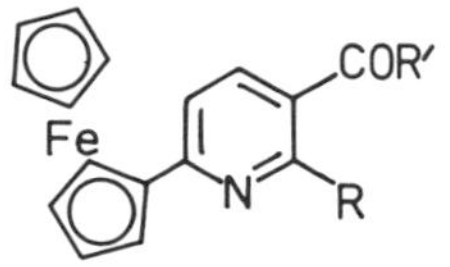   (496)

From *p*-phenylenebispyrylium salts one can obtain bispyridines with structure **497**[221,503–506]; for other bispyridines see Table XIX (Appendix, Section VII). In the case of the pyrylocyanine **320**, the stepwise replacement of oxygen by nitrogen atoms leads to the isolable intermediate **498** and thence to the trispyridine **499**.[156]

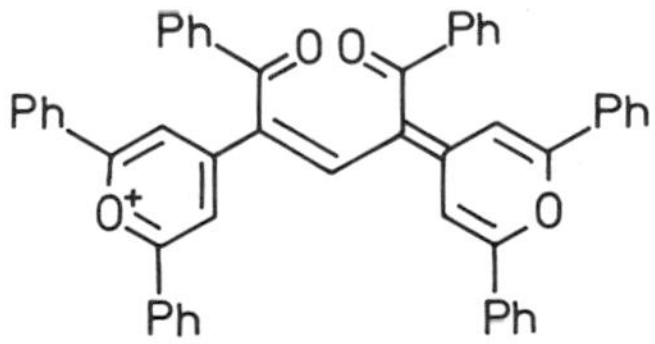

(320)

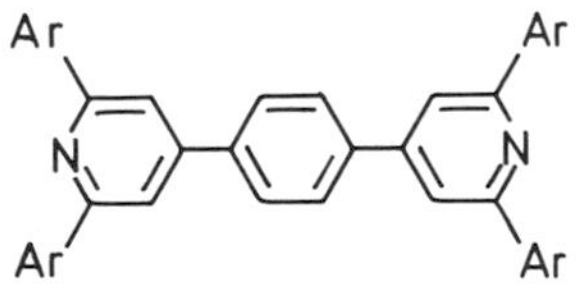

(497)

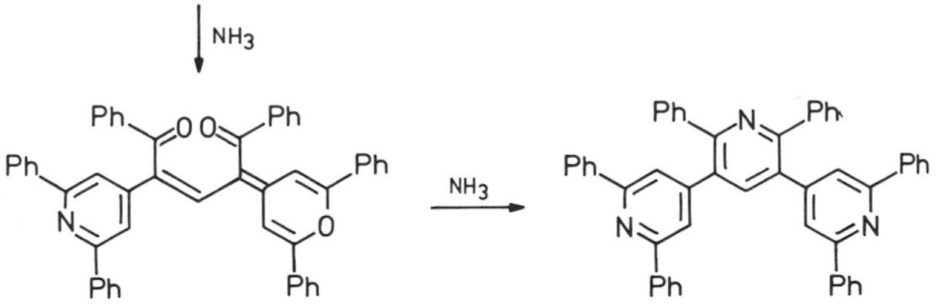

NH3

(498)   NH3   (499)

Analogously to Baeyer's preparation of 4-methoxy-2,6-lutidine,[29] one can obtain (from pyrylium salts with S-, N-, P-, and C-containing side chains with functional groups) pyridines of types **500** (R = Me, PhCH$_2$),[507] **501** (NR$_2'$ = piperidino, morpholino, indolino),[28,228,508] **502**,[65] and **503**.[146]

The O → N exchange of pyrylium salts for the synthesis of bi-, tri-, and polycylic pyridine derivatives has been widely used; cf. Tables XXII–XV (Appendix, Section VII). Balaban and co-workers succeeded in thus obtaining 3,4-tri- and tetramethylenepyridines **504** (n = 3, 4)[509,510] as well as 2,6-dimethyl-3,5-heptamethylenepyridine (**505**)[184] and 2,6-di-methyl-3,5-nonamethylenepyridine (**506**).[185,511] Compounds of type **504** were also independently obtained by Praill and Whitear.[172,173,512] In the ^{1}H-NMR spectrum of **505**, the CH$_2$ group in the middle of the saturated chain gives rise to a signal at δ = −0.08 ppm, whereas the corresponding ^{1}H-NMR signal of **506** is at δ = 1.00 (in CS$_2$ in both cases)[184] proving that the ring current of the pyridine exerts a considerable shielding of the CH$_2$ group of **505** held rigidly above the plane of the ring (see Section IV,A,2,a).

Pyridines of type **504** were also obtained by Dorofeenko and co-workers[190] without isolating the corresponding pyrylium salts in a "one-pot" reaction from cyclopentanols; in the same laboratories 3,4-hepta-methylenepyridines **507** were prepared.[215]

(500)   (501)   (502)   (503)

(504)   (505)   (506)   (507)

Whereas benzo[*b*]pyrylium salts (chromylium salts) whose oxygen heteroatom is in a "phenolic" position are unreactive toward ammonia,*

---

* However, benzo[*c*]pyrylium salts, whose oxygen heteroatom is not in a "phenolic" position, do react with ammonia affording isoquinolines by O → N exchange.[98,513–526] Also, [*c*]annelated heteroaromatic systems do not cancel the reactivity of pyrylium salts toward ammonia.[516,527–532] However, benzopyrylium systems or pyrylium rings condensed with het-eroaromatics are outside the scope of this review, and will not be discussed in more detail.

their 5,6,7,8-tetrahydro derivatives react readily, affording the corresponding 5,6,7,8-tetrahydroquinolines **508**.[2,3,117,533-535] From variously substituted 5,6,7,8-tetrahydrobenzo[c]pyrylium salts the corresponding tetrahydroisoquinolines **509** have been obtained.[99,190,206,213,214,479,509,536-540]

Compounds with structures **510** (n = 3, 5)[102,541] and **511**,[190] octahydroacridines **512**,[534,542-545] the dihydrobenzoquinolines **513**[543,544] and **514**[130] as well as the tetrahydrodibenzoacridines **516**[116,543,544] and **517**[130] are examples for tri- and pentacyclic systems which can be readily obtained from simple starting materials. This ready availability of such partly hydrogenated systems makes them attractive as starting materials for obtaining totally aromatic condensed heterocycles as demonstrated by Schroth, Fischer, and Rottmann[130] who dehydrogenated **514** to substituted benzo[f]quinolines **515**, and **517** to dibenzo[a,h]acridines **518**.

(508)   (509)   (510)   (511)

(512)   (513)   (514)   (515)

(516)   (517)   (518)

Pyrylium salts **519** with an α-methoxy group react with ammonia to give 2-pyridones **520**.[103] The pyrylium salt **521** leads on treatment with ammonia to 3,6-diphenylcopyrine (**522**).[231] Starting from pyrylium salts with a condensed steroid skeleton, Dorofeenko *et al.*[110,546] obtained pyridines of types **523–525** (cf. Tables XXIV and XXV, Appendix, Section VII). Also more complex systems like **526** and **527** which contain more

(519)   (520)

(521)   (522)

(523)   (524)

(526)

(525)

(527)

than one steroid unit in the same molecule can be obtained from the corresponding pyrylium or bispyrylium salts.[111,112]

In all of the above cases (except the deuterated pyridines) the intermediate pyrylium cation was the direct product of a two- or three-component synthesis. Pyrylium salts with modified side chains (cf. Section

III,A,2), e.g., by condensation with aldehydes to styrylpyrylium salts **528**, can also be converted to the corresponding pyridines.[547]

R	Me₂N	MeO	HO	HO
R′	H	H	MeO	HO

(528)

The O → N exchange in the pyrone series leading to pyridones[548] is as important as the Baeyer synthesis of pyridines from pyrylium salts. Thus many 2- or 4-pyrones[549-554] were converted by ammonia to the corresponding pyridones **529** and **530**, respectively. Analogous ring transformations are known for pyranthiones.[28,446,555-558] Classical examples for the formation of pyridones from pyrones are, among others, the reactions between ammonia and derivatives of coumalic acid (**531**),[559-562] comanic acid (**532a**),[563] comenic acid (**532b**),[564] oxycomenic acid (**532c**),[564] chelidonic acid (**532d**),[565] and kojic acid (**533**).[566,567] Pyridones are formed by treating not only pyrones but also acetoxypyrylium salts with ammonia.[568]

(529)   (530)

(531)   (532)

	R	R′
a	H	H
b	OH	H
c	OH	OH
d	H	COOH

(533)

The 4-methylenepyrans **534** which are related to 4-pyrones or to 4-alkoxypyrylium salts (cf. Section III,A,5,a) are converted by ammonia to pyridines **535**.[2,3,224,569] Formamide can act in this reaction as NH₃ donor.[570] However, Simalty and co-workers[402,403] showed that 4-(carbonylmethylene)pyrans **537** ring-opened to the intermediate **538** (or to tautomeric forms thereof) which does not cyclize normally by a 2,6-linkage to the expected 2,6-diarylpyridine **536** but cyclizes instead through a 2,4-[C₅ + N] transformation via **539** (or its tautomers) affording the isomeric α-methylpyridine **540**. This reaction constitutes additional support for the reaction mechanism of the Baeyer pyridine synthesis.

(534)   (535)   (536)

(537)   (538)   (539)   (540)

(ii.) Primary Amines. Analogously to the 2,6-[C_5 + N] transformation of pyrylium salts with ammonia, their reaction with primary amines
RNH_2 constitutes an important synthesis of 1-substituted pyridinium salts
544 which for R = Alk or CH_2Ar is an alternative to the quaternization
of pyridines. The R group can be alkyl, aralkyl, hetarylalkyl, aryl, or
hetaryl, as seen in Table XXVI (Appendix, Section VII). This reaction
which was also discovered by Baeyer and Piccard[44] allows a wide variation of pyrylium salt and primary amine structures. It has therefore
found as large a synthetic application as the O → N exchange with
ammonia.

In the reaction of aqueous methylamine with 2,4,6-triphenylpyrylium
perchlorate, Susan and Balaban[408] obtained carbon tetrachloride extracts
which showed in the IR spectrum carbonyl stretching bands which vanished in a few minutes; this evidence indicated the intermediate formation
of a vinylogous amide **542**. Toma and Balaban[407] had observed earlier
that the reaction stopped at this stage if the primary amine was *t*-bu-
tylamine, due to steric hindrance toward pyridinium formation.

More recently, Katritzky and co-workers (cf. Ref. 13), in connection
with their new method for converting primary amines to other functionally substituted compounds (see below), investigated in more detail the
factors which influence the conversion of pyrylium into pyridinium salts:
acid–base catalysis, solvent and substituent effects, etc. From kinetic
data obtained by ^{13}C-NMR[409] and UV methods[410] it follows that the formation of the acyclic intermediate (e.g., **542**) takes place rapidly and is
base-catalyzed, whereas the cyclization to **544** is acid-catalyzed and constitutes the rate-determining step. One may associate the base-catalyzed

step with the deprotonation of an intermediate (e.g., **541**) while the acid-catalyzed step is probably associated with the dehydration of the cyclized dihydropyrydinium ion **543**. The rate of the cyclization step is furthermore strongly influenced by the nature of the solvent (Me$_2$NCHO : MeCN : CH$_2$Cl$_2$ give relative rates 1 : 20 : 270; the divinylogous amide from triphenylpyrylium and cyclohexylamine does not cyclize at 20°C in DMSO at all, but cyclizes in chloroform or other such solvents), and the nature of the amine (RCH$_2$NH$_2$: RR'CHNH$_2$: PhNH$_2$: p-O$_2$NC$_6$H$_4$NH$_2$ have rates relative to n-BuNH$_2$ of 1.3–0.7 : 0.01–0.002 : 0.02 : 0.0007). Thus, optimum conditions for preparing pyridinium salts are as follows: 1 mol each of pyrylium salt, RNH$_2$, and NEt$_3$ are stirred for 5 min in CH$_2$Cl$_2$ or CHCl$_3$, then 2 mol of AcOH are added, and after another 15 min at room temperature diethyl ether is added to precipitate the pyridinium salt.[410]

In the following we shall draw attention to some specific data on the O → N exchange by primary amines, and to some subsequent reactions of preparative interest.

In the reaction of primary aliphatic amines with pyrylium salts possessing α-oriented ethyl or methyl side chains, along with O → N exchange reactions, a benzene ring closure is also possible leading to N-alkyl-3,5-xylidines[407,571] as will be seen below in the reaction of such pyrylium salts with secondary amines where the latter reaction becomes the main one.

After Sammes and Yip,[572] methylamine reacts with 4-methoxy-2,6-diphenylpyrylium perchlorate (**122**) affording not only the known pyridinium salt **545**, R = Me (α-attack)[28,507] and its subsequent product **546**, R = Me (α- and γ-attack),[28] but also the iminopyran salt **547**, R = Me (γ-attack). With p-substituted anilines as primary amines R'C$_6$H$_4$NH$_2$, the course of the reaction is strongly dependent on the nature of the R'

substituent[572]: electron acceptor groups like NO_2 and Ac yield exclusively
547 (R = R'C_6H_4), while in the series R' = Br, Cl, F, H, Me, MeO the
fraction of pyridinium salt **545** (R = R'C_6H_4) increases from 39 to 68%.
With excess aniline, the formation of cations **546** (R = Ph) is favored.[28,572]
2,6-Diphenyl-4-(N-piperidino)pyrylium perchlorate forms with methylamine the expected 1-methyl-2,6-diphenyl-4-(N-piperidino)pyridinium
perchlorate by α-attack.[228]

(122)   (545)   (546)   (547)

(548)   (549)   (550)   (551)

The incorporation of amino acids (or their derivatives) leading to *N*-
carboxyalkylpyridinium salts, e.g., **548–550**, was thoroughly investigated
in Balaban's[94,573,574] and Dorofeenko's[575–579] laboratories. Such reactions
lie at the basis for using pyrylium salts as selective reagents for the
chemical modification of the terminal amino groups in proteins.[580]

From primary aliphatic and aromatic diamines bispyridinium salts of
type **551** are formed; for details see Table XXVII (Appendix, Section
VII).

The conversion of pyrylium salts, e.g., **159**, to 1-substituted pyridinium
derivatives succeeds also when instead of primary aryl amines the corresponding azomethines with aromatic aldehydes are employed: the cations **552** and **553** (among other forms) have been postulated as intermediates to explain the elimination of PhCHO.[500,581] Analogous reactions
were observed with phenyl isothiocyanate PhN=C=S[500] and with sul-

finylanilines ArN=S=O.[582] On treating 2,4,6-triarylpyrylium salts **381** (R^1–R^3 = Ar), however, with arylsulfenylamides the primarily formed pyridinium salts **554** are not isolable but react with excess arylsulfenylamide to give pyridines **555** and diaryl disulfides.[241]

(159) (552) (553)

(381) (554) (555)

From the multitude of 1-hetaryl-substituted pyridinium salts obtained via pyrylium salts, compounds **556**,[583] **557**,[584] and **558**[578] represent examples of carbon-bonded hetaryl groups; compound **558** may deprotonate to the betaine **559**.[578] The reaction of pyrylium salts with N-amino nitrogen heterocycles, studied by Katritzky and Suwinski,[585,586] represents an elegant synthesis of N,N'-bonded bishetaryl monocations, e.g., **560–563** (see also Ref. 587).

As seen from X-ray diffraction data,[341,588] and from the atropisomerism[187] observed by means of chemical shift and nonequivalence of the two methyls in the isopropyl group of **564**, N-aryl groups of pyridinium salts are more or less tilted out of coplanarity, therefore they shield magnetically the protons bonded to the α-oriented carbons due to the ring current in the aryl group.[179,186,407] The difference of chemical shifts between α-and γ-methyl protons in the ¹H-NMR spectrum of 1-R-2,4,6-trimethylpyridinium salts **565** may therefore be used, after Balaban et al.[589] as a measure for the existence and magnitude of the ring current in the cyclic substituent R (cf. Section IV,A,2,a). Due to its ready formation from easily accessible 2,4,6-trimethylpyrylium perchlorate (**286**) and a primary amine RNH_2, system **565** offers advantages over system **566** which had been used earlier[590] for a similar evaluation of ring currents in groups R, since the synthesis of **566** is less simple.

(556)   (557)   (558) $-H^+$→   (559)

(560)   (561)   (562)   (563)

(564)   (286) $\boxed{R}-NH_2$→   (565)

(566)   (567)

As found by Balaban et al.,[591] the two α-methyl groups in pyridinium salts **567** (R = Me, Et, Ph; R′ = H, Ac, Ts) give rise to two distinct signals which coalesce reversibly on heating, indicating rotation barriers $\Delta G^{\ddagger}$ of 15.0 (R = Ph), 17.0 (R = Me), and 19.3 kcal/mol (R = Et), with little or no influence exerted by varying the R′ group. Of course, due to ring current effects, with R = Ph the chemical shift differences between the two methyl groups is much larger (0.80 ± 0.05 ppm) than with R = Me or Et (0.05–0.09 ppm).

Appropriately substituted 1-arylpyridinium salts are able to undergo interesting deprotonation reactions. Thus Dilthey et al.[592–595] obtained deeply colored anhydrobases of types 569 and 571 starting from 4- and 1-(p-hydroxyphenyl)pyridinium salts 568 and 570, respectively. Wizinger and Wenning[596] prepared from 572 the betaine 573 which is blue in benzene solution. As found by Dimroth et al.[597–602] pyridinium-N-phenolbetaines 571 present the strongest solvatochromy yet observed, extended over the whole visible spectrum, on varying the solvent polarity, and are therefore useful for the characterization of solvent polarities (the E_T empirical parameter is the energy corresponding to the electronic transition of the largest-wavelength band measured for 571, Ar = Ph, R = t-Bu or Ph in the appropriate solvent).

(568) (569) (570) (571).

(572) (573)

The pyridinium salts 574 (R = Me, Ph) obtained from indeno-[1,2-b]pyrylium perchlorate (474) were deprotonated by Boyd,[205] analogously to the thiopyrylium salt 475 (cf. Section III,C,3,b) yielding deeply colored pseudoazulenes 575 (cf. also Section III,A,7,a).

If both the pyrylium salt and the primary arylamine have adequate substituents in ortho positions, cyclizations may occur in the pyridinium cation after the O → N exchange reaction. Thus Dimroth and Odenwälder[603]

(474)   (574)   (575)

obtained in one step benzo[c]quinolizinium salts **577** from pyrylium salts
of type **576** and o-aminobenzaldehyde in acetic acid. Similarly, pyrylium
salts **578** (R = H, Me) afford the pyrido[1,2-a]quinoxalinium salt **580**
with o-phenylenediamine through intermediate **579**.[268] In analogy to the
reaction course **521** → **522**, the reaction of **521** with primary amines leads
to 3,6-diphenylcopyrinium salts **581** (R = Me, cyclohexyl).[231]

(576)   (577)

(578)   (579)   (580)

(521)   (581)

By a photochemically induced dehydrocyclization, Dorofeenko and
co-workers[604] converted pyridinium salts **582** (R = Me, Ph, COOH) to
tetracyclic cations **583**. When R = Ph, Katritzky et al.[605,605a] observed,
however, a double photocyclization leading to **584**, and the same authors
similarly prepared compounds **585** (X = CH, CMe, CCOO$^-$). 1-Pyrid-
2-ylpyridinium salts **586** undergo a one-side photocyclization to **587**.[605]

(582)   (583)   (584)

(585)   (586)   (587)

A large variety of reactions of 1-substituted pyridinium salts results from the ability of the pyridine moiety to function as a leaving group in nucleophilic substitutions. For 1-alkyl-substituted pyridinium salts, the first observation in this respect is due to Ziegler and Fries[606] who discovered that the thermolysis of 1-methyl-2,4,6-triphenylpyridinium chloride affords 2,4,6-triphenylpyridine and (supposedly) methyl chloride. By careful investigations involving thermogravimetric analysis of the above chloride and the corresponding iodide, and by chemical trapping of CH_3Hal as CH_3HgHal (Hal = Cl, I) Susan and Balaban[408] brought additional proofs and predicted the synthetic usefulness of this bond cleavage–bond forming reaction, e.g., for converting alkyl- and benzylamines to the corresponding halides, of interest, for instance, for obtaining isotopically labeled iodides $R^{14}CH_2I$ from $^{14}CN^-$ via $R^{14}CN$, $R^{14}CH_2NH_2$, and $R^{14}CH_2\overset{+}{P}yPh_3I^-$. Dinculescu and Balaban[574] noted that N-(p-methoxybenzyl)-2,4,6-triphenylpyridinium perchlorate decomposes in excess trifluoroacetic acid at 70°C to 2,4,6-triphenylpyridine and p-methoxybenzyl trifluoroacetate by first-order kinetics with a half-life of 7 min, and that the decomposition rate of N-p-substituted benzyl cogeners increases with increasing donor capability of the para substituent in the benzyl group.

In recent years, Katritzky and co-workers (cf. Ref. 13) extended this reaction to other halides (Br, F) as well as to other amines (aryl, hetaryl, hetarylalkyl), and they systematically generalized this reaction principle to oxygen, sulfur, phosphorus, and carbon nucleophiles. Table III presents an overview of the extent of this synthetic concept. Details (prep-

TABLE III

APPLICATION OF THE 2,6-[C₅ + N] TRANSFORMATION OF PYRYLIUM INTO PYRIDINIUM SALTS FOR
THE CONVERSION OF PRIMARY AMINES INTO COMPOUNDS WITH OTHER FUNCTIONAL GROUPS[13]

No.	Substituent R of primary amine	Nucleophile Y	Reaction product R-Y	Reference
1	Alk, Ar-CH₂, Ar, Hetaryl	I⁻	R-I	408, 607, 608
2	Alk, Ar-CH₂, Ph-CH₂CH₂	Br⁻	R-Br	609, 610
3	Alk, Ar-CH₂	Cl⁻	R-Cl	609, 611
4	Alk, Ar-CH₂	F⁻	R-F	612, 613
5	Alk, Ar-CH₂, Ph-CH₂CH₂, Hetaryl-CH₂	R'-COO⁻	R'-COOR	614, 615
6	Alk, Ar-CH₂	NO₃⁻	R-O-NO₂	616
7	Ph-CH₂, Hetaryl-CH₂	Ar-O⁻	R-O-Ar	617-619
8	Alk, Ar-CH₂, Hetaryl-CH₂		R'-CHO	620
9	Alk, Ar-CH₂, Ph-CH₂CH₂, Ar	SCN⁻	R-SCN	621-624

TABLE III (continued)

No.	Substituent R of primary amine	Nucleophile Y	Reaction product R–Y	Reference
10	Alk, Ar–CH$_2$	$S=C \begin{smallmatrix} OEt \\ S^- \end{smallmatrix}$	$S=C \begin{smallmatrix} OEt \\ S-R \end{smallmatrix}$	624
11	Alk, Ar–CH$_2$	$S=C(NH_2)_2$	$R-\overset{+}{S}=C(NH_2)_2$	625a
12	Ph–CH$_2$, Hetaryl–CH$_2$	Ar–S$^-$	R–S–Ar	617, 618, 626
13	Hetaryl–CH$_2$	Ar–SO$_2^-$	R–SO$_2$–Ar	617
14	Alk, Ar–CH$_2$	$\begin{smallmatrix} CO \\ CO \end{smallmatrix} N^-$	$R-N \begin{smallmatrix} CO \\ CO \end{smallmatrix}$	627
15	Alk, Ph–CH$_2$	PhSO$_2$NR′$^-$	$\begin{smallmatrix} R \\ R' \end{smallmatrix} N-SO_2Ph$	627
16	Ar–CH$_2$, Hetaryl–CH$_2$	HNR′$_2$	R–NR′$_2$	617, 618
17	Ar–CH$_2$, Hetaryl–CH$_2$	NR′$_3$	$R-\overset{+}{N}R'_3$	617, 618
18	Alk, Ar–CH$_2$, Ph–CH$_2$CH$_2$	N$_3^-$	R–N$_3$	627
19	Ar–CH$_2$, Hetaryl	PPh$_3$	R–PPh$_3$	617, 618
20	Ph–CH$_2$	(EtOOC)$_2$CH$^-$ (EtOOC)(CN)CH$^-$	R–CH(COOEt)$_2$ R–CH(COOEt)CN	625b 625b
21	Alk, Ar–CH$_2$	$\begin{smallmatrix} R' \\ O_2N \end{smallmatrix} CH^-$	$R-CH \begin{smallmatrix} R' \\ NO_2 \end{smallmatrix}$	628

aration of pyridinium salts, reaction conditions for the formation of the R—Y bond, and the sterically produced acceleration of this reaction by special substituents of the pyridinium system) may be found in the original literature listed in Table III and in Katritzky's review,[13] mentioned in the introduction (Section I). It is particularly useful to start from a primary amine as a synthon (and by means of a 2,4,6-triarylpyrylium salt to convert the NH_2 group to a 2,4,6-triarylpyridinium group which functions as a leaving group in nucleophilic substitutions) when primary amines are easily available (e.g., as natural products), more stable or less toxic than the corresponding halogen or tosyl derivatives (e.g., it is preferable to start from ω-picolylamines than from the unstable and more toxic ω-picolyl halides). A further advantage is the higher selectivity of such substitution reactions (e.g., selective conversion of secondary into tertiary amines, Table III, No. 16, without danger of quaternary salt formation). Finally, this method allows reactions which otherwise are impossible such as the C-alkylation of nitroalkane RNO_2 anions (R = Me, Et, i-Pr; cf. Table III, No. 21).

Closely related to the reactions included in Table III is the reductive deamination of primary amines to hydrocarbons, i.e., the replacement of the amino group by hydrogen. Starting from 2,4,6-triphenylpyrylium perchlorate **(159)** methyl-, allyl-, benzyl-, and heterylamines are first converted to the corresponding 1-substituted 2,4,6-triphenylpyridinium salts **588** which are readily reduced by sodium borohydride to 1,2-dihydropyridines **589**. Thermolysis of the latter yields 2,4,6-triphenylpyridine and hydrocarbons RCH_3 in yields useful for preparative purposes.[629,630] Initially an electrocyclic mechanism was assumed for the cleavage step,[629,630] which ought to lead, in the case of the deuterium-labeled benzyl derivative **590**, to a mixture of ring- and side-chain-labeled toluenes **592** and **593** through the intermediate **591**; however, experimentally only **593** was obtained, indicating a radical mechanism.[583] On the other hand, for the cleavage of 1-allyl-substituted 1,2-dihydropyridines an electrocyclic mechanism cannot be excluded.[631]

Amines which would afford less stable radicals (primary alkyl- and arylamines) require conversion to 2,3,5,6-tetraphenylpyridinium salts **594** whose selective reduction by sodium borohydride affords 1,4-dihydropyridines **595**; pyrolysis of the latter yields the hydrocarbons RH corresponding to the amines RNH_2.[583]

An example for converting primary amines to alkenes is provided by the reaction sequence leading to **597** via **596** (R = Ph). This reaction type ("deammoniation" of primary amines $RCH_2CH_2NH_2$ to $RCH=CH_2$ under the action of **159** followed by treatment with bases) represents a promising alternative to the Hofmann degradation.[13] 1-(2-Hydroxyethyl)-2,4,6-triphenylpyridinium fluoborate **(596, R = OH)** is cleaved pyrolytically in the presence of potassium hydroxide to 2,4,6-triphen-

ylpyridine and ethylene oxide,[632] while the 1-(2-chloroethyl) derivative
596 (R = Cl) under similar conditions (heating with potassium *t*-butoxide
in dimethyl sulfoxide) eliminates hydrogen chloride affording a 2,4,6-
triphenyl-1-vinylpyridinium salt (**598**).[574,632] Such salts cannot be obtained
by quaternization.

In contrast to the 2,6-linkage in all O → N exchange reactions of
pyrylium salts with primary amines described so far, the action of cyanamide on 2,4,6-triarylpyrylium salt in the presence of triethylamine
leads via intermediates **599** and **600** to α-amino-β-aroylpyridines **601**, and
hence represents a 2,5-linkage (2,5-[C_4 + NC] synthesis)[633]; the recyclization includes two atoms from cyanamide instead of the "normal"
reaction course which would have only included one nitrogen atom leading to the energetically unfavorable 1-cyanopyridinium salts.

In the 4-pyrone series **5**, the reaction with primary amines RNH_2 yields
by a normal O → N exchange 1-substituted 4-pyridones **602** (R = al-
kyl,[565–567,634–639] hydroxyalkyl,[567] dialkylaminoalkyl,[567,640] carboxyalkyl,[567,638,641]

(599)   (600)   (601)

and aryl[555,556,565,634,639,642–644]). Analogous reactions are known for the
2-pyrone series.[645] Pyridones of type **602** have proved to be valuable
starting materials in syntheses, because their alkylation to **603**[639] or their
halogenation to **604**[639,646,647] affords reactive reagents which can be converted to many other pyridine derivatives.

(5)   (602)   (603)   (604)

The action of primary amines on 4-methylenepyrans **534** leads in several instances to "normal" O → N exchange, proceeding through a 2,6-
linkage,[329a,648–652] but in other cases, analogously to the reaction sequence
537 → **540**, gives ring transformations including an α-methylene side-
chain carbon. Thus, for example, the reaction of the methylenepyran **605**

with alkylamines RNH_2 (R = Me, *n*-Bu, $PhCH_2$) leads, according to the group R and the reaction conditions, to different reactions products. Van Allan and co-workers[352] obtained at 100°C from all three amines indicated above a mixture of two pyridine methides **608** and **609**, the latter being formed by normal recyclization of the ring-opened intermediate **606** through a 2,6-linkage, and the former by 2,4-linkage via **607** incorporating a cyano carbon into the ring (2,4-[C_5 + N] transformation). At higher temperatures (150–180°C), for R = Me, only the 2-pyridone **610** results, probably by solvolytic cleavage of the pyridone methide **608** which is formed preferentially under these conditions; for R = $PhCH_2$ with excess benzylamine the pyridone methide **608** undergoes ring closure to the condensed bicyclic system **611**; on the other hand, *n*-butylamine (R = *n*-Bu) reacts with the pyrone methide **605** at 150°C, affording exclusively the pyridone methide **609**.

A 2,4-[C_5 + N] transformation corresponding to the reaction sequence **605** → **606** → **607** → **608** was also observed by Belsky, Dodiuk, and Shvo[653] who treated the methylenepyran **612** with *n*-butylamine to obtain the 2-pyridone derivative **613**.

(iii.) Secondary Amines. Diels and Alder[411] discovered that the reaction of 2-methylpyrylium salts **614** bearing various substituents R^2, R^3 in positions 4 and 6 with dialkylamines like dimethylamine or piperidine

(612)   (613)

yields 3,5-disubstituted *N,N*-dialkylanilines **616**.* This reaction parallels
the Baeyer phenol synthesis described in Section III,C,3,a. As mentioned
in the previous subsection, an analogous ring transformation also appears
as a side reaction on treating pyrylium salts with primary amines along
with the formation of 1-alkylpyridinium salts; Toma and Balaban[407]
showed that 2,4,6-trimethylpyrylium reacts with primary arylamines
yielding almost exclusively pyridinium salts, but with alkylamines a fair
yield of xylidines is obtained (15–40% with alkylamines RCH_2NH_2 where
pyridinium yields are 60–80%; with cyclohexylamine, however, the xy-
lidine is formed as the. main product in 75% yield and only a 7% yield

(614)   (615)   (616)

(122)   (209)   (210)   (617)

of pyridinium salt is obtained (cf. also Ref. 407). Dorofeenko and co-
workers extended this little used 2,6-[C_6] synthesis to vinylogously substituted pyrylium salts **614** [R^2 and/or R^3 = $Ar(CH=CH)_n$ where n =
1,2][222] and used also indoline[508] as secondary amine. Pyrylium salts of
type **405** (R = Me) react under simultaneous halogen substitution, yield-

---

* The reverse reaction **615** → **614** corresponds in principle to a pyrylium synthesis
described by Schroth and Fischer[128,129a] [acid-catalyzed cyclodeamination of ketovinylenamines which can be obtained from enamines and β-chlorovinyl ketones (cf. Section
II,B,2,b)].

ing *N,N*-dialkyl-3-(*N,N*-dialkylamino)-5-methylbenzylamines.[654] 4-Methoxy-2,6-dimethylpyrylium perchlorate (**122**) is converted through the isolable intermediates **209** and **210** (cf. Section III,A,6,c) to substituted *m*-phenyllendiamines **617**, where the two dialkylamino groups may be equal or different.[228]

The reaction of 2-(2-dimethylaminovinyl)-4,6-diphenylpyrylium perchlorate (**462**) with aqueous dimethylamine proceeds after Reynolds and Van Allan[230] as a 2,5-[C$_6$] synthesis over the nonisolable intermediates **618a**, **618b**, and **619** yielding 4-dimethylamino-2-phenylbenzophenone (**471**); the same compound was obtained along with other products on treating **462** with alkali hydroxides, as mentioned earlier in Section III,C,3,a; as seen in the scheme, the reaction involves an electrocyclic reaction of the acyclic intermediate **618a**. Analogous conversions of **462** were observed on treating with piperidine and morpholine.[230]

On refluxing piperidine with the methylenepyran **605**, Van Allan and co-workers[352] observed a mixture of the 2-(*N*-piperidino)pyridine **621** as main product, and of the enamine **622**. As a common intermediate an acyclic adduct **620** may be postulated. Its 2,4-[C$_5$ + N] reaction yields **621**, while its 4,6-[C$_5$ + N] transformation affording **622** is so far the only known example of a 4,6-linkage. Compound **621** may react with excess amine yielding the bicyclic compound **624**, while **622** gives by acid hydrolysis pyridone **623** which is able to undergo other cyclizations.[352]

(iv.) Hydroxylamine. The conversion of 2,4,6-trisubstituted pyrylium salts to pyridine *N*-oxides by reaction with hydroxylamine (a normal O → N exchange) was independently discovered by Schmitz[655] and by Balaban and Nenitzescu.[94] The latter authors found that with bulky α-groups like *i*-Pr and Ph the product is the pyridine rather than the pyridine *N*-oxide. For α-aryl-substituted pyrylium salts, this latter reduction as

well as the competing formation of isoxazolines which was described
earlier in Section III,C,2,c can be suppressed by careful selection of
reaction conditions especially in the case of highly aryl-substituted pyrylium salts. Thus, Pedersen, Harrit, and Buchardt[463] obtained, for instance, from 625 (R = H) and hydroxylamine in acetic acid (AcOH/
NaOAc) 51% pyridine N-oxide 626 (R = H), along with 21% 341 and
24% 627, while the pentaphenylpyrylium salt 625 (R = Ph) affords under
the same conditions almost quantitatively the N-oxide 626 (R = Ph). As
shown by comparing the yields, this synthesis of pyridine N-oxides by
O → N exchange of aryl-substituted pyrylium salts with hydroxylamine
gives better results than the N-oxidation of the corresponding pyridines.
Since the latter are ordinarily prepared from pyrylium salts, the direct
introduction of the NO group represents the most rational approach.
Table XXVI (Appendix, Section VII) includes pyridine N-oxides obtained from pyrylium salts in the form of their conjugated acids, i.e., N-
hydroxypyridinium salts.

The analogous reaction of 4-pyrones leading to 1-hydroxy-4-pyridones
have been known since the last century.[634,656,657] The 1-hydroxy-2-pyri-
done 629 obtained similarly[658] from 4,6-diphenyl-2-pyrone (628) reacts
(in the form of its sodium salt) with halides RCH$_2$Hal affording crystalline

1-alkoxypyridones **630*** which may be isolated and cleaved pyrolytically to aldehydes and 4,6-diphenyl-2-pyridone (**631**). This method developed by Katritzky and co-workers[659] allows the conversion of halides RCH_2Hal into carbonyl derivatives RCHO in neutral and nonoxidizing media, and provides therefore an attractive alternative to known methods for performing such conversions, especially in the case of sensitive R groups. The analogous use of the sodium salt of **629** for converting primary amines via pyridinium salts into aldehydes[620] has been mentioned above in Table III (No. 8).

(628)   (629)   (630)   (631)
                                                   +
                                                 R—CHO

(v.) Hydrazine Derivatives. Hydrazines can react with pyrylium salts not only bifunctionally, according to Section III,C,2,c, with ring contraction to a pyrazoline or, according to Section III,C,4,a, as will be seen further, with ring enlargement to a 1,2-diazepine, but also monofunctionally like a primary amine. The 1-aminopyridinium salts **633** thus formed by a normal O → N exchange via **632** were first isolated by Schneider and co-workers.[416,464] When they treated pyrylium salts with phenylhydrazine, the initially formed "α-pyranolhydrazide," whose structure was later proved[414] to be **632**, cyclized on refluxing in acetic acid to 1-anilinopyridinium salt. The structure of acyclic intermediates and alternative reaction pathways are described in more detail in Sections III,B,3,c and III,C,2,c.

(632)   (633)   (634)

Analogous reactions are known with variously substituted phenylhydrazines, monoalkylhydrazines, N,N-disubstituted hydrazines, acid hy-

---

* Compounds of this type can also be obtained in one step by treating pyrylium salts or pyrones with the corresponding alkoxyamines.[389]

drazides as well as hydrazine itself; 1-aminopyridinium salts of type **633** obtained in this way are included in Table XXVI (Appendix, Section VII). The separation of side products which are not salts (e.g., 2-pyrazolines, pyrazoles, 1,2-diazepines, or acyclic products) is easy because of their solubility in nonpolar solvents like ether, which do not dissolve pyridinium salts; the separation of pyridinium salts from unchanged pyrylium is avoided by using an excess of the hydrazine derivative; when the organic base is precious and one suspects traces of unchanged pyrylium salt, these traces are converted by ammonia to the corresponding pyridine which is soluble in acids or in nonpolar solvents.

By deprotonation of salts possessing structure **633** (R = Ar, R' = H) with alkali hydroxides, Schneider and co-workers[415,418,464] obtained deeply colored anhydrobases which are N^+,N^- betaines **634**.[660] Due to their polar character, similarly to the pyridinium-N-phenol betaines **571**, compounds **634** present a pronounced negative solvatochromy.[660] N^+,N^- Betaines of type **635** with ortho-oriented methyl groups undergo in solution on heating, more or less rapidly according to their substitution pattern, an electrocyclic rearrangement to 2-(2-aminobenzyl)pyridines **636**.[417,418,660]

(635)   (636)

An interesting reaction of such N^+,N^- pyridinium betaines is their photochemical conversion, through the intermediacy of diazanorcaradiene valence isomers **637**, to 1,2-diazepines **638**.[661-664] This reaction is reminiscent of the photochemical conversion of pyridine N-oxides to 1,2-oxazepines.[665]

3,4,5-Trichloropyrylium salts (**639**) yield, on treatment with 2,4-dinitrophenylhydrazine, after hydrolysis of the γ-chloro substituent, N-(2,4-dinitrophenylamino)-3,5-dichloro-4-pyridone [**640**, Ar = 2,4-$(NO_2)_2C_6H_3$].[666]

Dorofeenko et al.[500] showed that phenylhydrazine may be replaced in the reaction with 2,4,6-triphenylpyrylium perchlorate (**159**) by benzaldehyde phenylhydrazone yielding a 1-anilino derivative of type **633**. The same pyrylium salt with benzalazine, however, afforded 2,4,6-triphenylpyridine and benzonitrile; under the reaction conditions (heating in

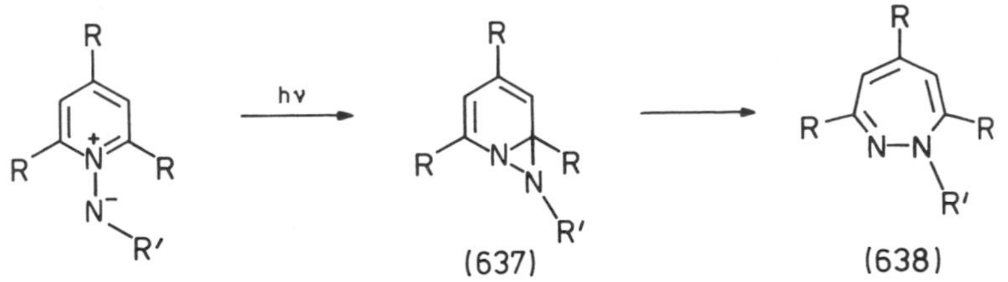

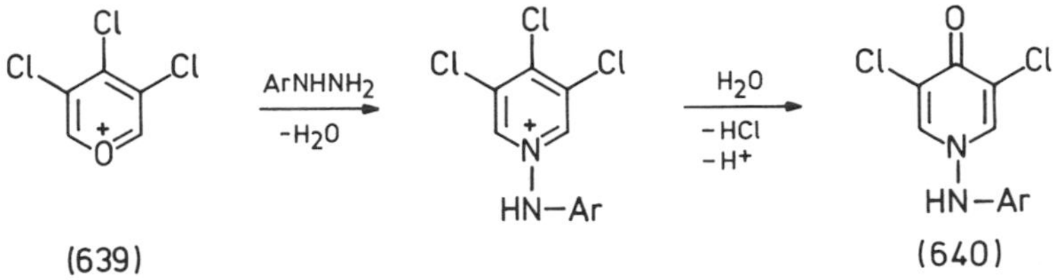

(637)   (638)

(639)   (640)

dimethylformamide), the intermediate nonisolated pyridinium salt **641** is easily cleaved.

A similar N—N bond cleavage was put to use by Katritzky and co-workers[667,668] for the synthesis of nitriles from aldehydes on a preparative scale: the 1-amino-4,6-diphenyl-2-pyridone (**643**) obtained from 4,6-diphenyl-2-pyrone or -thiopyrone (**642**, X = O, S) and hydrazine, was converted by aldehydes to the aldimine **644**, whose pyrolysis afforded in high yields nitriles along with 4,6-diphenyl-2-pyridone (**631**).

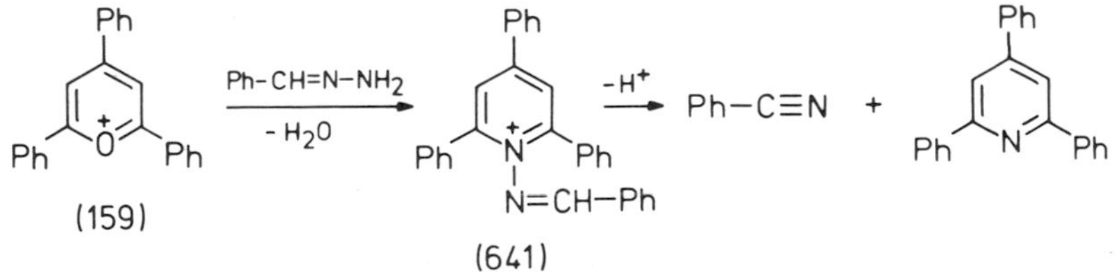

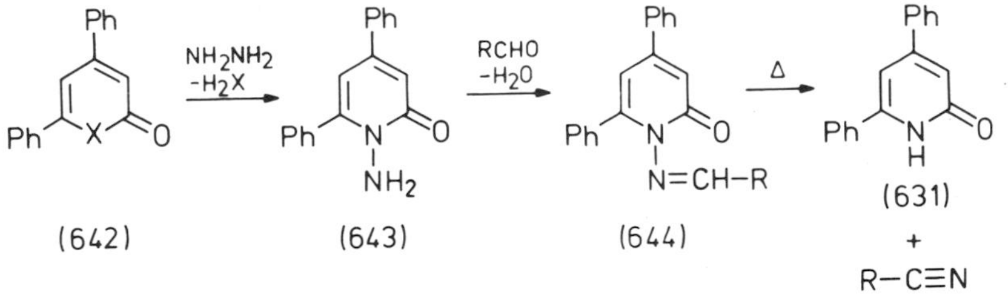

(159)   (641)

(642)   (643)   (644)   (631)

A further illustrative example for the synthetically useful coupling of ring transformations with subsequent pyrolytic reactions from Katritzky's laboratories[471,668-670] is provided by the reaction of 2,4,6-tri-phenylpyrylium perchlorate with carboxylic acid hyrazides or amidra-zones, yielding salts of type **646** (X = O, NAr), which can be deprotonated to acyl-*N*-imines **647** (X = O) and imidoyl-*N*-imines **647** (X = NAr), respectively. Pyrolysis of these compounds affords in high yields iso-cyanates **648** (X = O) and diarylcarbodiimides **648** (X = NAr), respec-tively. The intermediate pyridinium salts **646** can also be obtained by acylation of 1-amino-2,4,6-triphenylpyridinium perchlorate (**645**). This new method for converting acid chlorides or hydrazides into isocyanates represents an alternative to the Curtius reaction.

Similarly to pyrylium salts, pyrones can also react ambifunctionally with hydrazines, yielding either pyrazole derivatives (cf. Section III,C,2,c) or six-membered nitrogen heterocycles. Thus, for example, the reaction of kojic acid (**533**) with anhydrous hydrazine results in a mixture of the pyridazine derivative **651** and the pyrazole **652**.[671] The latter results by 2,4-linkage of the common acyclic intermediate **649** through a 2,4-[$C_3 + N_2$] reaction, the former via **650** through 2,5-linkage (2,5-[$C_4 + N_2$] synthesis). Analogous mixtures of pyrazoles and pyridazines result similarly from pyromeconic acid and its 6-methyl derivative (allomaltol).[671]

In the case of the methoxy derivative **653** of kojic acid a 2,5-linkage is no longer possible because the keto–enol equilibrium allowing the formation of a compound analogous to **650** is blocked; therefore in this case the formation of the pyrazole **654** is accompanied by a normal 2,6-[$C_5 + N$] reaction giving the 1-aminopyridine **655**.[671,672] Similarly, a nor-mal O → N exchange was observed in the reaction between hydrazine and 4-methylenepyrans of type **534** (R^1 = Me, R^2 = CN, R^3 = CN or COOEt).[650]

(vi.) *Further Nitrogen Nucleophiles.* A 2,4-[C$_3$ + NCN] synthesis was discovered by Zhdanova, Zvezdina, and Dorofeenko[673] when they treated 2,4,6-triphenylpyrylium perchlorate (**159**) with guanidine. The acyclic intermediate **656** recyclizes with 2,4-linkage to the cyclic intermediate **657**, which, in close analogy to the reactions of phenacyl-2-isoxazolines and phenacyl-2-pyrazolines described above in Section III,C,2,c, aromatizes by cleavage into acetophenone and 2-amino-4,6-diphenylpyrimidine (**658**) which reacts further with excess pyrylium affording the 2-pyrimidinyl-*N*-pyridinium salt **659** as isolated final product. If amidines[674] or alkylisothioureas[675] are used instead of guanidine the reaction yields, as expected, pyrimidines **660** (R = Me, Ph, MeS, PhCH$_2$S). According to the same principle the reaction of 2,4,6-triphenylpyrylium perchlorate (**159**) with 2-aminobenzimidazoles **661** yields pyrimido[1,2-*a*]benzimidazolium salts **662**.[676,677] In such syntheses the pyrylium salt functions as a synthesis equivalent of a 1,3-diketone (see also reactions with hydroxylamine, phenylhydrazine, and benzylmagnesium halides, Sections III,C,2,a, c, and III,C,3,e, respectively).

Under drastic conditions (250 h at 215°C) the pyrone derivative **663** reacts with benzonitrile yielding the pyridine derivative **664**.[678] Although the mechanism of this reaction is unknown, it can be interpreted as indicated in Scheme 10. Formally this reaction corresponds to a 2,5-[C_4 + NC] transformation with elimination of carbon dioxide.

The spontaneous rearrangement of the pyrylium salt **665** (on heating above 165°C or on dissolving in chloroform) which yields the 2-thiopyridone derivative **666**[679] represents also a 2,5-[C_4 + NC] transformation. Here the ring interconversion is initiated by nucleophilic α-attack of the thiocyanate anion.

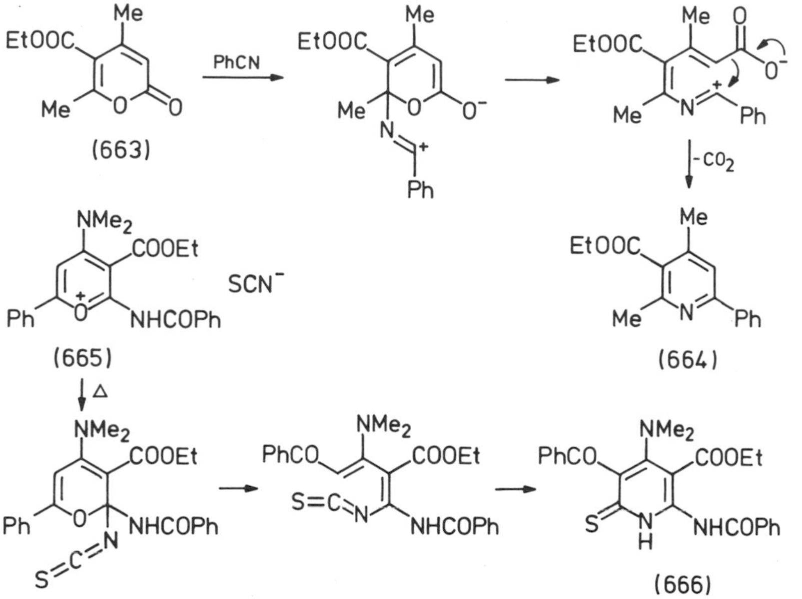

SCHEME 10

d. *Reactions with Phosphorus Nucleophiles.* Under certain definite reaction conditions, pyrylium salts react with phosphine derivatives after the 2,6-[C$_5$ + P] synthesis scheme leading to λ^3-phosphorins (phosphabenzenes). Thus by reaction of 2,4,6-triphenylpyrylium perchlorate with tris(hydroxymethyl)phosphine, P(CH$_2$OH)$_3$, in boiling pyridine, Märkl[680] obtained 2,4,6-triphenylphosphabenzene (**667**), the first representative of a new heterocycle with dicoordinated phosphorus. Under similar conditions a whole series of 2,4,6-tri- and higher substituted pyrylium salts[681,682] was converted by means of tris(hydroxymethyl)phosphine to λ^3-phosphorins which are listed in Table XXVIII (Appendix, Section VII).

When P(CH$_2$OH)$_3$ is employed as the phosphorus nucleophile, the irreversible step of the reaction sequence is the elimination of water, formaldehyde, and protons (these protons are trapped by the basic solvent), as shown in Scheme 11.

On the other hand, if the pyrylium salts are iodides and if the potential PH$_3$ is introduced as tris(trimethylsilyl)phosphine, P(SiMe$_3$)$_3$, then only Me$_3$SiI and (Me$_3$Si)$_2$O are eliminated and there are neither water nor protons to be trapped (Scheme 11). The yields in this variant of the method are higher than when P(CH$_2$OH)$_3$ was used.[683]

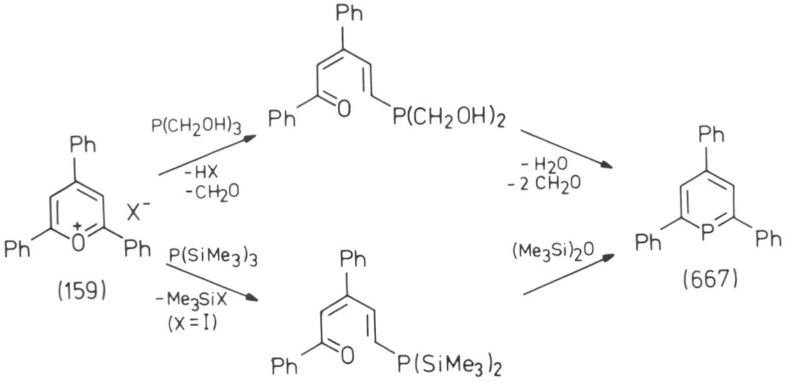

Ph

P(CH₂OH)₃

Ph   O   P(CH₂OH)₂

Ph

−HX
−CH₂O

−H₂O
−2 CH₂O

Ph

Ph   O   Ph
        X⁻

(159)

P(SiMe₃)₃

−Me₃SiX
(X = I)

Ph

(Me₃Si)₂O

Ph   P   Ph

(667)

Ph   O   P(SiMe₃)₂

SCHEME 11

Phosphine itself, PH₃, is too weak a nucleophile to attack the carbonyl group of the acyclic intermediates and give a spontaneous irreversible cyclization. Therefore earlier attempts by Balaban[389] and by Dimroth[684] were unsuccessful. However, if this step is catalyzed by acids, e.g., by carrying out the reaction with phosphonium iodide, PH₄I, λ³-phosphorins are obtained in preparatively useful yields.[308] Unlike the reactions with P(CH₂OH)₃ and P(SiMe₃)₃, which succeed only with α-aryl or bulky alkyl substituents on the pyrylium ring, the method employing PH₄I has the advantage that it can be applied also to pyrylium salts possessing α- or γ-methyl substituents. In the latter case, P(CH₂OH)₃ adds at the less hindered γ-position giving a γ-pyran which is unable to cyclize (cf. Section III,A,6,d), while in the case of pyrylium salts with α-methyl groups the liberated formaldehyde gives unwanted condensations with the methyl groups.

λ³-Phosphorins are much less basic than the corresponding pyridines, hence their inability to be protonated by nonoxidizing acids like trifluoroacetic acid, and the failure to isolate phosphorus analogs of N-alkylpyridinium ions (such ions, e.g., **673**, appear only as reaction intermediates). The lower electronegativity of phosphorus than that of nitrogen, the different orbital energies of λ³-phosphorins and pyridines, as well as the high tendency of phosphorus to become tetracoordinated leading to λ⁵-phosphorins, are the main features of λ³-phosphorins.

In Section III,C,3,e a 2,6-[C₅ + P] synthesis with subsequent 2,6-[C₅ + C] transformation will be discussed in more detail.

2,6-Diphenylpyrylium salts **668** with the 4-substituent H, MeO, COOR', etc. fail to undergo an O → P exchange but yield 2,2',6,6'-tetraphenyldipyrylene (**221**) instead[685] (cf. Section III,A,6,d). This compound results also from the easily reducible bipyrylium dication **669** under the action

of PR$_3'$ (R' = H, CH$_2$OH, SiMe$_3$). No trace of the λ^3-phosphorin derivative
670 was detected; its formation was, however, possible only through
partly hydrogenated precursors.[685] The somewhat less easily reducible
dication **671** is, however, accessible to an O → P exchange affording **672**
in low yield.[685]

(668)   (221)   (669)   (670)

(671)   (672)

The reaction of 2,4,6-triphenylpyrylium salts (**159**) with phenylphosphine in pyridine affords after Price *et al.*[686] as main product a hydrate
of the λ^5-phosphorin **674a** formed by addition of the water arising from
the reaction to the intermediate cation **673** (R = Ph). Märkl *et al.*[687] have
shown that in such reactions alcohols, phenols, or thiols may compete
successfully with water forming 1-alkoxy-, 1-aryloxy-, or 1-alkylmer-
capto-λ^5-phosphorins **674b** or **674c**, respectively. The yields can be increased if instead of the primary phosphine, RPH$_2$, one uses bis(hydroxymethyl)phosphines, RP(CH$_2$OH)$_2$, which can be easily prepared
from RPH$_2$.

An independent approach yielding λ^5-phosphorins, i.e., 1,1-disubsti-
tuted phosphabenzenes, starts from 2,4,6-triphenyl-λ^3-phosphorin **667**
which reacts with organolithium compounds yielding adduct **675**. This
may be oxidized by mercury(II) acetate to **673** which then results in
674.[687] Quaternization of **675** yields phosphonium salts **676**. These can
be deprotonated to 1,1-disubstituted λ^5-phosphorins **677**.[688] When **675** is
oxidized by hydrogen peroxide, the resulting cyclic phosphinoxide **678**
is, according to UV evidence, in equilibrium with 1-hydroxy-λ^5-phos-
phorin **674a**.[688]

Reviews on syntheses and properties of λ^3- and λ^5-phosphorins were
written by Märkl,[689] Dimroth[684] as well as Mel'nikov *et al.*[690]

Ph, Ph, Ph structures:

(159) → (673) → (674)

RPH₂ ; R'XH

	X	R'
a	O	H
b	O	Alk, Ar
c	S	Alk

Hg(OAc)₂ | −2e, −H⁺

(667) → (675) → (676) → (677)

a) LiR b) H⁺ ; MeI ; −H⁺

H₂O₂

(678) ⇌ (674a)

e. *Reactions with Carbon Nucleophiles.* As shown by the Baeyer phenol synthesis (Section III,C,3,a) and by the Diels–Alder dialkylaniline synthesis (Section III,C,3,c), the oxygen heteroatom of pyrylium salts can be replaced not only by heteroatoms such as sulfur, nitrogen, or phosphorus, but also by carbon, leading to the closure of benzene rings. In the above ring transformations, this carbon originates in an α-oriented methyl or methylene group so that the external nucleophilic reagents (OH⁻ and NHR₂, respectively) function only as ring-opening and condensation agents.

Analogous 2,6-[C₆] transformations can also be initiated by organometallic compounds and by other carbon nucleophiles. Thus, Köbrich and Wunder[423,424] obtained from 2,4,6-trimethylpyrylium perchlorate (286) and p-dimethylaminophenyllithium a mixture of the ketone 354 (cf. Section III,B,4,b) and the biphenyl derivative 355 formed through its cyclodehydration; at higher temperatures, the latter compound 355 becomes the main product. Gompper and Christmann[691] obtained by analogous reactions from 286 and Grignard reagents benzene derivatives 679 (R = Me, Ph, p-tolyl) while from 286 and malonitrile in the presence of triethylamine they obtained a similar benzene derivative 680. According to the same scheme 2-methyl-4,6-diphenylpyrylium perchlorate (114) reacts with 1-phenyl-3-methylpyrazol-5-one as a compound with an active methylene group in the presence of triethylamine to give compound 681.[330]

On treating 4,6-disubstituted 2-pyrones with Grignard reagents R'MgX the primary acyclic adduct 682 reacts with a second mole of R'MgX affording intermediate 683 which cyclizes to a benzene derivative 684 or 685 if the originally present R group, or the added R' group, is methyl.[692] In the former case (R = Me, R' = Ar) this ring transformation represents a 2,6-[C$_6$] synthesis, in the latter case (R = Ar, R' = Me) it represents a 2,6-[C$_5$ + C] synthesis. For the formation of 2H-pyrans via 683, see Section III,A,6,e.

According to Dimroth and co-workers[314,315,693] benzylmagnesium chloride and substituted derivatives thereof add to 2,4,6-trisubstituted pyrylium salts **381** forming isolable 4H-pyrans **686** (cf. Section III,A,6,e), which rearrange easily (e.g., on UV irradiation, see below) forming 2H-pyrans **687**. The latter are converted by bases to tetraarylbenzenes **688** ($R^1 = R^2 = R^3 = Ar$) in a 2,6-[C$_5$ + C] transformation. Under more drastic conditions (e.g., heating with calcium oxide), the 4H-pyrans **686** may also be converted directly to **688**. The above reactions can be extended to more highly substituted pyrylium and to thiopyrylium salts.

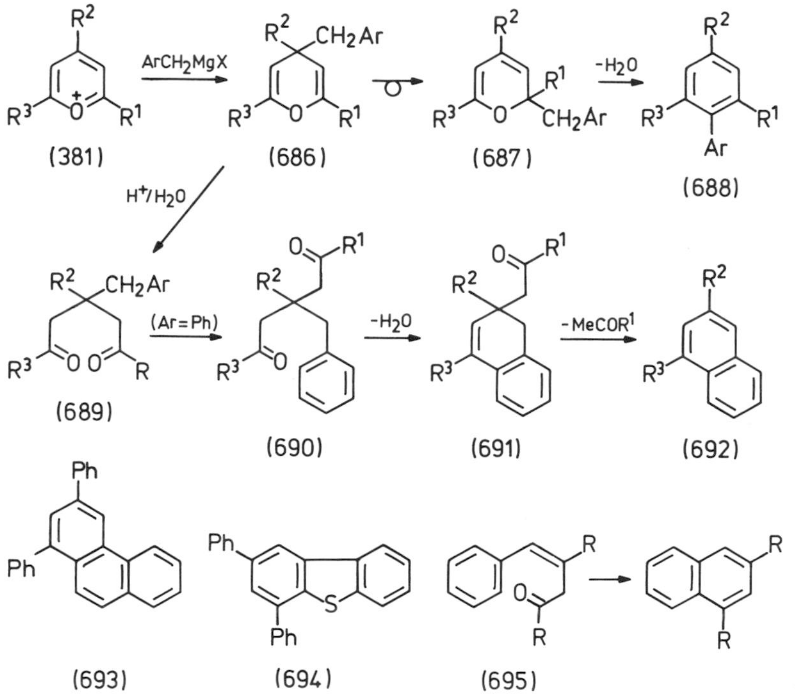

On treatment with 70% perchloric acid the 4H-pyrans **686** pass through the intermediate stages **689–691** and aromatize to naphthalene derivatives **692** by elimination of a methyl ketone. Analogously one may obtain, e.g., 1,3-diphenylphenanthrene (**693**) and 2,4-diphenyldibenzothiophene (**694**) from 2,4,6-triphenylpyrylium and the corresponding arylmethylmagnesium halide, by elimination of acetophenone as the methyl ketone.[648,694] 2,4,6-[2,6-[14]C$_2$]Trimethylpyrylium perchlorate leads to 1,3-[1-^{14}C]dimethylnaphthalene,[437] and the tri[D$_3$]methylpyrylium cation leads to 1,3-di[D$_3$]methylnaphthalene.[373] Since in this reaction the pyrylium cation functions as a potential 1,3-diketone, it was logical to develop a synthesis of 1,3-disubstituted naphthalenes from benzylmagnesium halides and 1,3-diketones. Balaban and Barabas[695,696] starting from this idea, and Canonne et al.,[697,698] starting from other considerations, independently developed such a synthesis; the monoaddition product of benzylmagnesium halide to a 1,3-diketone may be dehydrated to 4-phenyl-3-butenone derivative **695** which is then cyclized under acid catalysis.

The already mentioned photochemical rearrangement of 4H-pyrans **686** formed from 2,4,6-triarylpyrylium salts and benzylmagnesium halides to the corresponding 2H-pyrans **687** and the thermal conversion of the latter to benzene derivatives **688** was studied in more detail by Cuong, Fournier, and Basselier.[699] In the case of the 4H-pyran derivative **696** obtained from 2,3,4,6-tetraphenylpyrylium perchlorate (**312**) and benzylmagnesium chloride these authors found that on irradiation, besides the sigmatropic benzyl group migration, a photochemical linkage of the vicinal α- and β-phenyl groups takes place. The formed 2H-pyran **697** undergoes, on heating at 300°C, a normal 2,6-[C$_5$ + C] transformation, affording 1,2,4-triphenyltriphenylene (**698**).

An important preparative conversion of pyrylium salts to functionally substituted benzene derivatives was developed by Dimroth and co-workers[2-4] using as C-nucleophiles compounds with activated methyl(ene) groups bonded to electron-attracting substituents. 2,4,6-Trisubstituted pyrylium salts **248** react with nitromethane[700-702] in the presence of two equivalents of alkoxides (e.g., potassium t-butoxide) yielding, through nonisolable intermediates **699** and **700**, 2,4,6-trisubstituted nitrobenzene derivatives **701**. Thus, several difficultly accessible aromatic nitro compounds, and hence primary arylamines or other corresponding reaction and reduction products, may be obtained.[703-707] With the nitromethane residue an isotopically labeled carbon atom (e.g., ^{13}C) can be incorporated into the benzene ring in an exactly defined position.[708] The reaction succeeds also with pyrylium salts possessing more than three substituents,[2-4] or having an alkoxy,[228,329b] a methylmercapto,[228,329b] or a dialkylamino group[228] in the γ-position, as can be seen from Table XXIX (Appendix, Section VII). The analogous synthesis of [2,6-$^{14}C_2$]nitromesitylene from 2,4,6-[2,6-$^{14}C_2$]trimethylpyrylium perchlorate was described by Balaban and co-workers.[437]

Pyrylium salts with a secondary alkyl residue at the γ-position may undergo, in parallel to the ring transformation, a deprotonation reaction yielding a 4-methylenepyran. For example, from 4-cyclohexyl-2,6-diphenylpyrylium perchlorate with nitromethane under the above conditions one obtains 15% nitrobenzene derivative **702** and 80% γ-methylenepyran derivative **703**[324] (cf. Section III,A,7,b).

For certain combinations of substituents (e.g., R = p-C$_6$H$_4$CH$_2$OH, R' = Ph)[324] or with γ-unsubstituted pyrylium salts **248** (R' = H)[56,323] the nucleophilic attack occurs at the 4-position with formation of 4H-pyran derivatives **704** (cf. Section III,A,6,e), therefore 2,6-disubstituted pyrylium salts cannot be converted directly to nitrobenzene derivatives. However, 2,6-diphenylnitrobenzene (**708**) can be obtained from 4-(p-anisyl)-2,6-diphenylpyrylium fluoborate (**705**) through the corresponding nitrobenzene derivative **706**, followed by chromic acid oxidation to the corresponding carboxylic acid **707**, which is finally decarboxylated.[146] Analogously one can obtain 2- and 4-nitrobiphenyl and 2,4-diphenylnitrobenzene.[146]

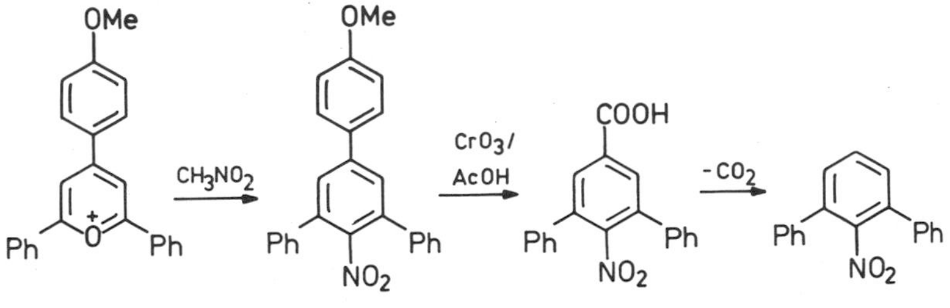

(705)   (706)   (707)   (708)

Dimroth et al.,[705,709] on attempting to perform the same base-catalyzed 2,6-[C_5 + C] transformation of 2,4,6-triaryl-substituted pyrylium salts **381** with phenylnitromethane instead of nitromethane, found that the intermediate product **709** (which can be isolated when $R^1 = R^3 = Ph$; $R^2 = p$-tolyl) aromatizes to a nitrobenzene by allylic migration of the nitro group. With triethylamine or one equivalent of potassium t-butoxide as base the main product is the nitrobenzene derivative of type **710**, while longer heating with potassium t-butoxide or performing the addition in the presence of one equivalent of ethyldiisopropylamine in o-dichlorobenzene or in a mixture of o-dichlorobenzene and toluene leads to tetraarylphenols **712** ($R^1 - R^2 = R^3 = Ar$) as main products. These are probably formed by hydrolysis of intermediately formed nitrite esters **711** which are formed by the alternative fixation of the bidentate migrating nitro group. An analogous 1,3-rearrangement of a nitro group to a 3-nitrite ester group leading finally to phenols **713** can also be observed with nitromethane if instead of excess base one introduces only one mole of base and if chlorobenzene, ethylene dichloride, or ethanol are used as solvents.[705,710]

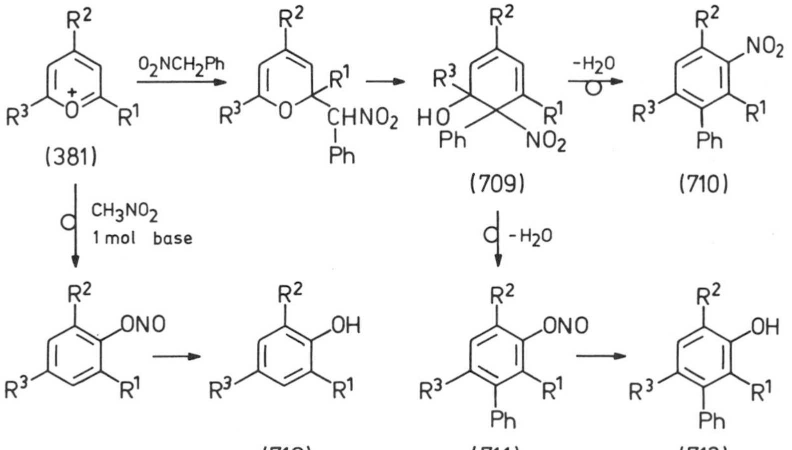

(381)

(709)   (710)

(713)   (711)   (712)

The same principle of a 2,6-[C$_5$ + C] synthesis, as in the reaction of pyrylium salts with nitromethane, can be extended to other compounds with active methylene groups. Thus acetylacetone, ethyl acetoacetate, and ethyl cyanoacetate react with 2,4,6-triphenylpyrylium fluoborate (159) in the presence of potassium *t*-butoxide via 714 and 715 affording benzene derivatives 716.[314,711] The recyclization to the aromatic benzene ring is accompanied here by elimination of a resonance-stabilized anion, R'O$^-$, namely AcO$^-$ for acetoacetate or for acetylacetone, and EtOCOO$^-$ for cyanoacetate. By performing this reaction with 2,4,6-[2,6-^{14}C$_2$] trimethylpyrylium perchlorate, the labeled compounds [2,6-^{14}C$_2$] mesitonitrile and 2,4,6-[2,6-^{14}C$_2$]trimethylacetophenone were obtained[437]; alternatively, starting from labeled cyanoacetate or acetoacetate the isotopically labeled carbon may be introduced into the 1-position.

A special case is the reaction of 2-*t*-butyl-4,6-diphenylpyrylium with acetylacetone. Here, after addition of the acetylacetonate anion at position 6, the two acetyl groups are eliminated successively, leading to 3,5-diphenyl-*t*-butylbenzene.[321]

Malonitrile and diethyl malonate behave differently, reacting with 2,4,6-triphenylpyrylium (159) by a 2,5-[C$_4$ + C$_2$] transformation (via non-isolable intermediates 717 and 719, respectively) and yielding 2-amino-

	R	R'
a	COOEt	COMe
b	COMe	COMe
c	CN	COOEt

3-cyano-4,6-diphenylbenzophenone (718) and 2-hydroxy-3-carbethoxy-4,6-diphenylbenzophenone (720), respectively.[314,323] In these reactions two carbon atoms of the nucleophile become incorporated into the benzene ring. On the other hand, 2,4,6-triphenylthiopyrylium perchlorate (721) reacts with malonitrile affording 2,4,6-triphenylbenzonitrile (716c) by a 2,6-[C$_5$ + C] transformation similar to the reaction with ethyl cyanoacetate; in agreement with the rearomatization step 722 → 716c, the thiocyanate ion which functions as a leaving group could be identified as a reaction product.[712]

Starting from 2,4-dimethoxypyrylium salts [accessible by alkylation of corresponding 2-pyrones (cf. Section II,A,1,a)] and using phosphonates (R'O)$_2$P(O)CH$_2$COOMe as active methylene compound and two equivalents of NaH as base, Griffin and Staunton[31] obtained in a "one-pot" reaction resorcylic acid derivatives of structure 723. In this 2,6-[C$_5$ + C] transformation the phosphate ion (R'O)$_2$PO$_2^-$ functions as a leaving group.

Recently Fischer, Zimmermann, and Weissenfels[301,307,322] showed that also 2H-pyrans of type 725 and 727, obtained in crystalline form by addition of sodium methoxide and dialkylamines, respectively, to 3-methyl-2,4,6-triphenylpyrylium perchlorate (724) (cf. Sections III,A,6,a and III,A,6,c), react with nitromethane to give the nitrobenzene deriv-

ative **728**. However, the reaction of **725** as well as of **727** with acetylacetone leads to the crystalline 2*H*-pyran derivative **726**, obtainable also
directly from **724** and acetylacetone in the presence of one equivalent
of triethylamine. On heating with potassium *t*-butoxide in *t*-butanol **726**
undergoes, analogously to the reaction sequence **714** → **716c**, the expected 2,6-[C_5 + C] transformation yielding 3-methyl-2,4,6-triphenyl-
acetophenone (**729**), while on treatment with aqueous sodium hydroxide
a 2,5-[C_4 + C_2] transformation occurs affording 3-acetyl-2,5-dimethyl-
4,6-diphenylbenzophenone (**730**).

2,6-Disubstituted 4-(*N*-piperidino)pyrylium salts **731** (R = Me, Ph)
react with malonitrile or with ethyl cyanoacetate under basic conditions
according to a 2,5-[C_4 + C_2] transformation, yielding benzene derivatives
732 and **733**, respectively; however, with cyanoacetamide or cyano-
acetanilide, the products are 2-pyridones of type **734**, formed by a 2,4-
[C_3 + C_2N] transformation.[228]

The examples presented above show how strongly the substituent
pattern on the pyrylium ring, the nature of the carbon nucleophile (active
methylene component), and the reaction conditions influence the out-

come of such ring interconversions. γ-Unsubstituted pyrylium salts do not undergo ring transformation reactions on treatment with active methylene compounds in the presence of bases, but add instead the nucleophile, yielding 4H-pyran derivatives as in the case of nitromethane where the product was **704** (cf. Section III,A,6,e).

Boyd and Dando[427] reported that 2,4,6-trisubstituted pyrylium salts can also be converted to benzene derivatives by reaction with carbon nucleophiles obtained from cyclic carbonyl compounds with active methylene groups. Thus the azlactone 2-phenyl-2-oxazolin-5-one leads in the presence of triethylamine to benzanilides of type **736** through a 2,6-[C_5 + C] transformation. In some cases the acyclic intermediates **735** could be isolated (cf. Section III,B,4,c). Their base-catalyzed recyclization to the aromatic benzanilide **736** proceeds with loss of carbon dioxide. Again in this reaction, 2,6-diphenylpyrylium perchlorate reacts differently, forming a 4-pyranylidene derivative (cf. Section III,A,6,e).

Another possibility of converting pyrylium salts **381** to benzene derivatives by a 2,6-[C_5 + C] transformation consists of using ylids as C-nucleophiles. Thus the reaction of various alkylidenetriphenylphosphoranes yields, after Märkl,[428] intermediate vinylogous acylmethylenephosphoranes **737** which, depending on substituents, may be isolated (e.g., **737**, $R^1 = R^2 = R^3 = Ph$; $R^4 = COOMe$), or cyclize spontaneously by an intramolecular Wittig reaction to a substituted benzene **738**. When the simple methylenetriphenylphosphorane Ph_3PCH_2 is employed, this reaction allows the conversion of the O^+ heteroatom in pyrylium to a CH group of a benzene ring.* The reaction can be extended to 2,3,4,6-tetraphenyl- and pentaphenylpyrylium salts.[713]

* Under special conditions, the same reaction between pyrylium salts and methylene-triphenylphosphorane affords azulenes (cf. Section III,C,4,b).

According to reaction scheme **739** (X = O⁺, S⁺) → **740**, the heteroatom of pyrylium or thiopyrylium salts can also be replaced by an unsubstituted CH group by means of another ylid, namely dimethylsulfoxonium methylide.[714] By contrast, the sulfonium benzoylylid, PhCOCHSMe₂, reacts with 2,4,6-triphenylpyrylium fluoborate (**159**) under cleavage of methyl benzoate and incorporation of the sulfur ylid to give 2,4,6-triphenyl-thioanisole (**741**).[333] A possible mechanism for this unexpected 2,6-[C₅ + C] transformation is indicated in Scheme 12. In the case of 2,6-di-phenylpyrylium fluoborate a γ-attack of the ylid takes place, yielding a phenacylidenepyran (cf. Section III,A,6,e). In Section III,C,2,d another reaction between sulfonium acylylids and pyrylium salts was described, leading to aroylvinylfurans **447**.

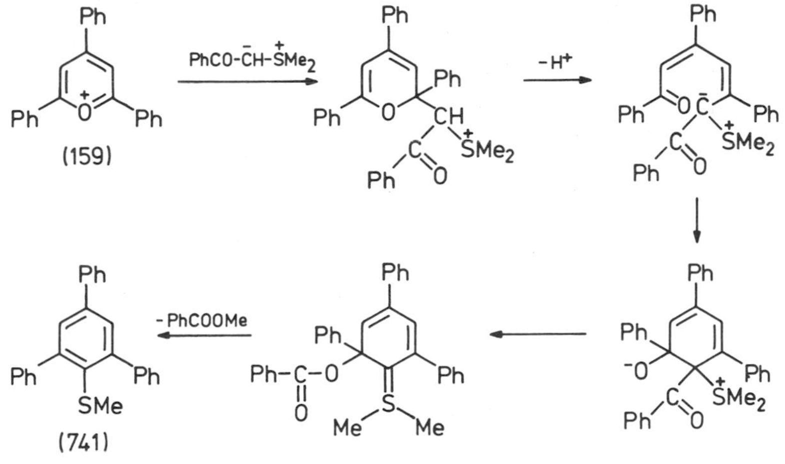

SCHEME 12

The reaction of two moles of 4-methylpyrylium salts **742** with one mole of phenylphosphine, leading to λ^5-phosphorins **745** possessing a substituted benzene ring in the 4-position, is explained as a 2,6-[C₅ + P] synthesis followed by a 2,6-[C₅ + C] transformation. The C-nucleophile is probably the ylid **744** (formed by deprotonation of the nonisolable λ^4-intermediate **743**) which reacts with another mole of pyrylium cation in the presence of alcohol R′OH, affording the final product **745**.[687]

2,4,6-Triphenylpyrylium perchlorate (**159**) reacts with pyridinium aroylylids after Katritzky and co-workers[475,476] via **746–748**, yielding 1-(3-benzoyl-2,4,6-triarylphenyl)pyridinium perchlorates with structure **749**. In contrast to the foregoing reactions with ylids, in this case the process is a 2,5-[C₄ + C₂] transformation.

(742)   (743)   (744)   (745)

(159)   (746)   (747)

(749)   (748)

Another 2,5-$[C_4 + C_2]$ transformation with multilateral applications allowing the conversion of pyrylium salts **381** to benzocycloalkanes was found by Märkl and Baier[715] when they treated these salts with cyclic enamines. The acyclic intermediate **750** cyclizes by an electrocyclic process to the cyclohexadiene derivative **751**; depending on the nature of the substituents, this eliminates either an amine or an amide undergoing thereby stabilization to an aromatic benzene ring: pyrrolidinoenamines afford almost exclusively aryl ketones **752**, while morpholinoenamines with six- or eight-membered rings (n = 4 or 6, respectively) yield aromatic hydrocarbons **753**, however, seven-membered morpholinoenamines (n = 3) give rise to a mixture of **752** and **753**. Analogous reactions were described for pyrrolidino- and morpholinoenamines derived from

1-phenylphosphorinan-4-one and its 1-oxide, yielding phosphorus heterocycles of type **754** and **755**.[716]

(381)　　　　　　　　(750)　　　　　　　　(751)

(754)　　　(755)　　　(752)　　　(753)

Finally, still another 2,5-[C_4 + C_2] transformation is found in the 1,3-
dipolar cycloaddition of 1,2-disubstituted acetylenes RC≡CR (e.g., R
= Ph, COOMe) to the betaine pyrylium 3-oxide **395**, followed by thermolysis of the isolable adducts **756**: thermal rearrangement gives the
cyclohexadienones **757**, whose alkaline cleavage leads to phenols **758**.[717]

(395)　　　　　(756)　　　　　(757)　　　　　(758)

4. Formation of Seven-Membered Rings

a. *Reactions with Nitrogen Nucleophiles.* In Sections III,C,2,c and
III,C,3,c the reactions of pyrylium salts with various hydrazines H_2NNHR
(R = H, Alk, R'CO, R'SO₂) were reviewed and it was shown that such
reactions can lead to five-membered and six-membered nitrogen heterocycles, respectively. So far, preparatively useful ring enlargements of

pyrylium salts to seven-membered nitrogen heterocycles with neighboring heteroatoms succeeded only with unsubstituted hydrazine, N_2H_4, which converts 2,4,6-triarylpyrylium (or trialkylpyrylium salts with bulky substituents such as t-Bu in α-positions*) to 3,5,7-trisubstituted $4H$-1,2-diazepines **760** in high yields.[414,719] On brief treatment of 2,4,6-triphenylpyrylium perchlorate with hydrazine Balaban[420] could isolate in the solid state the acyclic intermediate hydrazone **759** (R = R' = Ph; cf. Sections III,B,3,c and III,C,2,c) which in solution dehydrated easily to the diazepine (half-life in CDCl₃ at 40°C about one hour as indicated by ¹H-NMR spectra). This reaction represents a 2,6-[C₅ + N₂] transformation. An analogous ring enlargement was observed with 2,4,6-triphenylthiopyrylium perchlorate (**761**, R = R' = Ph).[719,720]

From appropriately substituted pyrylium salts, $4H$-1,2-diazepines **762**,[721] **763**,[722] and **764**[723,724] were prepared. All these compounds were $4H$-diazepines, but the related product obtained from 2,6-di-t-butyl-4-(1-methylindol-3-yl)pyrylium perchlorate, according to IR and UV spectra, was reported[723] to possess the 1H-form **765** (this structural assignment should be checked by NMR methods). It is certain that a $1H$-diazepine structure is present in 1-methyl-3,5,7-triphenyl-$1H$-diazepine [obtained in 4% yield along with the main product (a pyrazole derivative; cf. Section III,C,2,c) in the reaction of 2,4,6-triphenylpyrylium with methylhydrazine in benzene at 0–25°C] as shown by Snieckus and Kan.[467] The conversion of

---

* The structure of the resulting 3,5-di-t-butyl-5-methyl-$4H$-1,2-diazepine was investigated by IR, UV, ¹H-NMR, ¹³C-NMR, as well as mass spectroscopy and by lanthanide shift reagents with the above compound and its methyl-deuterated congener.[718]

4-methoxy-2,6-diphenylpyrylium perchlorate to 3,5-diphenyl-5-hydra-zino-4H-1,2-diazepine was described by Zhungietu et al.[725]

Buchardt, Pedersen, Balaban, et al.[719] reported that the temperature-dependent ^{1}H-NMR spectra observed for the diazepine **760** (R = R' = Ph) in various nonpolar solvents is due to ring inversion, and not to a possible valence tautomerization to a diazanorcaradiene form which would possess an energetically unfavorable azo structure. The free energy of activation for the ring flip between the two degenerate boat conformations is $\Delta G^{\ddagger} \cong 17.5$ kcal/mol (90°C), a larger value than in other hetero analogs of cycloheptatriene[726]; however, in trifluoroacetic acid the energy barrier is appreciably lower (~ 12 kcal/mol at -12°C) probably due to protonation of **760**. Interestingly, the intense M − 28 peak in the mass spectrum of 3,4,7-triaryl-1,2-diazepines due possibly to loss of an N_2 molecule[719] is absent when the 3,7-groups are t-butyl.[718] It exists in the mass spectrum of 3,7-diisopropyl-, -diethyl-, or -dimethyl-1,2-diaze-pines as $M - C_2H_4$.

Another possibility of converting tetra- and pentaphenylpyrylium salts **766** (R = H, Ph) to seven-membered ring systems (this time, however, with nonadjacent heteroatoms) consists of treating them with sodium azide in acetonitrile.[727] At -30°C an addition product of the $N_3{}^-$ ion can be isolated in crystalline form. At room temperature, for the 2,3,5,6-tetraphenyl derivative, this adduct decomposes to nitrogen and a crystalline ketoazirine **767** (R = H) in 72% yield. This azirine on heating at 100°C, or the azide adduct from pentaphenylpyrylium at room temperature afford in 80–100% yield crystalline compounds for which the French authors indicate a 1,3-oxazepine structure **768**. This reaction would represent a 2,6-[C_6O + N] transformation. The oxazepines referred to were also obtained by an independent photochemical rearrangement of the corresponding pyridine N-oxides. Thiopyrylium salts react differently,

(766)   (767)

(768)

the azide adducts being much more stable and leading at more elevated temperatures to pyridines and to thiophenes.

b. *Reactions with Carbon Nucleophiles.* Suitably 2,4,6-trisubstituted pyrylium salts 248 react after Hafner and Kaiser[728-730] with sodium cyclopentadienide in tetrahydrofuran at room temperature yielding 4,6,8-trisubstituted azulenes 771. This elegant azulene synthesis represents a 2,6-[C_5 + C_2] transformation proceeding through the nonisolable intermediate 769 and its tautomer 770, differing by the position of the double bonds in the five-membered ring. Satisfactory preparative yields were obtained from pyrylium salts with 2,4,6-trialkyl substituents, as well as with 2,6-dialkyl-4-methoxy- or 2,6-dialkyl-4-phenylpyrylium salts, whereas pyrylium salts with α-aryl groups react in lower yields. Table XXX (Appendix, Section VII) presents the azulenes so far obtained by the Hafner synthesis. Compounds 772,[372] 773,[731] and 774[731] represent examples for special applications of this reaction. An isotopically labeled azulene, 4,6,8-[4,8-$^{14}C_2$]trimethylazulene was prepared.[437] The condensation products of 2,4,6-trimethylpyrylium perchlorate with various aldehydes lead likewise to corresponding azulenes.[369b] The preparation of 2,4,6,8-tetramethylazulene 775 indicates that also substituted cyclopentadienes are capable of this reaction.[729]

Not only do 2,4,6-triaryl substituted pyrylium salts give poor results, but pyrylium salts with unsubstituted α-position(s) (such as 4-methoxypyrylium, 2-phenyl-4-methoxypyrylium, and the unsubstituted pyrylium perchlorate) yield no azulenes in this reaction. However, the preparation of 6-methoxyazulene (**777**) starting from 4-methoxypyrylium perchlorate (**338**) succeeds if prior to reaction with sodium cyclopentadienide the ring opening is performed with *N*-methylaniline leading to the more resistant pentamethinecyanine **339** (cf. Section III,B,3,b). This reacts with sodium cyclopentadienide forming the fulvene **776**, which in agreement with another, earlier, Hafner azulene synthesis[732] (from cyclopentadiene and Zincke's aldehyde, i.e., 1-(*N*-methylanilino)penta-1,3-dien-5-al) cyclizes with elimination of *N*-methylaniline to the 6-methoxyazulene **777**.[413] Another possibility is to introduce the cyclopentadiene as (C_5H_5)CuPBu$_3$ in the reaction with pyrylium salts.[733]

If the reaction between aryl-substituted pyrylium salts **248** (R = R' = Ar) with methylenetriphenylphosphorane is carried out in methylene dichloride or in dilute acetonitrile solution (instead of *t*-butanol or concentrated acetonitrile solution, where the product is a benzene derivative as discussed above in Section III,C,3,e), then, after Dimroth and co-workers,[713] azulenes of type **780** are formed. Under these modified conditions, the primary adduct gives rise to a new ylid **778** which reacts with another mole of pyrylium salt faster than it is able to undergo the electrocyclic ring opening, followed by the subsequent intramolecular reaction ending in a benzene derivative. Thus the phosphonium salt **779** produced from one mole of ylid and two moles of pyrylium salt, which is the probable intermediate, affords a 1,3,4,6,8-pentaaryl-5-aroylazulene

780 by a combination of a ring enlargement (2,5-[C₄ + C₃] transformation) with a ring contraction (2,5-[C₄ + C] transformation).

Cycloadducts of the betaine 2,4,6-triphenylpyrylium 3-oxide (**395**) with diphenylacetylene can be converted not only to six-membered ring systems (cf. Section III,C,3,e) but also to seven-membered carbocycles. Thus the catalytic hydrogenation of the cycloadduct **781** gives a ketone with structure **782**, which on reduction with lithium aluminum hydride yields a mixture of the pentaphenylcycloheptadiene **783** and the alcohol **784**. The latter on heating with *p*-toluenesulfonic acid in toluene is converted to pentaphenylcycloheptatriene (**785**).[717] These reactions represent a 2,6-[C₅ + C₂] transformation.

D. Special Reactions of Pyrylium Salts

1. Photochemistry

a. *Photochemistry of Pyrones and Hydroxypyrylium Salts.*   The photochemical conversion of 2-pyrone to the bicyclic system **786** analogous to Dewar benzene was reported[734]; later investigations demonstrated also the formation of the tricyclic system **787** analogous to benzvalene.[735,736]

Following Paterno's observation that 2,6-dimethyl-4-pyrone (**788**, R = Me) affords a photodimer,[737] and an early incorrect structure assignment,[738] Yates *et al.* reported the correct head-to-tail cage structure **789** of this dimer[739,740] and found[741] that at low concentration a monomeric product results, namely 4,5-dimethylfurfuraldehyde (**795**), suggesting a dimethylcyclopentenone epoxide intermediate **793**. Padwa and Hartman[742] and Yates *et al.*[743] investigated 2,6-diphenyl- (**788**, R = Ph) and 2,6-diethyl-4-pyrone (**788**, R = Et) finding analogous photodimers **789**. It was shown[742] that furfuraldehydes result from acid-catalyzed isomer-

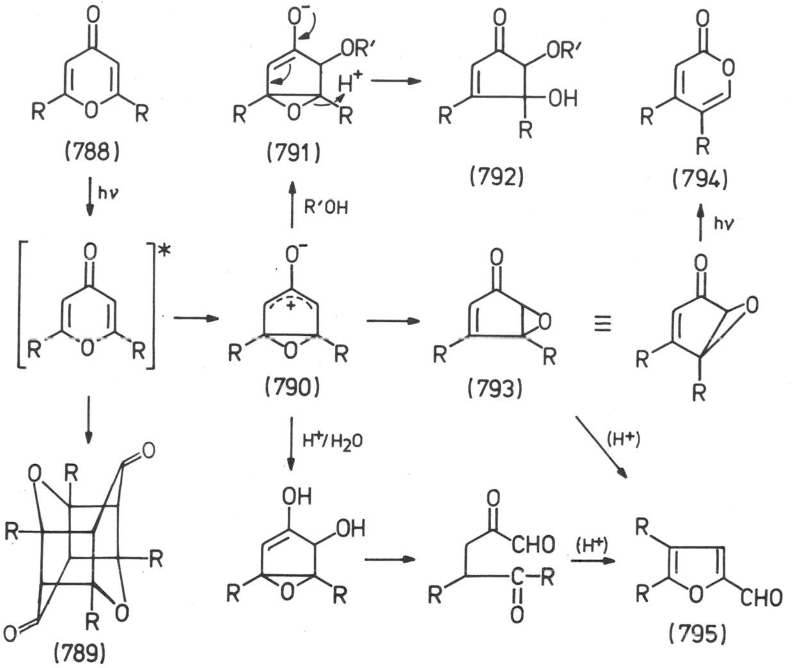

SCHEME 13

ization of epoxides. The true monomeric photoproducts of 4-pyrones are 2-pyrones **794** formed through photorearrangement (two electron pairs are involved) of the same epoxide **793**.[744–746] This epoxide **793** results from a zwitterionic intermediate **790** by an "oxygen walk"; proofs for this intermediate result from (i) the isolation[747,748] of solvent adducts **792** when alcohols R'OH are used as solvents, along with 2-pyrones **794**; (ii) in case of 3-hydroxy-4-pyrones **796**, hydroxycyclopentenediones **797** are formed[749]; (iii) photorearrangement of the isolated cyclopentadienone epoxides **793** affords 2-pyrones **794**.[750] The results are summarized in Scheme 13 for 2,6-disubstituted 4-pyrones in nonprotonating solvents.

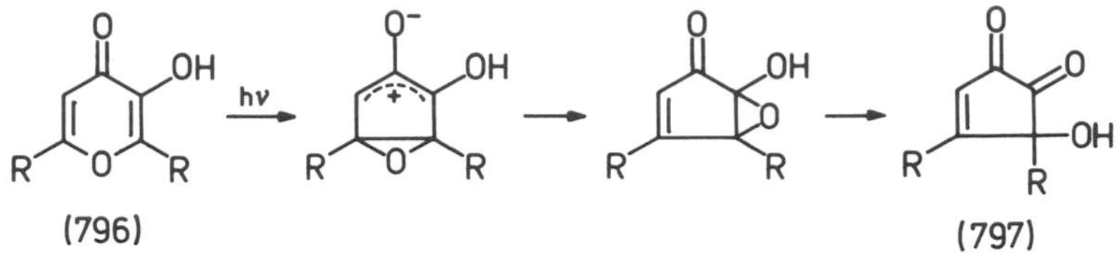

(796)   (797)

For 3,5-disubstituted 4-pyrones **798** (R = H, Me), Barltrop, Day, and Samuel[750] proposed Scheme 14. In trifluoroethanol as solvent, two photodimers, **804** and **805**, result from **798** along with a 1,3-cyclopentenedione

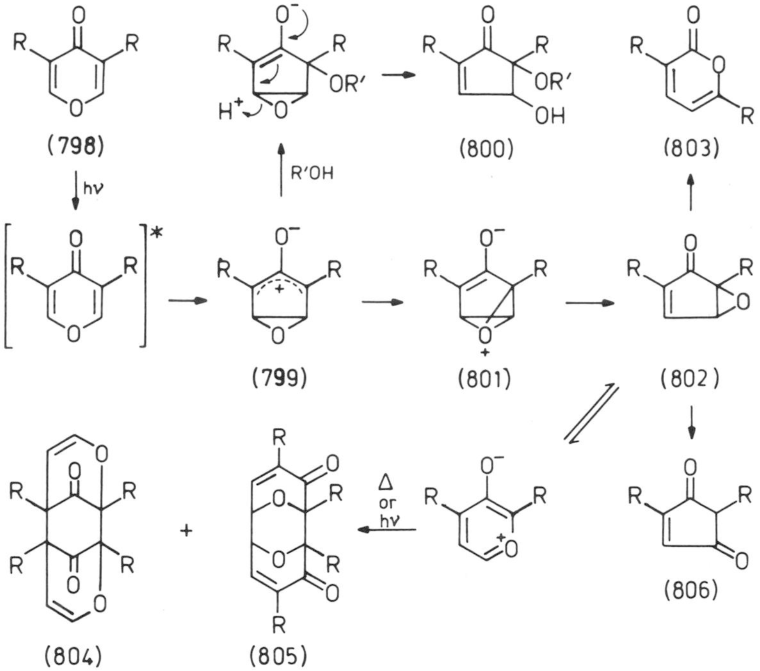

(798)   (800)   (803)

(799)   (801)   (802)

(804)   (805)   (806)

SCHEME 14

806, the 3,6-dimethyl-2-pyrone (**803**, R = Me), and the solvent adduct **800**.[750,751] In furan as solvent, the zwitterion **799** was trapped, yielding the 2 + 2 photoadduct **807** and the 3 + 2 photoadduct **808**.[750]

The rearrangement of the zwitterion **799** into the cyclopentenone epoxide **802** probably involves an oxoniabenzvalene zwitterion **801**. The photorearrangement of 2-hydroxypyrylium cations **809** to the isomeric cations **810** involves oxonia-Dewar benzene intermediates.[752]

Pavlik and Clennan[752] investigated the photochemistry of the 2,6-dimethyl-4-hydroxypyrylium cation (**811**) by irradiating 2,6-dimethyl-4-

(798) (807) (808)

(809)

(810)

pyrone (11) in 96% sulfuric acid. The proposed reaction mechanism accounting for the isomerization to 4,5-dimethyl-2-hydroxypyrylium [813, i.e., protonated 4,5-dimethyl-2-pyrone (814)] and its subsequent photochemical conversion to 5,6-dimethyl-2-hydroxypyrylium [815, i.e., protonated 5,6-dimethyl-2-pyrone (816)] is depicted on Scheme 15.

2,6-Dimethyl-3,5-diphenyl-4-hydroxypyrylium (817) yields on photolysis not only a 2-hydroxypyrylium derivative but also the 2,4-diphenyl-5,6-dimethyl-3-hydroxypyrylium cation (818).[744]

The photorearrangement of 4-hydroxypyrylium salts 819 (i.e., 4-pyrones in sulfuric acid)[751-753] to the corresponding 2-hydroxy isomers 821 proceeds similarly to the photochemical reactions of 4-pyrones and involves oxoniabenzvalene intermediates. The authors[753] succeeded in demonstrating the intermediate formation of the cyclic sulfate 820, which in the presence of water affords the diol 822 (R = Me).

The pattern of ring atom permutations[753] is represented by one of the twelve possible such permutations, namely P_4, which is derived from the oxoniabenzvalene intermediate, with a minor contribution of the ring permutation P_8 derived from a Dewar benzene intermediate. These pathways were established by deuterium or substituent (methyl and/or ethyl) labeling of the various positions in 4-hydroxypyrylium (i.e., protonated 4-pyrones), whereas protonated 2-pyrones rearrange mainly via P_8 intermediates (Scheme 16).

As side products in photoreactions of 4-hydroxypyrylium cations, 2-acylfurans 823 (R = R' = Me, Et) were isolated.[751,753] Their relative

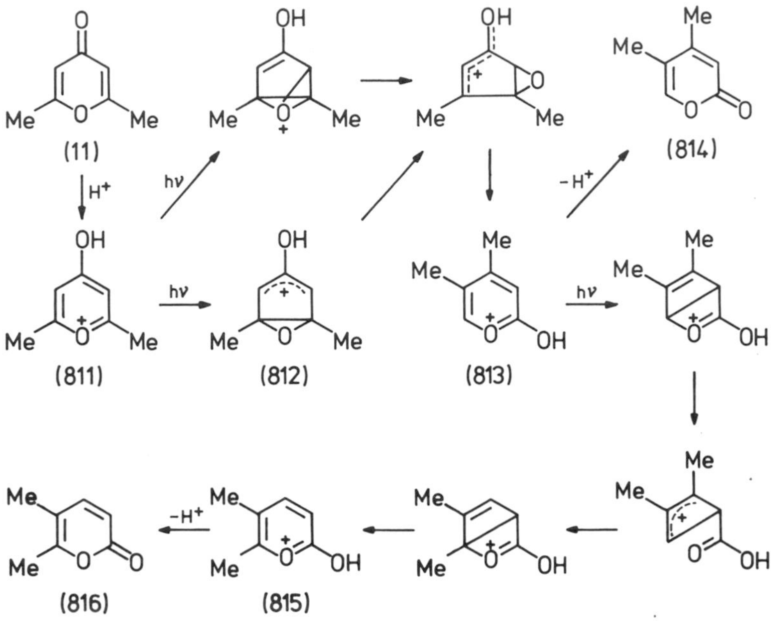

SCHEME 15

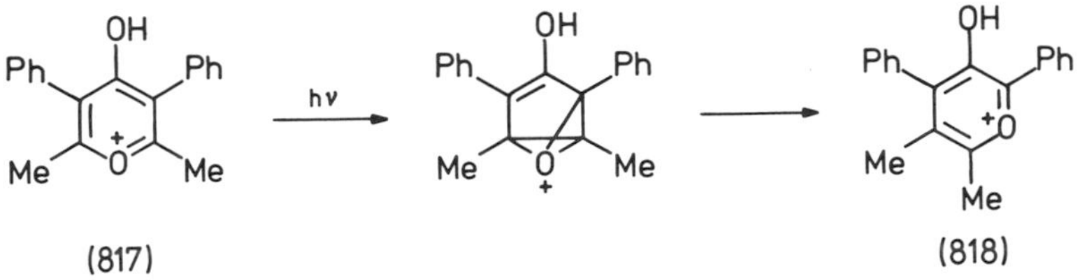

amount decreases with increasing sulfuric acid concentration: main products are furans in 90% H_2SO_4, but 2-pyrones are the exclusive products in oleum, and mixtures result at intermediate 90–100% H_2SO_4 concentrations.

b. *Photochemistry of Alkyl- or Aryl-Substituted Pyrylium Salts.* Barltrop, Day, and co-workers[754,755] investigated the photochemistry of 2,4,6-trialkylpyrylium salts 824 with γ-oriented methyl, ethyl, and isopropyl groups. The postulated intermediate, an oxoniabenzvalene 825, may rearomatize to form an α-unsubstituted pyrylium salt 827 (which could not, however, be isolated on performing the irradiation in anhydrous acetonitrile) or undergo ring opening directly to a ketoaldehyde 828; with R = Me, R' = H, the structure proof for 828 ruled out alternative mechanisms. The isolation of 5-alkylidene-2,3-dimethylcyclo-

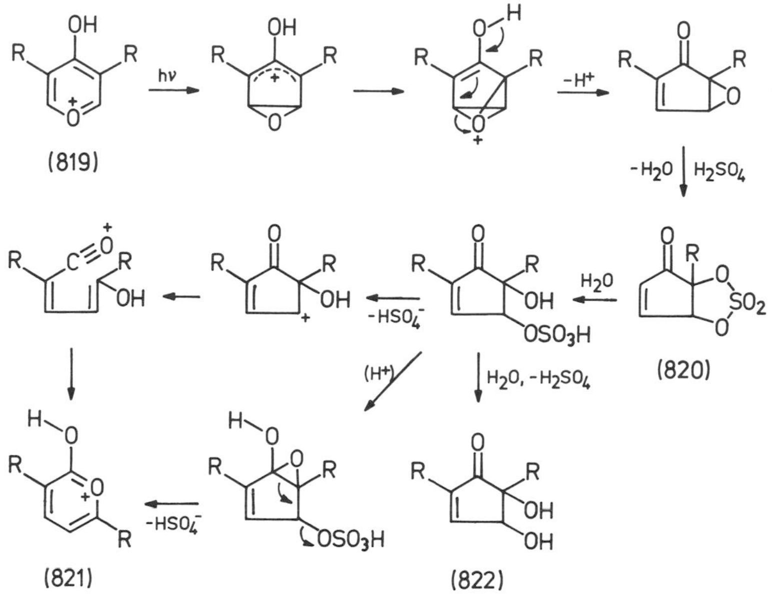

(819)

(820)

(821)

(822)

pent-2-enones **826** as side products is a strong argument in favor of the oxonia-benzvalene intermediate **825**.

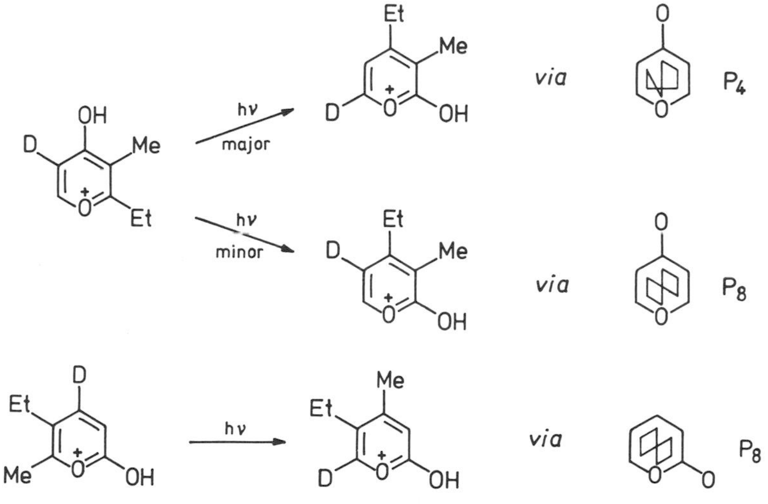

SCHEME 16

On the other hand, with a γ-oriented *t*-butyl group the products of the irradiation are quite different[756]: the methylenecyclopentenediol **829**, 1,1-bisacetonyl-2,2-dimethylcyclopropane (**830**), and the cyclopentenetriols

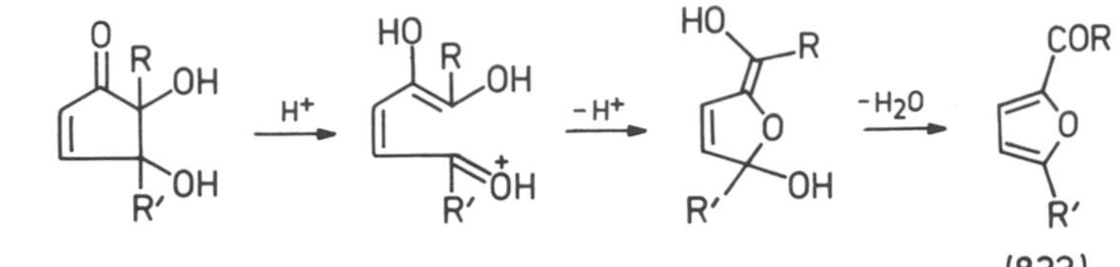

(823)

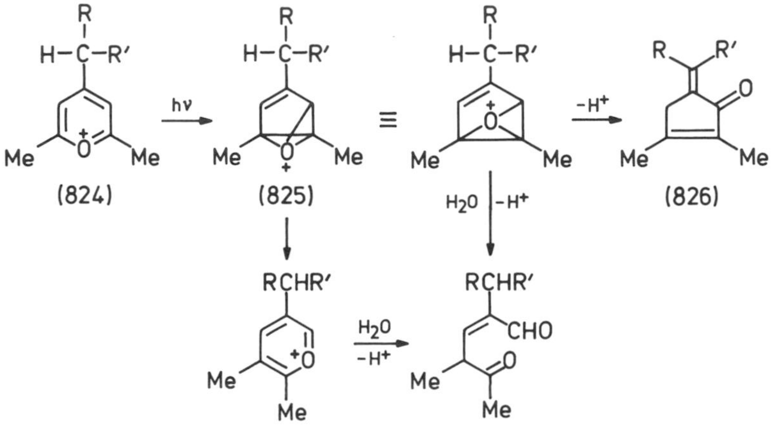

(824)   (825)   (826)

(827)   (828)

831 (two stereoisomers) can all be derived from an oxoniabenzvalene. In view of the above data analogous mechanisms were proposed to account for the photochemistry of 3-hydroxy-4-pyrones[757] and of 2,6-disubstituted 4-pyrones.[741]

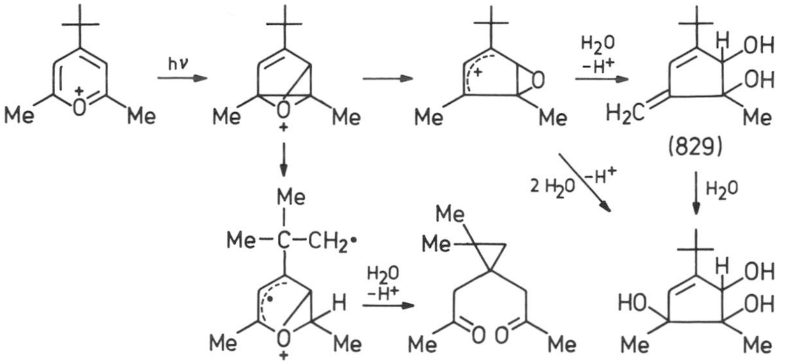

(829)

(830)   (831)

Graph-theoretical methods are useful for a complete analysis of all possible valence isomers which can result in such photochemical rearrangements.[758,759] The novel approach[760–762] of "ring permutations" to account for aromatic phototranspositions also has a graph-theoretical basis: the numbers of topologically distinct ways in which an n-membered ring can be "twisted" are: 1 ($n = 3$), 2 ($n = 4$), 4 ($n = 5$), 12 ($n = 6$), 39 ($n = 7$), . . . , 83435 ($n = 11$), . . . , 9223092 ($n = 13$), etc.

The photochemical isomerizations of 4H-pyrans to 2H-pyrans reported by Dimroth et al. were mentioned in Section III,C,3,e.

c. *Photochemistry of Pyrylium 3-Oxides.* On irradiating pyrylium 3-oxides 832 (R = H or Ph) with UV light, Ullman and Henderson[763,764] changed the stationary concentrations of valence isomers (cf. Section III,D,2). The cyclopentenone epoxide valence isomer 833 (R = Ph) is isomerized by irradiation to a 2-pyrone 834 (R = Ph) which on prolonged irradiation decarboxylates and affords 1,2,4,7-tetraphenylcyclooctatetraene. This product cleaves on further irradiation to diphenylacetylene and a fragment which cyclizes to p- terphenyl.[763-766]

(832) (833) (834)

2. Valence Isomerizations

The rigorous definition of valence isomers (as molecules whose constitutional graphs have the same partition of vertex degrees)[758,759] has allowed Balaban to find by graph-theoretical techniques all possible valence isomers of the pyran ring: there are 17 "planar graphs" (i.e., valence isomers; the word "planar" has a topological and not a geometrical connotation) depicted in Ref. 759. Photochemical valence isomerizations are discussed in the preceding section and involve several of these valence isomers.

The present section will discuss in more detail only the thermally allowed valence isomerizations of 2H-pyrans 377 to pentadienones 378 (cf. Ref. 387, p. 188). This is an important reaction because most of the nucleophilic additions to the pyrylium ring occur at the α-position leading to a 2H-pyran, which after valence isomerization to the acyclic pentadienone may undergo an intramolecular ring closure involving the nucleophile or an α- or γ-oriented side chain (cf. Section III,C).

(377) (378)

The first convincing evidence for this valence isomerization was provided by Marvell, Gosink *et al.*[430,432] on the basis of ^{1}H-NMR data. UV radiation of *trans*-β-ionone (**835**) had been reported[325] to afford the pyran **837** instead of the *cis*-β-ionone (**836**), but the former authors showed that in fact an equilibrium mixture results. Table IV indicates the equilibrium

(835)   (836)   (837)

and the rate constants. It may be seen that at room temperature the pyran **837** predominates, but at more elevated temperatures appreciable amounts of dienone **836** coexist at equilibrium.

TABLE IV

EQUILIBRIUM AND RATE CONSTANTS FOR THE PROCESS **836** ⇌ **837** AT VARIOUS TEMPERATURES IN TETRACHLOROETHYLENE

Temperature [°C]	K	$10^4 \cdot k_1 \ [s^{-1}]$	$10^4 \cdot k_{-1} \ [s^{-1}]$
0	0.054	0.086	1.58
8	0.070	0.25	3.57
18	0.094	1.31	13.9
54	0.217		
113	0.658		

With other 2*H*-pyrans, the equilibrium mixture contains insufficient amounts of dienone to obtain reliable estimates by ^{1}H-NMR for the thermodynamic and kinetic data.[432] In these cases indirect kinetic measurements were performed by reduction with excess borohydride or alanate (MH); provided that k_r [MH] $> k_{-1}$, the rate of disappearance of the pyran **838** (which can be measured by UV spectrometry) equals k_1; assuming that $K = k_1/k_{-1} > 100$ (when in NMR no bands of the diene **839** appear) this allows the estimation of a lower limit for k_{-1}.

While 2,2-disubstituted 2*H*-pyrans exist mainly in this form, the reverse is true for 2-monosubstituted 2*H*-pyrans which valence isomerize almost completely to cis dienones. Indeed, on reinvestigating the reduction of

(838)   (839)

(840)   (841)

2,4,6-trimethylpyrylium perchlorate with sodium borohydride (described in Section III,B,5) Marvell and Gosink[767] could demonstrate (along with the γ-pyran) the intermediate formation of the α-pyran **840** which valence isomerizes to the cis dienone **841**. The rates for the ring-opening reactions are for

$$\mathbf{840} \underset{k_{-1}}{\overset{k_1}{\rightleftharpoons}} \mathbf{841},\ t = 13°C,\ k_1 = 3 \times 10^{-3}\ \text{sec}^{-1};\ K \gg 1$$

whereas for

$$\mathbf{838} \underset{k_{-1}}{\overset{k_1}{\rightleftharpoons}} \mathbf{839},\ t = 14.6°C,\ k_1 = 1.6 \times 10^{-4}\ \text{sec}^{-1};\ K \ll 1\ (R^1\text{–}R^3 = Me)$$

The decrease of the ring-opening reaction rate on replacing the hydrogen in **840** by a methyl group appears reasonable for this electrocyclic process; the rate of the cyclization, however, is decreased much more by this substitution, and this explains why the equilibrium is shifted to the opposite direction. Only in compounds **836** and **837** are the rates matched.

Another case leading to equilibrium mixtures containing both valence isomers in comparable amounts involves the esters **842** and **843**.[768] A nearly equal relationship is demonstrated by a ¹H-NMR study of the equilibrium between the butadienylcarboxylic ester PhC(OEt)=CHC(NR₂)=C(CN)COOEt and its electrocyclic valence isomer (involving the car-

(842)   (843)

boxylic carbonyl group), where the acyclic structure dominates with increasing temperature.[769]

Attempts to add selectively one hydrogen molecule to 1-phenyl-3-(cyclohexen-1-yl)-2-pyropyne (844) failed because of the low selectivity of the palladium catalyst, but the 2H-pyran derivative 846 could be demonstrated in the reaction mixture by its UV and NMR spectra.[433] Its formation must involve a valence isomerization of the dienone 845.

(844)   (845)   (846)

However, Schiess and Chia[770,771] obtained by an analogous reduction of 847 2-vinyl-3,4,5,6-tetrahydrobenzaldehyde (848). All spectral data agree with structure 848, which could either mean that the equilibrium 848 ⇌ 849 has a high energy barrier, or that the equilibrium is completely displaced toward 848. To find which alternative was true the above authors synthesized the cis-dideuterated aldehyde 850. On heating this aldehyde at 70°C in ethanol for three hours, an almost statistical cis/trans ratio was found. Therefore the equilibration takes place, though the low amount of 851 is not detectable by spectral means; it can, however, be trapped by cycloaddition to tetracyanoethylene.

(847)   (848)   (849)

(850)   (851)

Schiess et al.[770,772] followed by UV spectrophotometry the kinetics of cis–trans isomerization of aldehydes 853, $n = 1$ or 2 (obtained by semi-hydrogenation of a triple bond) finding $\Delta H^{\ddagger} \sim 22$ kcal/mol and $\Delta S^{\ddagger} \sim -12$ e.u. between 60 and 70°C. The negative entropy indicates a six-membered transition state; in all probability, the cis–trans isomerization takes place through the intermediate formation of the 2H-pyran 852, which could be trapped for $n = 2$ as tetracyanoethylene adduct (for $n = 1$ no adduct was isolated, possibly because in this case its equilibrium concentration was too low because of steric strain). No spectral evidence for **852** could be obtained, the equilibrium favors **853**.

Photorearrangement of **853** does not lead to increased concentrations of **852** but to a 1,5-hydrogen shift leading to ketenes **854**.[773]

Schiess[774] showed that a cis–trans isomerization **855** ⇌ **856** takes place readily in the dark at 35°C. The negative entropy and the lack of dependence on the solvent polarity indicate that the formation of the 2*H*-pyran **857** is the rate determining step. The equilibrium concentration of **857** is too low to allow its spectral identification, but it can be trapped to yield **858** by [4 + 2] cycloaddition with tetracyanoethylene at 60°C.

The probable valence isomerism between polychloropentadienals or polychloropentadienones and the corresponding pyrylium chlorides was discussed by Roedig, Märkl, *et al.* in a series of papers quoted in Part I,[1] Section II,B,2,e. Newer data on the thermal rearrangement of perchloropentadienthioates **859** (R = Me, Et, or Ph) to 5-thio-substituted acyl chlorides **861** show that the probable course involves 2*H*-pyrans **860** and **861**. In the equilibrium mixtures **862** predominates.[775]

As will be mentioned in Section IV,A,2,b, the methoxide adducts of 2,4,6-triarylpyrylium cations were proved by ¹³C-NMR spectroscopy to possess a 2*H*-pyran structure.[297] No chemical reactions (e.g., borohydride or alanate reductions) have yet been performed to investigate their possible valence isomerization to dienones.

2,6-Diphenylpyrylium adds a methoxide ion in position 4 in a kinetically controlled reaction while the thermodynamically controlled product is (Z,Z)-PhCOCH=CHCH=CPhOMe formed by valence isomerization of the 2H-pyran adduct.[294,300] 4-Methoxy-2,6-diphenylpyrylium also adds a methoxide anion at the γ-position in methanol, but in methanol/acetonitrile (1 : 9) comparable amounts of 2H- and 4H-pyran are formed, and the 2H-pyran does not undergo ring opening[300] (cf. Section III,A,6,a). Oestenson et al.[776,777] demonstrated that 4H-pyrans 863 (R = H or Ph) isomerize to 2H-pyrans 864 which isomerize to dienones 865 in hot acetic acid or by intermolecular hydride transfer under the catalytic influence of the corresponding pyrylium salt.

Potts et al.[717] obtained from 2,4,6-triphenylpyrylium 3-oxide (395) and maleic anhydride or other dipolarophiles (e.g., methyl maleate and fumarate, fumaronitrile, ethyl vinyl ether, norbornene, norbornadiene)

[4 + 2] cycloadducts **867** demonstrating the valence isomerization **395** ⇌ **866** to a cyclopentadienone epoxide. Adducts analogously obtained from **395** and acetylenes (diphenylacetylene, methyl acetylenedicarboxylate) isomerize on heating and afford cyclohexadienone derivatives as described in Section III,C,3,e.

The photochemically induced valence isomerization **395** ⇌ **866** and subsequent conversions of **866** to 2-pyrone under UV irradiation in polar solvents was studied by Ullman and by other authors, and is discussed in Section III,D,1,c.

The thermochromic spiropyrans will not be discussed in detail (most of them are valence-isomeric with benzopyrylium zwitterions) but two reviews[778,779] will be mentioned.

3. Complexes Based on Pyrylium Salts

The first attempts to synthesize eight-membered chelate complexes of transition metals (Cu, Co, Ni) with pyrylium salts or the products of their hydrolysis, 1,5-pentenediones, failed.[780] The supposedly eight-membered chelate complexes of pyrylium pseudobases with 1,3,2-benzodioxaborole[781,782] were proved by ^{1}H-NMR to be pyrylium salts having a bis(pyrocatechol)spiroborate anion[783] (cf. Section IV,A,1,c). Pyrylium cations with boron-containing substituents were mentioned in Section II,A,1,c; a fair number of carboranylpyrylium salts have been synthesized in recent years.[74]

According to Ukhin et al.,[784] 2,6-disubstituted pyrylium salts **238** (R = t-Bu, Ph) react with palladium chloride in water-containing organic solvents (EtOH, MeOH, 50%AcOH) to give complexes **868** containing 1,3-bisacyl-π-allyl ligands.

(238)   (868)

The reaction is of interest in that pyrylium ring opening usually occurs under the action of bases which is obviously not the case here. The formation of the Pd—C bond is believed to be the driving force of the process.[784] On the other hand, basic conditions or the use of pyrylium pseudobases (e.g., 1,5-diphenylpentene-1,5-dione) rather than pyrylium salts themselves seem to favor the formation of π-complexes with PdCl$_2$.[784–786]

2-Ferrocenylpyrylium salts have been obtained by synthesis from FcCOCH$_3$ via FcCCl=CHCHO.[140] The formation of 2-ferrocenylpyrylium (by decarboxylation in the presence of ferrocene) was mentioned in Section III,A,4 and the synthesis of pyrylium salts with cyclopentadienyl manganesetricarbonyl (cymantrenyl) groups in Section II,B,2,b. The reaction of di- and trisubstituted pyrylium iodides and Fe$_2$(CO)$_9$ leading to 4,4′-dipyranyls (for 2,6-diphenyl substitution) or 2,2′-dipyranyls (for 2,4,6-triphenyl substitution) complexed with Fe(CO)$_3$ at each diene system, was reported.[787]

The reaction between 2,6-dicymantrenylpyrylium and PdCl$_2$ in the presence of aqueous sodium carbonate leads to a bimetallic π-allyl complex **869**[788] which undergoes conversion to the cyclopentadienyl derivative **870** under the action of cyclopentadienylthallium in acetonitrile at room temperature.[785]

(869)   (870)

On treating 2,6-di-t-butyl-4-methylpyrylium perchlorate with PdCl$_2$, deprotonation occurred and a PdCl$_2$ complex of the corresponding γ-methylenepyran resulted. A similar reaction occurred with 2-methyl-4,6-diphenylpyrylium perchlorate.[789]

2-Phenyl-3-benzoyl-6-p-nitrophenylpyrylium perchlorate (**871**, Ar = p-O$_2$NC$_6$H$_4$) reacts with copper(II) acetate in refluxing ethanol in the presence of Na$_2$CO$_3$·10 H$_2$O to give the β-ketopyranolate **872** in 44% yield.[790] Analogous complexes were obtained with 2-phenyl-3-benzoyl-6-cymantrenylpyrylium and copper or cobalt acetate. The complexes described above are less stable than similar complexes with pseudoaromatic metallocycles. They decompose under the action of donor solvents (pyridine, dimethylformamide) and regenerate the initial pyrylium salts when treated with 70% HClO$_4$ in acetic acid. Formic acid reacts with these complexes to give pyranyl esters.[141]

(871)   (872)   (873)

The isolation of stable pyrylium complexes **873** (R = MeO, EtO, Ph) with metal carbonyl (chromium and molybdenum pentacarbonyls) was reported.[791]

IV. Physical Properties of Pyrylium Salts

A. SPECTRAL PROPERTIES

1. Optical Spectroscopy

a. *Electronic Absorption Spectra.* The earliest papers on UV absorption spectra of pyrylium salts were published by Hantzsch. During the time when the distinction between ionic and covalent bonds was not yet clear, and when the site of methylation of γ-pyrones was still debatable, the similarity of UV spectra led to the conclusion that 2,4,6-trimethylpyrylium and the methiodide of 2,6-dimethyl-γ-pyrone must have a similar structure. Indeed, the methiodide is now known to be the 2,6-dimethyl-4-methoxypyrylium iodide.[792] In 1930, when the ketone and dipolar formulas of 2,6-dimethyl-γ-pyrone were not yet regarded as mesomeric (resonance formulas), ultraviolet spectra in solutions of various acidities sought to clarify this point, and used also 2,4,6-trimethylpyrylium as a standard compound.[793]

Wizinger and co-workers[225,487,794-796] investigated the positions of the longest-wavelength absorption maxima of triarylpyrylium salts, most of which had auxochromic (NH₂, OR) groups in para positions.

The first systematic studies of a large series of pyrylium salts were published in 1960 by Balaban, Sahini, and Keplinger[797]; the absorption bands of this system were correlated to those of benzene, pyridine, and pyridinium salts on the basis of the gradual trends observed when one CH group in benzene is replaced by a heteroatom X of increasing electronegativity: $CH < N < NMe^+ < O^+$. This sequence formed the basis of a classification of aromatic heterocycles and of the definition of "aromaticity constants" by Balaban and co-workers.[798–800] The results are presented for the unsubstituted systems I[801–803] and the 2,4,6-trimethyl substituted systems II[797] in Table V. The first two bands in the spectrum are denoted in Table V as the x-band and y-band, respectively, in order to emphasize their similarity to the corresponding bands in benzene, polarized according to the axes shown under formulas I and II. The same notation was used by G. N. Lewis. However, Platt's notation or the group-theoretic notation introduced by Maria Goeppert-Mayer and Sklar, are more widely used; the correspondence between various notation systems is presented in Table VI. From the factors used in assigning UV absorption bands to electronic transitions (position and sequence of bands, effect of substituents, absorption intensities, effect of temperature, effect of solvent polarity, and vibrational structure)[804–807] the last one is not detectable because pyrylium salts are soluble only in polar solvents which obscure the vibrational structure.

The first electronic transition (x-, or ^1L_b-band) becomes less and less forbidden with increasing electronegativity of the heteroatom X. The absorption intensities increase steeply in the sequence **a–d**. The vibrational structure, very marked for **a** (X = CH) vanished completely for **d** (X = O^+). The energy levels which are degenerate in benzene are no longer so in pyrylium (cf. Fig. 7).

The effect of replacing in 2,4,6-trimethylpyrylium the methyl groups by aryl substituents acts differently on the x- and y-bands: upon increasing the conjugative capacity (phenyl < p-tolyl < p-anisyl) of the α-oriented aryl groups, the x- and x'-bands are considerably affected (bathochromic and hyperchromic effects), while the γ-band (where this may be detected) is less affected (cf. Table VII). This is in agreement with the projections of the substituents on the x-axis (cos 30° = 0.87) and y-axis (sin 30° = 0.5), respectively. On the other hand, on affecting the same replacement for the γ-oriented group, the x-band is practically constant, while the y-band is strongly displaced bathochromically and hyperchromically. In fact, the y-band appears at longer wavelength than the x-band in the case of 4-aryl-2,6-dimethylpyrylium salts.

Simalty and co-workers[216,220,823–825] provided ample and conclusive experimental evidence for the different effects of α- versus γ-substituents in pyrylium salts, corroborating the above data. The most interesting results are those concerning styrylpyrylium salts, easily formed from aromatic aldehydes and pyrylium salts possessing α- or γ-methyl(ene) groups (cf. Section III,A,2,b). A styryl substituent at an α- or γ-position exerts an effect slightly higher than a p-biphenylyl group[824] as indicated

TABLE V

ABSORPTION MAXIMA (nm) AND EXTINCTION COEFFICIENTS (IN BRACKETS) OF MONOCYCLIC SIX-MEMBERED AROMATICS OF TYPE I AND II

X	
a	CH
b	N
c	NMe⁺
d	O⁺

Compound	Absorption Bands X-Band Unsubstituted	2,4,6-Trimethyl	Y-Band Unsubstituted	2,4,6-Trimethyl
Benzene (Ia) Mesitylene (IIa)	255 (250)	265 (220)	198 (8000)	215 (7400)
Pyridine (Ib) sym-Collidine (IIb)	250 (2000)	267 (4000)	195 (7500)	216 (6900)
N-Methylpyridinium ClO₄⁻ (Ic) 1,2,4,6-Trimethyl-pyridinium ClO₄⁻ (IIc)	259 (4700)	268 (7340)	–	221 (5100)
Pyrylium ClO₄⁻ (Id) 2,4,6-Trimethyl-pyrylium ClO₄⁻ (IId)	269 (8800)	285 (12000)	219 (2100)	230 (4550)

TABLE VI

NOMENCLATURE OF ELECTRONIC ABSORPTION SPECTRA OF BENZOID AROMATICS[808]

Benzene nm (lg ε)	Lewis 809–812	Platt 813–815	Mayer-Sklar 816 D_{6h}	Mayer-Sklar 816 C_{2v}	Moffit 817	Doub-Vanderbelt 818	Braude 819	Clar 820,821	Gillam-Stern 805
–	Y'	^1B_a	–	$^1B_2$	Y	–	–	β'	–
183 (46000)	X'	^1B_b	$^1E_{1u}$	$^1A_1$	X	Second primary	Group I	β	–
198 203 (7400) 207	Y	^1L_a	$^1B_{1u}$	$^1A_1$	U	Primary	Group II	para	K
229, 234 238, 244 249, 255 261, 289 (220)	X	^1L_b	$^1B_{2u}$	$^1B_2$	V	Secondary	Group III	α	B

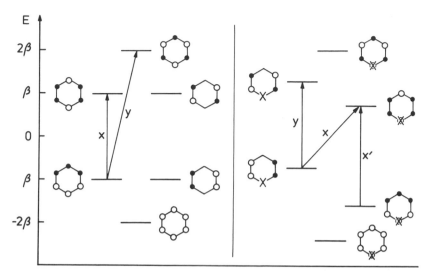

FIG. 7.   Energy levels, signs of atomic orbitals (LCAO–MO), and transitions for elec-tronic absorptions of benzene (left-hand side) and of six-membered heterocycles with heteroatom X (right-hand side).[753,822]

TABLE VII

ABSORPTION MAXIMA (nm) AND ABSORPTION INTENSITIES (lgε IN BRACKETS) OF 2,4,6-SUBSTITUTED PYRYLIUM PERCHLORATES OR CHLOROALUMINATES IN ACIDIFIED WATER (TO = p-TOLYL, AN = p-ANISYL)[361,826]

Substituent			Absorption Bands			
2	4	6	X	Y	X'	Y'
Me	Me	Me	284 (3.74)	233 (3.90)		
Ph	Me	Ph	386 (2.70)	273 (1.90)	236 (1.62)	219 (1.84)
To	Me	To	405 (3.00)	285 (1.96)	241 (1.45)	223 (1.95)
An	Me	An	438 (3.62)	306 (1.94)	267 (1.74)	231 (2.18)
Me	Ph	Me	294 (1.06)	326 (2.34)	231 (0.18)	
Me	To	Me	299 (1.16)	346 (2.64)	260 (0.20)	
Me	An	Me	295 (1.10)	378 (3.38)	255 (0.38)	235 (0.38)

TABLE VIII

ABSORPTION MAXIMA (nm) OF BIPHENYLYL (BI) AND STYRYL (ST) SUBSTITUTED PYRYLIUM
SALTS.[220] FOR COMPARISON, THE THIRD AND LAST LINES INDICATE THE MAXIMA OF PHENYL-
SUBSTITUTED PYRYLIUM SALTS

Substituent			Absorption Bands			
2	4	6	X	Y	X'	Y'
Ph	H	St	446	306	249	
Ph	H	Bi	429	301	264	
Ph	H	Ph	415	283	243	
Ph	Ph	St	452	366	306	256
Ph	Ph	Bi	435	363	303	258
Ph	St	Ph		418	277	240
Ph	Bi	Ph		408	278	240
Ph	Ph	Ph	408	361	278	

in Table VIII. In the series where only one α-, and γ-, or the two α-substituents were systematically varied, the y-band was sensitive only to the variation of the γ-substituent. In Table VIII it may be seen that with a γ-styryl or γ-biphenylyl group, the y-band is bathochromically shifted so that it overlaps the x-band. A definite analogy exists between triarylmethyl dyes [malachite green (874) and crystal violet (875)], 2,4,6-triarylpyryliun salts 876 and 877, and 2,4,6-tristyrylpyrylium salts 878 and 879, respectively (λ_{max} values in square brackets).[220]

As indicated on the formulas, the well-known hypsochromic effect when introducing the extra auxochrome (874 → 875) is due to the equivalence of the x- and y-axes in the triarylmethyl series, and to the fact that 875 has additional symmetry. The pyrylium ring has C_{2v} symmetry, therefore the x- and y-axes can never be equivalent. However, the central carbon in triarylmethyl cations and the pyrylium ring in 2,4,6-trisubstituted pyrylium salts have a definite analogy which leads to a comparable hypsochromic effect on introducing an auxiliary auxochrome, e.g., 876 → 877 or 878 → 879.

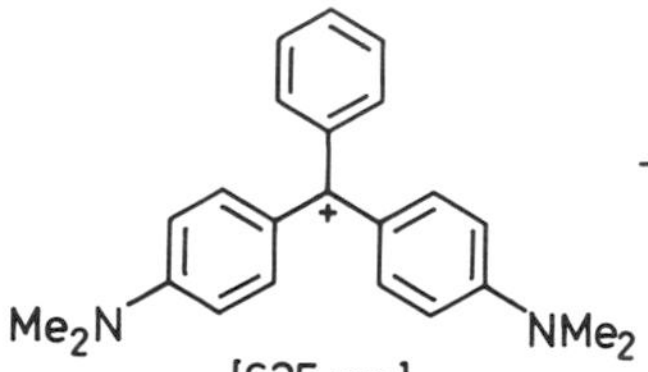

[625 nm]
(874)

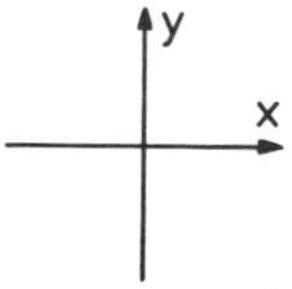

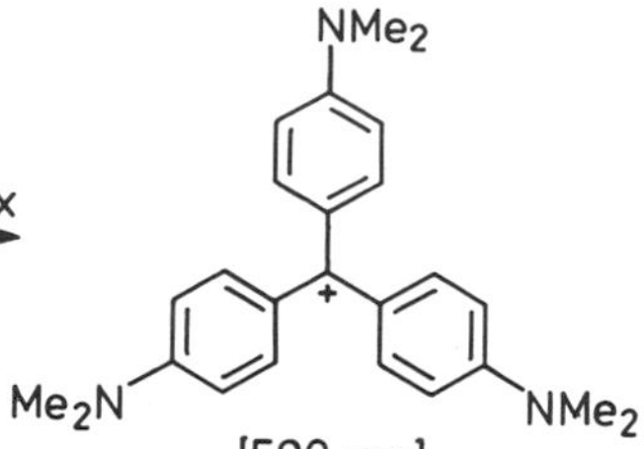

[590 nm]
(875)

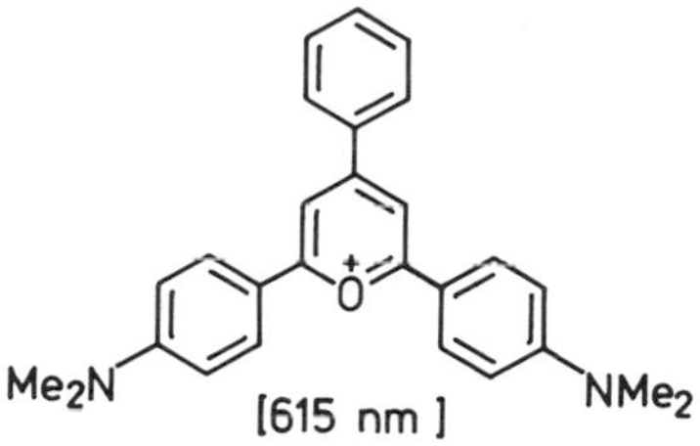

[615 nm]
(876)

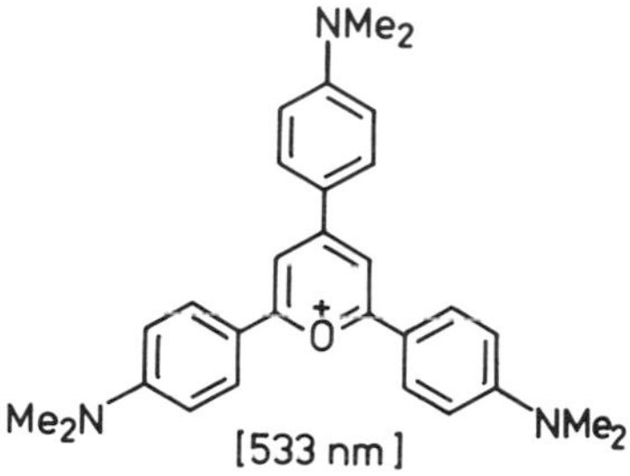

[533 nm]
(877)

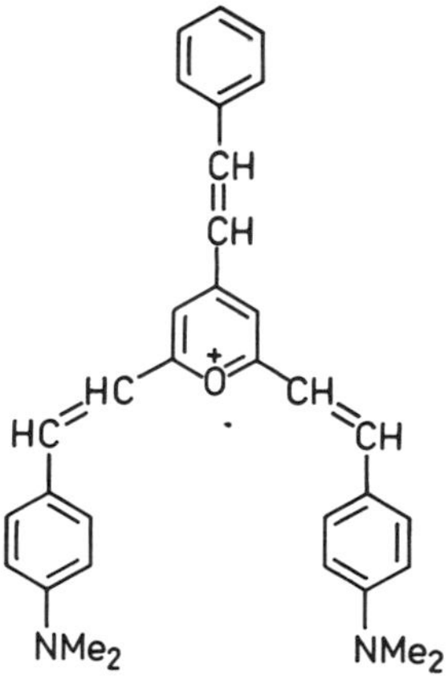

[725 nm]
(878)

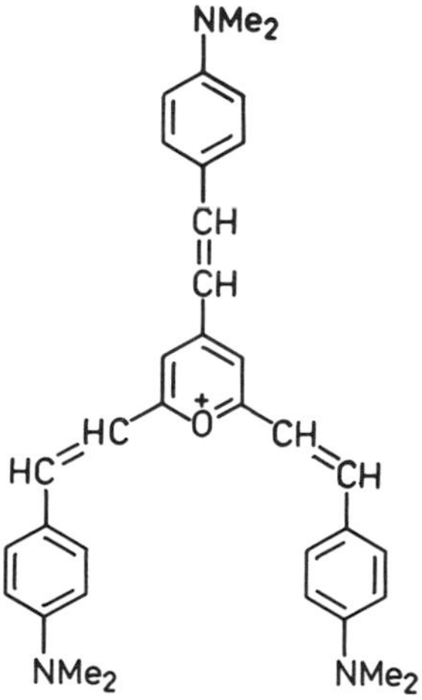

[682 nm]
(879)

No low-energy $n \rightarrow \pi^*$ transitions were observed in pyrylium salts, despite careful search which would have demonstrated an extinction coefficient as low as $\varepsilon = 1\ M^{-1}cm^{-1}$. Therefore it was inferred[827] that the lowest energy transition is of type $\pi \rightarrow \pi^*$. This fact was rationalized, in contrast with pyridine, as being due to the higher electronegativity of the oxygen, and to its positive charge.

b. *Emission (Fluorescence and Phosphorescence) Spectra.* The fluorescence of many aryl-substituted pyrylium salts in dilute solution is so intense that they can be detected in minute amounts; Kostanecki and Rossbach[828] noted in 1896 the strong green fluorescence of 1,3,5-triphenylpentane-1,5-dione in sulfuric acid, but they failed to isolate 2,4,6-triphenylpyrylium which caused the fluorescence [the oxidizing agent in the reaction is sulfuric acid (cf. Section II,A,2,b)]. Dilthey in 1917 characterized 2,4,6-triarylpyrylium salts[398] and later[829] reported that these salts fluoresce much more strongly than the corresponding pyridines, and that the pseudobases do not fluoresce.

An early paper[830] mentioned that 2,4,6-triphenylpyrylium chloride adsorbed on evacuated silica gel at -80 to $-180°C$ changes its yellow-green phosphorescence on introduction of oxygen, with a burst of bright emission of blue fluorescent light. This observation was interpreted as indicating that phosphorescent (triplet) triphenylpyrylium, with an energy r_2 lower than the energy r_1 of the (singlet) state leading to fluorescent emission, acts as sensitizer for oxygen leading to a quenched pyrylium state of energy r:

$$r_2 + O_2 \rightarrow O_2^* + (r_2 - r)$$

The activated O_2^* can be trapped by adsorbed oxygen acceptors such as carbon disulfide or allylthiourea without influencing the fluorescent light burst of the pyrylium adsorbate, therefore the energy of this burst originates neither in oxidations nor in O_2^*. To raise the energy from r_2 to r_1, i.e., to cause fluorescence, the concentration of pyrylium and the light intensity must be high enough, otherwise oxygen acts only as quencher; therefore the mechanism involves interaction between two phosphorescent molecules and O_2:

$$r_2 \rightarrow h\nu_2 \text{ (phosphorescence)}$$

$$2\,r_2 + O_2 \rightarrow r_1 + O_2^*$$

$$r_1 \rightarrow h\nu_1 \text{ (fluorescence)}$$

From a study of absorption and fluorescence spectra for a series of pyrylium and pyridinium perchlorates it was reported[831] that at least one of the 2,4,6-substituents must be phenyl for detectable fluorescence (this condition is contradicted by other data[832]) and it was suggested that 2,4,6-triarylpyrylium and 2,4,6-triaryl-*N*-methylpyridinium salts cannot be used as phosphors. A systematic study of the luminescence (at 77 and 293°K) of pyrylium salts in methylene dichloride, acetonitrile, or acetic acid was effected.[833]

An increase of planarity of the π-system by fixing the phenyl group positions with respect to the pyrylium ring by CH_2 chains results in a decrease of the Stokes shift and an increase of the fluorescence quantum yields. The length of the π-electron system, its topology, the existence and nature of substituents allow a prediction of the spectral properties.[833]

On the basis of the fluorescence spectrum structural assignments may be made[834] (tentative structural suppositions in Ref. 834 ought to be rechecked). Correlations between chemical structure and fluorescence spectrum were discussed for several sensitizers, among which were pyrylocyanines.[835]

The fluorescent properties of aryl-substituted pyrylium salts have been used in dye lasers: 2,4,6-triphenylpyrylium fluoborate (laser wavelength 485 nm in methanol at a concentration of 1.7 mmol/liter)[836]; compounds **880, 881,** and **882** have been used as Q-switches for neodymium lasers in acetonitrile (λ_{max} in square brackets), the compounds **883** and **884** as Q-switches for ruby lasers in the same solvent.[837]

[1026 nm]

(880)

[1026 nm]

(881)

[1062 nm]

(882)

[652 nm]

(883)

[685 nm]

(884)

A patent[838] for dye lasers describes the use of 2,4,6-triarylpyrylium (where the aryl is a *p*-alkoxyphenyl group with the alkoxy group containing 1–12 carbon atoms) or flavylium salts (which may contain other condensed benzenoid rings). As example, 4-(*p*-amyloxyphenyl)-2,6-di-(*p*-ethoxyphenyl)pyrylium perchlorate in 1,2-dichloroethane (10^{-2}–10^{-4} mol/liter) was pumped optically with a ruby laser yielding light of 347 nm; the emitted light had 559 nm with a band half-width of 20–35 nm. Other examples for benzopyrylium or xanthylium salts are indicated in the same patent affording emission in the range 500–600 nm. The solution can be circulated to avoid overheating.

    c. *Charge-Transfer Spectra.*   Charge-transfer absorption bands are now well understood.[839–844] The pyrylium cation may function as an electron acceptor in photochemically excited states leading to the appearance of low-energy absorption bands. The donor may be the anion or a neutral electron-rich molecule. It had been observed for a long time that crystalline pyrylium iodides are more deeply colored than salts of the same cations with other anions which cannot donate electrons, such as perchlorate or fluoborate; in ethanolic or aqueous solutions all anions lead to the same spectra, but in nonpolar solvents such as CH_2Cl_2 or CH_2ClCH_2Cl long-wavelength bands appear in iodides. The first rationalization of the CT absorption of 2,4,6-trimethylpyrylium iodide by Feldman and Winstein[845] was immediately afterwards[846] completed by an extensive study made by Balaban *et al.* of various pyrylium iodides.[207] Even with the hygroscopic 2,4,6-trimethylpyrylium bromide, a charge-transfer band appears as a shoulder, but with the chloride ion the CT band is submerged under the first $\pi \rightarrow \pi^*$ absorption (x-band).[207] Though electronic absorption bands of salts in general do not obey the Lambert–Beer law, especially for compounds which like pyrylium salts are fluorescent, the CT band of iodides deviates much more strongly from this law, i.e., its extinction coefficient is appreciably concentration-dependent.

2,4,6-Trimethylpyrylium iodide presents in CH_2ClCH_2Cl two bands (which are absent with anions other than iodide) at 360 and 450 nm. The separation between these bands agrees with that observed with pyridinium iodide[847,848] and indicates that the two bands are due to electronic transitions from the highest occupied MO of the iodide anion to the vacant MOs of the aromatic cation, rather than the involvement of different excitation states of iodine atoms.

In most cases, e.g., with aryl-substituted pyrylium salts, only one CT band may be observed. However, with 2,6-diaryl-4-methylpyrylium no CT band is visible because the electronic absorption x-band is much more strongly displaced bathochromically than the CT band.

A correlation was observed between the longest wavelength CT band and the y-band in the electronic absorption spectrum.[207]

Other anions causing the appearance of CT bands in pyrylium salts are the pseudohalides (thiocyanate and selenocyanate) for which a systematic study was made,[849,850] boron-containing anions (bispyrocate-cholborate, tetraphenylborate),[783] tricyanomethide,[491] 1,1,3,3-tetracyano-propenide[491,851–853] [with this anion, photo- and semiconducting properties were discovered[853] and the crystal structure of the salt with 2,4,6-tri-phenylpyrylium was investigated[853] (cf. Section IV,B)], 2-aza-1,1,3,3-tetracyanopropenide,[854] or pentacarbomethoxycyclopentadienyl.[855] Indirect evidence for CT complexes between the hydroxide ion or water and a pyrylium ring was obtained by studying the kinetics of hydrolysis of 4-ethoxy-2,6-dimethylpyrylium.[292] 7,7,8,8-Tetracyano-p-quinodimethane affords unidimensional conducting radical-ion salts with a variety of cations, among which 2,4,6-triphenylpyrylium was mentioned[856] (cf. Section III,A,1). Modeled after the tetracyanoquinodimethane + tetrathiaful-valene unidimensional conducting charge-transfer complexes, the complex between 2,2',6,6'-tetramethyl-4,4'-bipyrylene and hexacyanoben-zene was studied at various temperatures; the conductivity and magnetic susceptibility suggest a chain structure.[857]

Lastly, neutral molecules like secondary or tertiary amines, pyridine N-oxide, or anthracene, present CT absorption bands when admixed with solutions of 2,4,6-trimethylpyrylium salts; the data were rationalized by calculating the energy of the LUMO of the acceptor cations, and compared to analogous data for the tropylium cation.[858] It was concluded,[858] as had been done earlier,[207] that the electron-accepting capacity of pyrylium is lower than that of tropylium, but higher than that of pyridinium. Briegleb *et al.*[859] investigated the fluorescence and phosphorescence spectra of CT complexes formed by 2,4,6-trimethylpyrylium with anthracene, pyrene, and naphthalene at $-170°C$ in glassy solution.

d. *Infrared Spectra.* Vibrational spectra give information about the ground states of molecules.[363,860–862] More than other physical methods, infrared absorption spectra demonstrate the close parallelism which exists between the pyrylium cations, other six-membered heterocyclic aromatic systems (pyridine, the pyridinium cation, the borabenzene anion) and benzene or its anion. These six-membered monocyclic aromatic systems form a regular series (Scheme 17) where definite trends may be observed, as was emphasized in other sections of this review. Though certain trends are also apparent in the IR spectra, the most striking feature of these spectra is, however, their close similarity over the whole series.[363]

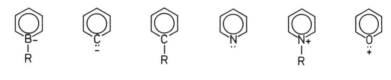

SCHEME 17

The first systematic investigation of pyrylium salts, including the unsubstituted pyrylium cation, was reported by Balaban *et al.*[363] in 1962. For the unsubstituted pyrylium cation, deuterated analogs were investigated[863]: $2\text{-}D_1$, $3\text{-}D_1$, $1,2,3,4,5\text{-}D_5$. These experimental data were supplemented later by a normal coordinate analysis for the unsubstituted pyrylium salt,[864] together with thiopyrylium. The high electronegativity of the oxygen heteroatom makes the force constants in pyrylium (which has a pronounced carbocation character) more different from benzene than in thiopyrylium; however, in thiopyrylium the mass effect of the sulfur heteroatom is detectable.

As a result of these factors, intense CH stretching vibrations at ~ 3100 cm^{-1} and B_1 ring modes at $1610–1640\ cm^{-1}$ appear, which serve as useful diagnostic data for the presence of a pyrylium ring, especially the latter band. The reduced symmetry relative to benzene leads to enhanced intensity of these vibrations. In non-, mono-, or pentaalkyl-substituted pyrylium systems, the highest-frequency band appears around 1620 cm^{-1}, while in di-, tri-, or tetraalkylpyrylium salts it appears around 1640 cm^{-1}. In aryl-substituted compounds which are not sterically distorted, it appears around $1630\ cm^{-1}$, but when overcrowding is present (e.g.,

2,3,4,6-tetraphenylpyrylium) the frequency is lowered to 1610–1620 cm^{-1}.[363] Chloro substituents lower the frequency of these ring vibrations.[865]

Detailed studies of 1,3-disubstituted 5,6,7,8-tetrahydrobenzo[c]pyrylium salts **885**[866] or substituted tetrahydrobenzo[b]pyrylium salts, e.g., **886** and **887**,[867] demonstrated common bands similar to those reported for alkyl-substituted pyrylium salts.[866] Styryl-substituted pyrylium salts present a common enhanced band for the pyrylium and C=C vibrations at 1620–1630 cm^{-1}.[824]

(885)   (886)   (887)

2. Nuclear Magnetic Resonance Spectra

a. ^{1}H-NMR Spectra.   Since the O$^+$ heteroatom constitutes the strongest single perturbation which can be introduced into a benzene ring, it is natural to expect pronounced effects in the NMR spectra.

The simplest, unsubstituted pyrylium cation affords a rather complicated A$_2$B$_2$C ^{1}H-NMR spectrum: the first extensive ^{1}H-NMR study indicated approximate resonance positions of α-, γ-, and β-protons (in increasing field order).[868] Degani et al.[869,870] using 60 MHz instruments, and more recently Radics and Kardos[871] with 60 and 100 MHz instruments, and with INDOR techniques refined the previous assignments and obtained the coupling constants. The evolution of technique is well illustrated in Table IX indicating the above assignments of pyrylium ^{1}H-NMR peaks. Table IX also includes recent data for pyridine.

Methyl-substituted pyrylium salts have methyl peaks appearing at the same order of increasing magnetic fields: α < γ < β. This order was evident from the first study on ^{1}H-NMR spectra of pyrylium salts by Balaban, Bedford, and Katritzky[868] which showed that α-methyl groups absorb at δ 2.85–3.00 in alkyl-substituted pyrylium salts, and at δ 3.00–3.15 for pyrylium salts possessing additional phenyl groups acting only as substituents which modify the charge density of the pyrylium ring (for shielding due to ring current effects, cf. below). β-Methyl groups absorb at δ 2.40–2.50, and γ-methyl groups at δ 2.70–2.83 for compounds devoid of phenyl groups. In pyrylium salts with phenyl groups which are not adjacent to the γ-methyl substituent, low field shifts to δ 2.85–2.95

TABLE IX

REFINEMENT OF ¹H-NMR ASSIGNMENTS FOR PYRYLIUM PERCHLORATE AND PYRIDINE
(CHEMICAL SHIFTS AS δ VALUES, J IN Hz)

Compound	Solvent	α	β	γ	$^3J_{23}$	$^4J_{24}$	$^5J_{25}$	$^4J_{26}$	$^3J_{34}$	$^4J_{35}$	$^1J_{2CH}$	$^1J_{3CH}$	$^1J_{4CH}$
Pyrylium 868	SO_2	~9.6	~8.5	~9.3									
Pyrylium 869	70% $HClO_4$	9.59	8.40	9.20	3.5	2.0	0.0	1.5	8.0	0.0			
Pyrylium 870	CF_3COOD	9.220	8.077	8.906	4.24	1.83	0.97	0.30	8.13	1.49	218	180	180
Pyridine 872,873	—	8.608	7.123	7.502	4.950	1.824	1.019	-0.043	7.627	1.466	178.4	162.3	162.0

are exerted on γ-methyl groups. The original paper[868] should be consulted for ethyl and other alkyl groups; in that paper all spectra were recorded in liquid sulfur dioxide. More recent experience shows that trifluoroacetic acid is more convenient, both regarding solubilities and ease of manipulation.

Calculations of charge densities in unsubstituted and substituted pyrylium salts fully substantiate the experimental findings. The simple HMO method[874] gave a fair correlation between π-electron charge densities and chemical shifts, both for the unsubstituted and for methyl-substituted pyrylium salts (the standard deviation for ten methyl-substituted pyrylium salts resulting in 18 points on the diagram is $\sigma = 0.077$ δ). The correlation is significantly improved on using a self-consistent technique, i.e., varying Coulomb integrals α according to charges, and resonance integrals β according to bond orders until the α,β-matrix converged for each element to within 0.0005: the ωβ-variant gives a standard deviation $\sigma = 0.0073$ δ but the ω'β-technique gives a considerable improvement, $\sigma = 0.065$ δ (all chemical shift δ values in ppm).

The same order of chemical shifts in increasing magnetic field $\alpha < \gamma < \beta$ is encountered in pyridines and pyridinium salts, with the exception of N-aryl-2,4,6-trimethylpyridinium salts[407]: due to the ring current of the N-aryl group which is almost perpendicular to the pyridinium ring[875] (this leads to atropisomerism of suitably substituted N-arylpyridinium salts[876]), the α-oriented methyl groups are shielded so much that they appear at higher field than the γ-methyl group.[407] This effect has recently been employed for demonstrating the presence and magnitude of ring currents in the R group of 2,4,6-trimethyl-N-R-pyridinium salts obtained readily from RNH_2 and 2,4,6-trimethylpyrylium[589] (cf. Section III,C,3,c).

2,4,6-Trimethyl-3-phenylpyrylium presents[179] three separate methyl peaks (in trifluoroacetic acid) at δ 2.98, 2.74, and 2.52, assigned to the methyls in positions 6, 2, and 4, respectively. The shielding of 0.2–0.3 ppm of the last two peaks is due to the ring current of the phenyl group which is tilted relative to the pyrylium ring. As a confirmation of this assignment, the peak at δ 2.52 disappears fastest on deuteration, as γ-methyl peaks do.

^{1}H-NMR spectra of substituted pyrylium, thiopyrylium, and selenopyrylium salts were also discussed by Tolmachev and co-workers.[877] The same authors[878] investigated pyrylium salts which had coplanar or noncoplanar α-phenyl groups; in the latter case, ortho, meta, and para protons appear as one multiplet, whereas coplanar phenyl groups have ortho protons resonating at lower fields.

An uncommonly strong shielding exerted by a pyrylium ring on side-chain protons was observed in a pyrylophanium salt (and in the corresponding pyridine or pyridinium salts): diacetylation of cyclodecene affords the pyrylium salt **888** ($X = O^+$) where the indicated methylene protons resonate at δ 0.49; the same group in the corresponding pyridinium salt **888** ($X = NH^+$ or NMe^+) resonates at δ 0.2–0.3, while in the respective pyridine **888** ($X = N$) the methylene protons resonate at δ -0.08.[184] The ^{13}C-NMR spectra of **888** indicate ring strain in the polymethylene bridge which obscures any shielding of the specified methylene carbon; in the ^{1}H- or ^{13}C-NMR spectra of the diacetylation product of cyclododecene **889** ($X = O^+$)[184-186] no such anomalies were observed.

(888)   (889)   (890)   (891)

(287)   (288)   (892)

The diagnostic value of ^{1}H-NMR spectra for elucidating structures of new pyrylium salts may be illustrated by several examples. The diacetylation of diisobutene yielded a pyrylium salt which could possess the structure 2,6-dimethyl-4-neopentylpyrylium (**890**), or 2,3,6-trimethyl-4-*t*-butylpyrylium (**891**); the ^{1}H-NMR spectrum showed that the former structure **890** is correct.[184] *t*-Butyl chloride and aluminum chloride can extract a hydride ion from 1,5-diketones such as 1,5-diphenyl-1,5-pentanedione, yielding pyrylium salts; however, with 1,3,5-triphenyl-1,5-pentanedione and 3-(2-oxocyclohexyl)-1,3-diphenylpropane-1-one, in addition to the expected pyrylium salts, mono- or di-*t*-butylated products **287**, **288**, and **892** were also obtained (cf. Section III,A,8,a), whose structure was elucidated by ^{1}H-NMR spectra.[264]

Another application of ^{1}H-NMR for structural studies of pyrylium salts refers to Kostanecki's compound, a tautomer of 1,2,3-tribenzoylpropene which in the solid state is probably **893**, and in solution a mixture of **894–897** in fast equilibrium.[879]

(893)   (894)   (895)   (896)   (897)

The triacetylation product of isobutene has the resonance structure **898A** ↔ **898B**[569] with an s-cis configuration of the α,β-unsaturated ketone moiety, as proved by[341] (i) decoupling experiments leading to coupling constants J = (α-Me − H) = 1 Hz; (ii) the very low field resonance of one β-pyran proton indicating that it is deshielded by a neighboring carbonyl group; (iii) ASIS and LIS data. An s-trans configuration **899A** ↔ **899B** is excluded by a computer-simulated comparison of LIS data[341]; such a configuration **899** has a higher electronic energy than configuration **898** and a higher steric repulsion between the β-pyran hydrogen and the methyl group.

(898)   (899)

(900)   (901)   (902)

If the ¹H-NMR spectra are recorded in trifluoroacetic acid, the tautomeric pyrylium salts **900** ⇌ **901** are evident; introduction of acetone into this mixture increases the contents of **901** in the reaction mixture due to hydrogen-bond formation.[341]

Analogous conclusions were reached by Oestensen and Undheim[328,880] for the condensation products of 2,6-dimethoxycarbonylpyrylium with methyl ketones, e.g., **902**. Despite the absence of couplings with the β-

pyran protons in these products, the s-cis configuration **902** was established on the basis of the strong deshielding of one β-pyran proton.

Khedija, Strzelecka, and Simalty[240] studied the ^{1}H-NMR spectra of
pyranylidenesulfonium salts **903** where X is H, Me, Ph, or COPh: the
β-pyran protons appear distinct because of the cis–trans relationships.
In the case of the benzoylpyranylidenesulfonium salt, two isomers appear
in the ^{1}H-NMR spectrum: one with distinct methyl peaks at δ 2.53 and
1.91 ppm (20%), the other with a degenerate methyl peak at δ 3.26 ppm
(80%). They were interpreted as **903** (X = PhCO, 20%) and **904** (X =
PhCO, 80%).[240] An alternative explanation would be geometric isomerism
905A, **905B** ⇌ **906A**, **906B**, since rotation around the C—C partial double
bond is expected to be slow (structures **905A** and **906B**). When X is Ph,
^{1}H-NMR spectra indicate a nonplanar structure of **904** because the SMe$_2$
peak is shielded.

(903)   (904)

A   (905)   B   A   (906)   B

Pyranylideneimonium salts **907A** ↔ **907B** possess a trans structure as
indicated by the coupling constant of 12 Hz when R = H. When R =
Ph, this phenyl group is again tilted out of coplanarity. In trifluoroacetic
acid, tautomeric dications **908** ⇌ **909** are indicated by the ^{1}H-NMR
spectra, in proportions which can be influenced by added solvents.[240]

A comparison between the substituent increments of the pyrylium ring
and other groups as substituents of aromatic rings showed in the case
of α-arylpyrylium salts, e.g., **910**, that an α-pyrylium ring is about as
electron-attracting as a COOR, COCl, or a CONH$_2$ group; a γ-pyrylium
ring exerts on the ^{1}H-NMR spectra a weaker electron-attracting effect,
about equal to that of a phenyl or vinyl group.[867]

A (907) B

(908) (909)

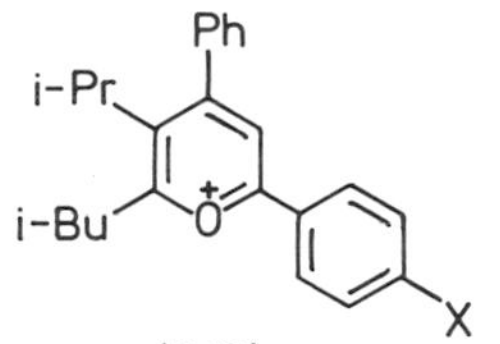

(910) 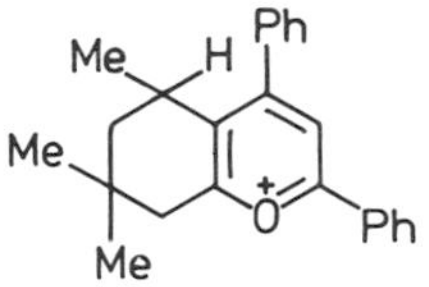 (886)

In 5,6,7,8-tetrahydrobenzo[b]pyrylium salts with the structure **886**, the 5-methyl group is quasi-equatorial, and the twisted 4-phenyl group exerts on this methyl a detectable shielding.[867]

By means of ^{1}H-NMR spectra and IR spectra[363] it was discovered that on heating in D_2O in buffered media, pyrylium salts exchange alkyl protons from the α- and γ-benzylic positions. NMR spectra are, however, much more convenient because they discriminate between α-, β-, and γ-side-chain and ring protons. It was thus possible to establish that the γ-side-chain exchange proceeds about ten times faster than the α-exchange allowing selective deuteration as discussed in detail in Section III,A,7,b.

The kinetic parameters of deuterations and dedeuterations of methyl and ethyl side chains were determined by direct integration of the respective proton signals (CH_3 and CH_2, respectively). However, for isopropyl groups the direct method for the isotopic exchange of the CH is too imprecise, therefore an indirect method was adopted. Since $CHMe_2$ groups present a widely spaced doublet ($J = 7$ Hz) for the methyl groups, while $CDMe_2$ groups indicate a closely spaced triplet ($J = 1$ Hz), integration of the methyl portion of the spectra allows an indirect measurement of the deuteration degree.

Balaban[783] showed that europium chelates cause downfield induced shifts of pyrylium and pyridinium protons. Possibly these cations behave as π-donors toward the lanthanide shift reagents.

b. *^{13}C-NMR Spectra.* So far, only a few papers have appeared on ^{13}C-NMR spectra of pyrylium salts. Balaban and Wray[302,881] investigated a large number of unsubstituted, alkyl- and/or aryl-substituted pyrylium salts, using as solvent a mixture of F_3CCOOH and CD_2Cl_2 (4 : 1). A comparison of the ^{13}C-NMR spectra of pyrylium with those of benzene, pyridine, and pyridinium reveals that charge density associated with the introduction of a heteroatom into the aromatic ring determines primarily the ^{13}C-NMR chemical shifts. The monotonous variations of the charge densities for the unsubstituted and the 2,4,6-trimethyl-substituted systems at the α- and γ-carbons when X = CH, N, NH$^+$ or O$^+$ is well reproduced by the experimental ^{13}C-NMR data (with the exception of the α-carbons of pyridine and pyridinium); the calculated total charge densities for the β-carbons (C-3) present in the above series a non-monotonous variation which is perfectly mirrored by the experimental ^{13}C-NMR data [302] (Table X).

TABLE X

CHARGE DENSITIES (INDO-MO) IN THE UPPER ROW, AND ^{13}C-NMR SHIFTS (ppm) IN THE LOWER ROW OF UNSUBSTITUTED AND TRIMETHYL-SUBSTITUTED BENZENOID AROMATICS[302]

X	C-2	C-3	C-4	C-2	C-3	C-4
CH	3.977	3.977	3.977	3.960	4.009	3.960
	128.7	128.7	128.7	137.69	127.00	137.69
N	3.839	4.029	3.924	3.834	4.059	3.915
	150.4	124.1	136.1	157.43	121.14	147.37
NH$^+$	3.821	3.993	3.857	3.802	4.046	3.842
	142.6	129.1	148.5	153.66	126.69	162.24
NMe$^+$				156.67	129.93	160.98
O$^+$	3.698	4.026	3.791	3.682	4.084	3.784
	169.32	127.74	161.21	180.15	124.88	177.20

The chemical shifts of methyl carbons bonded to the pyrylium ring also appear at characteristic fields: α-Me at 19–20 ppm, β-Me at ~17 ppm, γ-Me at 23–25 ppm. When a β-alkyl group is present the respective β-pyrylium carbon atom has a chemical shift of 133–135 ppm.[302] It may be seen from Table X that in pyridine and pyrylium the deshielding increases in the following order for the ring carbons: β < γ < α, whereas in pyridinium (both NH$^+$ and NMe$^+$) the order is β < α < γ. The side-chain deuteration rates show a similar inversion: for 2,4,6-trialkylpy-rylium they increase in the order α < γ, whereas for 1,2,4,6-tetraalkyl-pyridinium they increase in the order γ < α.

Table X also shows that, in agreement with ^{1}H-NMR and side-chain deuteration rate studies, the pyrylium ring is the six-membered ring with the highest possible single perturbation: there is no heteroatom or sub-stituent of higher electronegativity than an O$^+$ heteroatom, therefore the deshielding of α-ring carbons in pyrylium is the strongest yet observed in such six-membered aromatic rings. These considerations also explain why the resonance energy of pyrylium is the lowest, and why the py-rylium ring is so easily opened by nucleophilic attack; $^+$N—CN, $^+$N—Py, $^+$N—NO$_2$, or $^+$N—C$_6$H$_3$-2,4-(NO$_2$)$_2$ heteroatoms instead of O$^+$ ap-proach the electronegativity of O$^+$ and give rise to pyridinium salts able to afford ring opening on nucleophilic attack. It will be interesting to study ^{1}H-NMR, ^{13}C-NMR, and side-chain deuteration rates of such pyr-idinium salts and compare the results with those of pyrylium salts.

Since, especially for α-carbons in pyrylium salts, and for carbons of alkyl side chains, charge densities are not the only governing factors, empirical correlations of ring carbon chemical shifts in alkylpyrylium with those in alkylbenzenes (δ_B) were found[302]:

2-Alkylpyrylium	α—C	$0.67\,\delta_B + 88.19$
	β—C	$0.85\,\delta_B + 14.79$
	γ—C	no correlation
4-Alkylpyrylium	α—C	no correlation
	β—C	$0.87\,\delta_B + 12.11$
	γ—C	$0.82\,\delta_B + 64.64.$

Analogous substituent chemical shift correlations were found for the ^{13}C-chemical shift changes of the α-carbons in alkyl side chains (Me, Et, i-Pr, t-Bu) on comparing alkylbenzenes with alkylpyrylium salts. For phenyl substituents on pyrylium rings in the absence of steric hindrance there exists considerable electronic delocalization of the positive charge in the para position of the phenyl ring (138 ± 1 ppm); when because of steric overcrowding the phenyl groups are tilted out of coplanarity with the pyrylium ring, the para carbon resonates at 132–135 ppm (in biphenyl the para carbon resonates at 128 ppm). Similarly, a low frequency shift of the para carbon for the 2-phenyl relative to the same carbon in the 6-phenyl in 3,4-dimethyl-2,6-diphenylpyrylium (911) indicates that the methyl group at C-3 causes a significant out-of-plane twist for the 2-phenyl group.[302]

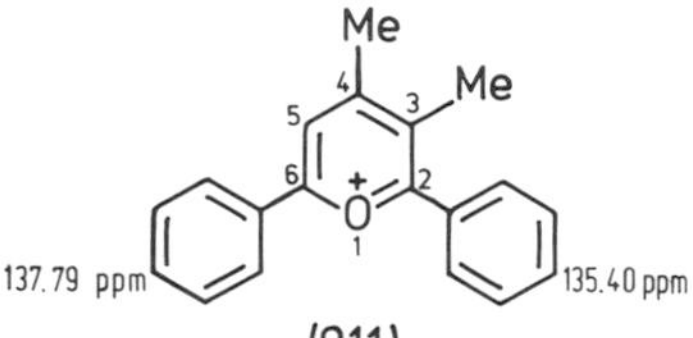

(911)

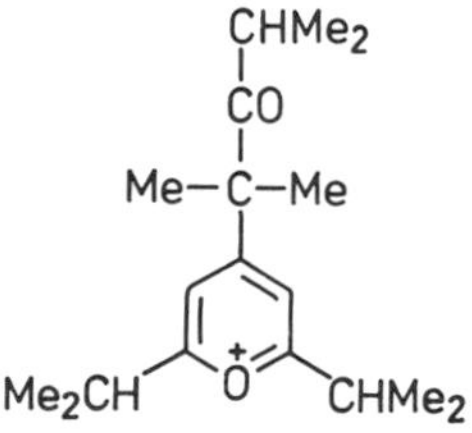

(912) 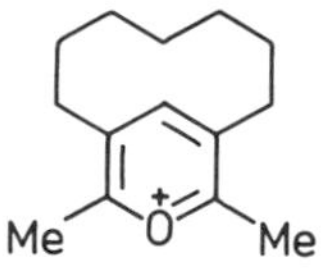   (108) 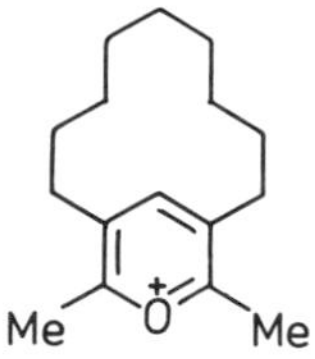   (913)

Similar ^{13}C-NMR chemical shifts appear in pyrylium salts with more complicated structures, e.g., 912,[180] 108,[510] and 913.[510] It may be confidently inferred that in the near future ^{13}C-NMR spectroscopy will become as useful a tool for structural investigations of pyrylium salts as ^{1}H-NMR spectroscopy.

Unpublished data,[882] making use of α-deuterated analogs, indicate chemical shifts for octahydroxanthylium in CD_2Cl_2–CF_3COOH (1 : 9) (letters indicate multiplicities for off-resonance proton decoupling) as shown in Scheme 18.

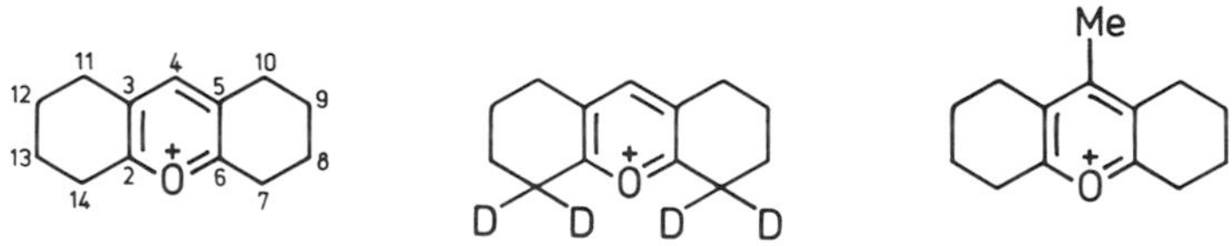

2,6 :	179.86	s	179.57	broad	179.94	
3,5 :	135.70	s	135.44		134.24	
4 :	158.63	d	158.44		176.66	
7,14:	30.56	t	29.77	quintet	30.87	
10,11:	27.92	t	27.74		26.41	
9,12:	22.01	t	21.83		22.38	
8,13:	22.01	t	21.65		21.81	
Me :	—		—		17.83	

SCHEME 18

In connection with an investigation of pyran versus dienone structures for the adducts of 2,4,6-triarylpyrylium salts with methoxide, Katritzky and co-workers[297] studied the ^{13}C-NMR spectra of 2,4,6-triarylpyrylium salts where the aryl was phenyl, p-tolyl, and p-fluorophenyl using [D$_6$]DMSO as solvent. The latter substitution proved extremely useful for assigning unambiguously the phenyl ring carbons because of the distance-dependent C–F coupling ($J = 256$ Hz for p-carbon, 22 Hz for m-carbon, 8–10 Hz for o-carbon, and 2.4 Hz for ipso-carbon). Then the phenyl and p-tolyl-substituted cations were completely assigned on substituent chemical shifts (SCS) considerations using the SCS in para-nitro compounds as a model. The agreement with the previous assignments is very good.

Chenon, Sib, and Simalty[883] have made a detailed study of rotation barriers around the ring carbon side chain C—N bond in 2-N,N-dimethylaminopyrylium cations by using ^{13}C-NMR spectroscopy. This study was triggered by the observation that whereas 4-alkoxypyrylium salts react readily with alcohols, exchanging the alkoxy groups, the introduction of another donor group like dialkylamino in position 2 suppresses this exchange by diminishing the susceptibility of the pyrylium ring toward nucleophilic attack. Chemical shifts and carbon–hydrogen coupling constants are given for two series of pyrylium salts in nitromethane. Assignments were made on the basis of previous data[302] and of coupling constants. Dynamic ^{13}C-NMR spectroscopy allowed the determination of $\Delta G^{\ddagger}$, $\Delta S^{\ddagger}$, and $\Delta H^{\ddagger}$ values for the internal rotation around the CN group by line shape analysis for the dimethylamino carbon peaks. Results from Table XI indicate that on decreasing the donor ability of the other substituents (Me$_2$N > OMe > Me > Ph), the rotation barrier increases for the α-Me$_2$N group, in agreement with the expected increasing conjugation between this group and the pyrylium ring. Rotation barriers are much higher for γ-NMe$_2$ groups than for α-NMe$_2$ groups: for 2,4-bis(dimethylamino)-6-methylpyrylium (**914a**) there is a difference of about 5 kcal/mol between the free enthalpies of activation and for compound **915** the coalescence temperature could not even be reached. This finding indicates that though the electron deficit is lower at γ than at α, a γ-NMe$_2$ group conjugates better than an α-NMe$_2$ group. Again, this conclusion is consistent with side-chain deuterium exchange studies which also indicate that a γ-methylenepyran is favored over an α-methylenepyran, which explains why the γ-methyl exchanges faster than the α-methyls of 2,4,6-trimethylpyrylium in deuterium oxide.

3. Electronic Spin Resonance Spectra

It had been noted by Balaban et al.[379] that zinc reduction of 2,4,6-trisubstituted pyrylium salts leads to dimers (4,4'-bi-4H-pyrans) and

TABLE XI

Free Activation Enthalpies $\Delta G^{\ddagger}$, Coalescence Temperatures t_C and Chemical Shift Differences $\Delta\nu$ for Internal Rotation of Dimethylamino groups Bonded to α or γ Positions of Pyrylium Salts Determined by ^{13}C-NMR Spectroscopy[883]

	R
a	NMe$_2$
b	OMe
c	Me
d	Ph

Rotating α-Me$_2$N Group				Rotating γ-Me$_2$N Group			
Compound	$\Delta G^{\ddagger}$ (kcal/mol)	t_C (°C)	$\Delta\nu$ (Hz)	Compound	$\Delta G^{\ddagger}$ (kcal/mol)	t_C (°C)	$\Delta\nu$ (Hz)
914a	12.6	-26	19.4	914a	~17.3	47	6.6
914b	17.0	49	14.2	915	~21.8	>110	2.9
914c	18.6	81	16.2				
914d	19.1	82	14.2				

therefore free pyran-4-yl radicals were postulated as intermediates. Much earlier, Conant et al.[381,884] had also speculated about stable pyranyl free radicals.

Palchkov, Zhdanov, and Dorofeenko[380] reported that zinc powder or other metals reduce 2,4,6-triphenylpyrylium salts in organic solvents to a stable free radical 916a demonstrated through ESR spectroscopy. Various other metals (K, Na, Hg, Cu, Mg) reduce 2,4,6-triphenylpyrylium perchlorate in solvents such as tetrahydrofuran to the same pyranyl free radical.[885] Degani and co-workers[886,887] prepared the radical 916a and its deuterated congeners 916b–916c in cyclohexane.

	Ar	Ar'
a	Ph	Ph
b	Ph-d$_5$	Ph-d$_5$
c	Ph-d$_5$	Ph
d	Ph	Ph-d$_5$

(916)

The well-resolved ESR spectra were deciphered by simulation and proton hyperfine coupling constants were assigned by comparison with calculated spin densities (Fig. 8). These calculations were performed by

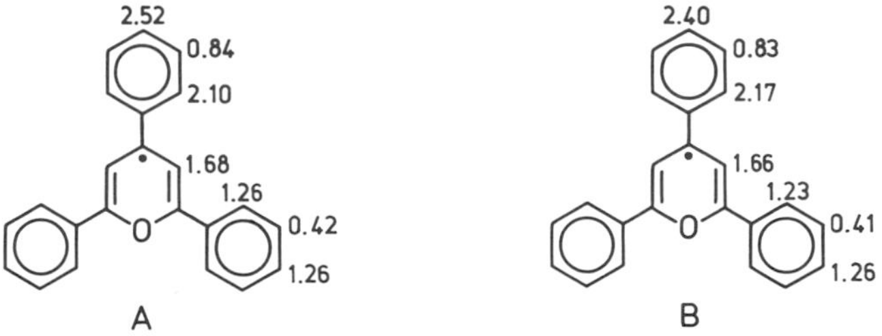

FIG. 8. Experimental coupling constants of the 2,4,6-triphenylpyranyl radical (in Gauss, 1 G = 0.1 mT): A after Ref. 887, B after Ref. 888.

the simple Hückel LCAO–MO approach (with poor results) and by the McLachlan method in two geometries: the planar geometry (unsatisfactory correlation) and with phenyl groups twisted out of the heterocyclic plane. The best correlation between experimental and calculated spin densities, using the latter approach, was found for a twist angle of 42° for the α-phenyl groups and of 28° for the γ-phenyl group. Though X-ray data of 2,4,6-triphenylpyrylium cations show that the γ-phenyl is more twisted than α-phenyl groups (cf. Section IV.B), in the radical it is reasonable to assume the reverse twisting trend: the spin density is known to be highest in the γ-position of the pyran ring (as indicated by the odd electron in formula 916), hence the γ-phenyl (which possesses the higher spin density) will probably be less twisted than the α-phenyls.

The reversible dimerization of two 2,4,6-triphenylpyranyl radicals in various organic solvents to the diamagnetic 4,4'-bi-4H-pyranyl dimer was studied by Okhlobystin and co-workers.[889]

Farcasiu and Farcasiu[890] demonstrated that the donor–acceptor complex obtained on dissolving aryl-substituted pyrylium perchlorates or hexachloroantimonates in anhydrous pyridine present ESR spectra identical to those obtained from the same salts on reduction with zinc powder. Other efficient one-electron donors are 2,6-di-t-butylphenol or N,N,N'N'-tetramethylphenylenediamine.[49]

A thorough experimental and theoretical investigation by Hacquard and Rassat[891] allowed a complete understanding of spin densities in the 2,4,6-tri-t-butylpyran-4-yl radical which was independently reported by Nekhoroshev and Okhlobystin.[73] The experimental coupling constants indicate high spin density at the γ-, lower at the β-, and undetectably low at α-positions: $a_{H-\beta} = 1.85$ Gauss, $a_{H-\gamma-t-Bu} = 0.30$ Gauss, $a_{\gamma-^{13}C} = 10.9$ Gauss; theoretical spin densities were determined by the Hückel, McLachlan and INDO methods, and are in agreement with experiment.[891,892]

Two interesting structural combinations of aroxyl and pyranyl radicals were studied by Okhlobystin et al.[893] In radicals 917 where the γ-pyran position is substituted by an aroxyl system the unpaired electron was mainly delocalized into the heterocyclic ring. However, in radical **918**, there is no delocalization on the benzopyrylium ring; the authors ascribed this lack of delocalization to lack of coplanarity between the two systems. A more probable rationalization should also take into account the less favorable position α of the aroxyl ring relative to the oxygen heteroatom, and the electronic differences between the pyrylium and benzopyrylium systems.

R = Ph, t-Bu

(917)   (918)

Radical-cations **920** with two linked pyran rings are formed by several reactions and may be detected by ESR spectroscopy: (i) dehydrogenation of 2,2′,6,6′-tetrasubstituted 4,4′-bi-4*H*-pyrans **43** with triphenylmethyl salts ending in bipyrylium salts **44** and producing cation-radicals as intermediates[885] (cf. Section II,A,1,c); (ii) reduction of 2,2′,6,6′-tetra-*t*-butylbipyrylium salts **44** (R = *t*-Bu) with electron donors such as ferrocene, *N,N*-dimethylaniline, triphenylphosphine, *N,N,N′,N′*-tetramethyl-*p*-phenylenediamine; (iii) oxidation of 4,4′-bipyranylidenes **919** with iodine, mercury(II) bromide,[49] or tetracyano-*p*-quinodimethane[209,894]; (iv) combination of a bipyranylidene **919** with a bipyrylium salt **44**[311]; (v) decarboxylation of 2,4-diphenyl-4-carboxypyrylium perchlorate **33** (R = Ph) catalyzed by Vaska's compound, Ir(CO)(PPh₃)₂Cl, under reflux in acetonitrile yielding the cation-radical **920** (R = Ph)[267] (cf. Section III,A,4).

Electrochemical reductions of bispyrylium salts of type **921** also lead to stable cation-radicals **922** in the first step, and these may undergo further reduction to corresponding bispyranylidenes **923**.[895]

Other instances in which electrochemical processes involve stable free radicals are mentioned in Section IV,C,3.

Tamamura *et al.*[852] observed an enhanced ESR spectrum on irradiating with the wavelength of the CT band the complexes formed by 2,4,6-triarylpyrylium cations with aromatic electron donors like pyridine. This effect is probably due to photodissociation of the CT complex.

4. *Mössbauer Spectra*

One paper mentions Mössbauer spectra for the tetrachloroferrates of: 2-methyl-4,6-diphenyl-, 2,6-diphenyl-, 2-*t*-butyl-4,6-diphenyl-, 2-styryl-4,6-diphenyl-, 2,4,6-triphenyl-, and 2,4,6-trimethylpyrylium.[896] The determination were carried out at 80–440°K. For the first two compounds the anions are weakly distorted $FeCl_4^-$ tetrahedra, but the other four compounds present anomalous Mössbauer signals indicative of octahedral coordination around the Fe(III) atom. This anomaly has not been explained.

5. *Mass Spectra*

The formation of pyrylium rings in electronic-impact produced fragment cations was advocated frequently in mass spectra of 2-alkylfurans[897,898] and of unsaturated esters.[899]

The first mass spectrometric study by Duffield, Djerassi, and Balaban[900] made use of pyrylium halides (iodides, bromides) or fluoborates. The base peak for 2,4,6-triphenylpyrylium salts corresponds to the molecular weight of the cation, but whenever α- or γ-methyl groups are present, the base peak corresponds to the loss of one hydrogen from the cation (as hydrogen halide). Fragmentation schemes (substantiated by studying metastable peaks) were worked out for several 2,4,6-trimethyl- and/or phenyl-substituted pyrylium salts of type **824** (R = Me, Ph) and **924**. Whereas pyrylium salts with α- or γ-methyl eliminate CH$_3$, a 4-phenyl is not eliminated; however, α-phenyls are eliminated as such or as benzoyl cations.

A subsequent study by Hvistendahl, Gyorösi, and Undheim,[901] involving also the determination of appearance potentials, indicated that 2,4,6-triphenylpyrylium bromide, iodide, and fluoborate undergo a thermal reaction affording the corresponding triphenylpyrylium free radical during evaporation; the fluoborate also gives rise to a small intensity peak (5% of the base peak) with the elemental composition of an adduct between the cation and a fluoride ion, which then may decompose into a fluorine atom and the pyrylium free radical. In the case of pyrylium salts with α- or γ-methyl groups the appearance potentials indicate that on evaporation anhydrobases (methylenepyrans) are formed. Perchlorate ions behave differently from other anions in mass spectra, oxidizing the organic fragments to a base peak (M + O − H) which differs from base peaks obtained with other anions.

Unlike pyrylium salts, N-methylpyridinium halides undergo thermally induced demethylation to methyl halide and pyridine; however, N-phenylpyridinium salts behave like pyrylium salts, i.e., give redox processes on evaporation in the mass spectrometer.[902]

B. STRUCTURAL DATA (X-RAY INVESTIGATIONS)

In the interionic charge-transfer complex 2,4,6-triphenylpyrylium 1,1,3,3-tetracyanopropenide,[853] the anion is planar while in the cation the

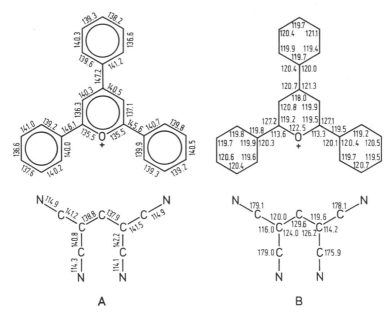

FIG. 9. X-ray diffraction results for 2,4,6-triphenylpyrylium 1,1,3,3-tetracyanopropenide. A: Bond lengths (in pm); B: Bond angles (in °C).[853]

γ-phenyl ring is rotated 18.0° and the α-phenyl rings are rotated 10.4° and 2.3° relative to the pyrylium ring. The interplanar spacing is 3.31 Å, and the crystal is built up from infinite columns stacked by ion pairs. Interatomic lengths (pm) and bond angles are presented in Fig. 9.

Treatment of acetylacetone (acacH) with $WOCl_4$ afforded a crystalline ionic compound $C_{10}H_{13}O_2{}^+C_5H_7Cl_2O_4W^-$ whose cation was 3-acetyl-2,4,6-trimethylpyrylium and whose anion was $[WO_2Cl_2acac]^-$; the cation is planar and has the interatomic bond distances (in pm) and bond angles shown in Fig. 10.[903]

A comparison between the two molecular structures presented in the Figs. 9 and 10 shows that the C—O bonds in the planar pyrylium ring

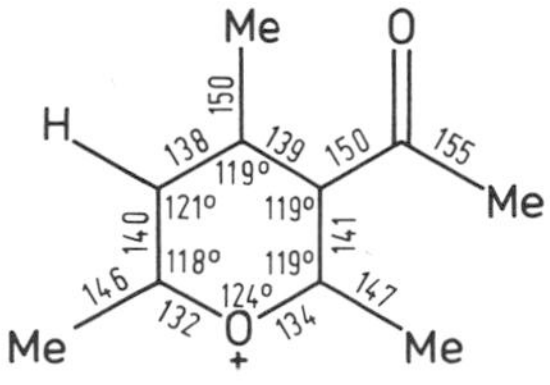

FIG. 10. Bond lengths (in pm) and bond angles for the 3-acetyl-2,4,6-trimethylpyrylium cation.[903]

are shorter than the other ring bonds, and that the C—O—C ring angle must therefore be larger than 120°. The α-CH$_3$C$_{ring}$ C—C bonds are shorter than the γ-CH$_3$C$_{ring}$ C—C bond.

An X-ray determination was necessary to elucidate the chemical (and at the same time the molecular) structure of a novel dimer of 2,4-pentanedione obtained from thallium(I) acetylacetonate and dichlorodimethylsilane; it proved to be 6-(2-hydroxyprop-1-enyl)-2,4-dimethylpyrylium chloride (**925**). The unusual enolic structure in the crystal is reminiscent of the keto–enol forms observed[341] by ^{1}H-NMR for solutions of 4-(2-hydroxyprop-1-enyl)-2,6-dimethylpyrylium: in trifluoroacetic acid both forms coexist, while in [D$_6$]acetone/trifluoroacetic acid (1 : 4) only the enol is observed (cf. Section IV,A,2,a). Since a 4,6-dimethylpyrylium-2 ring is a stronger electron attracting substituent than a 2,6-dimethylpyrylium-4 ring, it is logical to expect a higher stability of an enol with the former substituent.

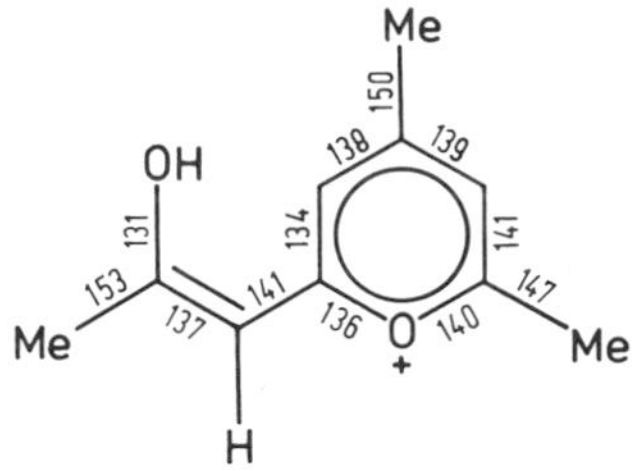

FIG. 11.   Bond lengths (in pm) for the 6-(2-hydroxy-prop-1-enyl)-2,4-dimethylpyrylium cation.[904]

Interatomic bond lengths are shown in Fig. 11 for the 6-(2-hydroxyprop-1-enyl)-2,4-dimethylpyrylium cation. In this case too the α-CH$_3$C$_{ring}$ C—C bond is shorter than the γ-CH$_3$C$_{ring}$ C—C bond, but the bond lengths in the pyrylium ring are difficult to rationalize. The very short C—C bond (141 pm) between the pyrylium ring and the vinyl carbon proves that the methylenepyran structure makes a large contribution to the resonance hybrid **925A** ↔ **925B**. Additional proof is provided by the short C—OH bond distance.

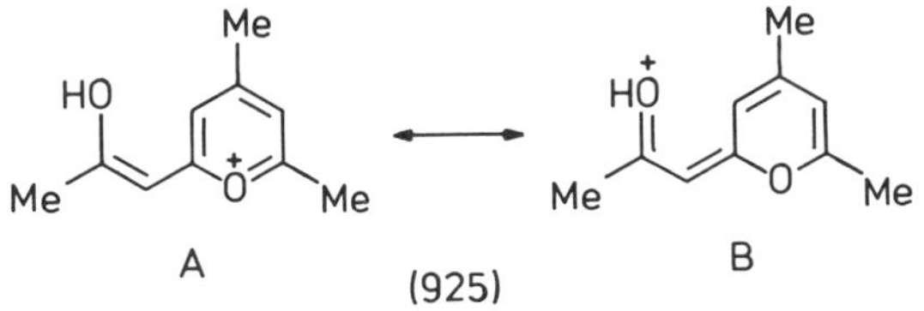

The crystal and molecular structure of 2,4,6-trimethylpyrylium tetra-chloroferrate was also determined.[896] Also the crystal structure of the palladium complexes **868** mentioned in Section III,D,3 was determined by X-ray analysis.[784] The crystal structure determination of the product formed from 2,6-diphenylpyrylium with aqueous bases was mentioned in Section III,B,2,a.

C. THERMO-, MAGNETO-, AND ELECTROCHEMICAL PROPERTIES

1. Thermochemical Properties

A calorimetric determination of the heat of combustion for 2,4,6-tri-methylpyrylium perchlorate afforded the value $\Delta H = -1092.6 \pm 0.7$ kcal/mol.[905] From the determination the standard heat of formation was calculated to be $\Delta H = -41 \pm 2$ kcal/mol at 21.2°C. This value includes the heat of formation of the anion.

Another relevant thermochemical paper[906] determined the enthalpy differences between methyltropic isomers. One such isomer pair is con-stituted by 2-methoxy-6-methyl-4-pyrone (**926**) and 4-methoxy-6-methyl-2-pyrone (**172**), while the equilibration catalyst was the common dimeth-

(926)   (171)   (172)

ylated product, 2,4-dimethoxy-6-methylpyrylium fluoborate (**171**). The α-pyrone is more stable, and the enthalpy of the conversion **926** → **172** at 115°C was determined calorimetrically to be -5.7 kcal/mol in the liquid and -8.8 kcal/mol in the gas phase. A similar energy difference (-10.6 kcal/mol) was found for the conversion of a γ- to an α-pyridone (2-methoxy-1,6-dimethyl-4-pyridone to 4-methoxy-1,6-dimethyl-2-pyri-done). In a refined localized model using the values of Benson,[907] an energy difference of 8.0 kcal/mol between the γ- and the α-pyrone is predicted, and the two isomers are therefore considered to possess es-sentially the same stabilization energies. In the subsequent discussion, Beak *et al.*[908] point out that magnetic susceptibility anisotropies have indicated that pyrones are nonaromatic (as they have also done for tro-pone), in agreement with the thermochemical data which also indicate absence of π-electron delocalization.

2. Magnetic Properties

Haberditzl[909,910] reported diamagnetic increments for several aromatic systems, among which 2,4,6-trimethylpyrylium perchlorate ($\chi = -0.497$) had the smallest negative exaltation relative to a localized structure ($\chi = -5.0$), in agreement with reduced aromaticity.

The free radical **916a** obtained by reduction of 2,4,6-triphenylpyrylium perchlorate, together with other persistent free radicals like 1,1-diphenyl-2-picrylhydrazyl, was studied in octafluoronaphthalene solution by multifield nuclear-electron double resonance.[911] The spin–spin coupling between fluorine nuclei and the electrons of dissolved free radicals results in field-dependent enhancement of nuclear relaxation. The 2,4,6-triphenylpyryl free radical gave the highest positive low-field enhancement of the dynamic polarization parameter (this enhancement decreases with increasing field), whereas 2,4,6-tri-*t*-butylphenoxyl presented the reverse effect (largest negative enhancement which becomes more positive with increasing field). This indicates that for the former radical the scalar coupling dominates low-field nuclear relaxation. This was rationalized in terms of steric and electronic factors, namely, of the peripheral spin delocalization of free radicals like 2,4,6-triphenylpyryl or tetrachlorosemiquinone, whereas *t*-butyl groups shield sterically the odd electron in tri-*t*-butylphenoxyl or galvinoxyl, thus preventing spin delocalization over the solvent during the radical–solvent encounter.

3. Electrochemical Properties

Ion pair dissociation equilibria of 2,4,6-trimethylpyrylium hexachloroantimonate in methylene dichloride at 0°C and −45°C were investigated by Ledwith and co-workers,[912] who calculated from the experimental data the equivalent conductance at infinite dilution, the Stokes radius, the interatomic distance, and the thermodynamic heat and entropy terms for the dissociation equilibrium. The behavior of the pyrylium salt parallels that of tropylium or triphenylmethyl salts.

Schwarzenbach and Lutz[913] were the first to determine the pK_a of a pyrylium, namely 2,4,6-trimethylpyrylium perchlorate (**286**), while performing acidity measurements of unstable substances like glutacondialdehyde. By titration with sodium hydroxide in a flow system it was observed that with ratios x = NaOH/trimethylpyrylium lower than 1, on interrupting the flow, the glass electrode potential varies rapidly, indicating that the solution becomes more acid. For $x > 1$ the potential stays constant. These electrochemical observations are paralleled by color changes [colorless in acid solution, yellow in alkaline solution; on adding hydroxide ($x < 1$) the solution becomes immediately yellow, then the color fades rapidly; for $x > 1$ the yellow color remains unmodified]. This indicates that the enolate **928** is yellow and that one of the mentioned processes proceeds with a measurably rapid rate. The pK_a of 4-methyl-3-hepten-2,6-dione (**929**) or of its enolic form **927** is 11.2 ± 0.2.

More recent and precise determinations of the pK_a values for three 2,4,6-trisubstituted pyrylium salts are due to Williams,[82] who performed kinetic measurements in various buffers and used UV measurements. The pK_a' values at 25°C in 0.1 M ionic concentration are: for 2,4,6-trimethylpyrylium in water, 6.7; item, in deuterium oxide, 7.6; for 2-methyl-4,6-diphenylpyrylium, 6.2 (for the reverse reaction, 4.4); for 2,4,6-triphenylpyrylium, 5.0 (for the reverse reaction, 3.3). These pK_a' values are apparent ionization constants; noting π = pyrylium cation, P = α-hydroxypyran, E = 1,5-enedione, we have $K' = [H_3O^+][E]/[\pi] = K_{R^+}[E]/[P]$. The true ionization constant, of course, is $K_{R^+} = [H_3O^+][P]/[\pi]$, but P is not stable. This indicates that the 2,4,6-tri-methylpyrylium cation is an appreciably stronger acid than acetic acid, and this is why a convenient technique for obtaining pseudobases is to treat aqueous or ethanolic solutions of pyrylium salts (perchlorate, halide, tetrachloroferrate) with an aqueous solution of sodium acetate (which forms a soluble complex with FeCl$_3$). It was noted by Balaban[783] that 2,4,6-triphenylpyrylium pseudobase affords a trifluoroacetate or tri-chloroacetate which are crystalline stable double salts (similar to the halides, i.e., with hydrogen-bonded anions HX_2^-), which fluoresce in solution, that the respective dichloroacetate does fluoresce but could not be obtained crystalline, while the same pseudobase does not afford fluorescent solutions with chloroacetic or acetic acids.[783]

Several studies have been made investigating the electroreduction of pyrylium salts. The first one, by Gârd and Balaban,[914] demonstrated that the first half-wave reduction potential of 2,4,6-trisubstituted pyrylium salts with alkyl and/or aryl substituents correlate with the frequencies

v_x of the longest wavelength absorption band (the so-called x-band; cf. Section IV,A,1,a) in the electronic spectra of the same pyrylium salts. Actually the theory predicts a correlation between $E_{1/2}$ and LUMO, whereas v_x equals Δ(LUMO − HOMO). The height of the first polarographic reduction wave in acid buffers depends linearly on concentration and on the square root of the mercury pressure and has a normal temperature coefficient; this wave may be considered to be diffusion-controlled.

In a subsequent paper,[379] 2,4,6-trimethylpyrylium salts (perchlorate, tetrachloroferrate, and iodide) were investigated in more detail: the first polarographic wave is a one-electron reduction process. It was ascertained (through the independence of $E_{1/2}$ from the pH of the buffer in the range pH = 0.5–4.0) that no proton is involved in the electroreduction, and the product of the reaction was isolated and shown to be identical to **931**, the hexasubstituted 4,4′-bi-4H-pyran, which can be prepared in quantity by reducing pyrylium salts **290** (R = Me or Ph) with zinc dust in water–ether or in ethanol (cf. Section III,A,8,b). In electrochemical reduction, 2,4,6-trimethylpyrylium behaves very much like tropylium.

The reaction involves a free radical which is stable if R = Ph. The reversibility of the bipyran formation even with R = Me was demonstrated both by chemical oxidation (**931** → **290**) with chromium trioxide and perchloric acid and by electrooxidation on a rotating platinum electrode in acetonitrile with potassium perchlorate as electrolyte. The clean 4,4′-dimerization (indicated by the simple ^{1}H-NMR spectrum of **931** for R = Me), without any detectable 2,2′- or 2,4′-dimers indicates the high spin density in the 4-position of the free radical **930**.

Feldman and Winstein[845] obtained a similar value for the electroreduction half-wave potential of **290** (R = Me).

Bratu and Balaban[340] investigated the electrooxidation (in acetonitrile in the presence of LiClO$_4$ or HClO$_4$ on a rotating platinum electrode) of 1,5-pentanediones and of 4,4′-bi-4H-pyrans (2,2′,4,4′-tetrasubstituted by methyl or phenyl groups). The reaction products are pyrylium salts, as demonstrated by a preparative electrooxidation of 1,3,5-triphenylpentane-1,5-dione. The electrooxidation potentials show an inverse correlation with the yields in hydride transfer reactions to Ph$_3$C$^+$ClO$_4^-$, leading to the same pyrylium salts and to triphenylmethane. The electrooxidation is apparently a two-electron process.

Unlike 2,4,6-trimethylpyrylium which does not give a persistent free radical **930**, 2,4,6-triphenylpyrylium (**290**, R = Ph) gives a pyranyl radical **930** which coexists, in concentrations detectable by ESR, in equilibrium with its dimer **931** (R = Ph). Pragst[915] showed that homogeneous electron transfer between radicals **930** and the rubrene cation-radical results in electrochemical luminescence through the rubrene triplet state. The half-wave electroreduction potential of 2,4,6-tri-*p*-anisylpyrylium perchlorate and the UV absorption spectrum of the corresponding free radical **930** were also obtained. In a subsequent paper,[916] the cathodic dimerization of 2,6-phenylpyrylium perchlorate was studied in acetonitrile on platinum electrodes, by voltametry, oscillovoltametry on the rotating electrode, cyclic voltametry, and potentiostatic coulometry. The cathodically formed 2,6-diphenylpyran-4-yl radical dimerizes irreversibly to a 4,4'-bi-4*H*-pyran which can be oxidized anodically to 2,2',6,6'-tetraphenyl-4,4'-bi-pyranylidene. The latter compound is formed also by chemical reduction with zinc in acetonitrile,[14,66] but the structure of the product and the mechanism of its formation (involving hydride transfer to 2,6-diphenyl-pyrylium, which is converted to a 4*H*-pyran that can hydrolyze to 1,5-diphenylpentane-1,5-dione) was elucidated later.[209]

The two-electron reduction of 2,4,6-triarylpyrylium salts in the presence of alkyl halides RX leads to 4-alkyl-2,4,6-triaryl-4*H*-pyrans. The alkylation occurs by nucleophilic substitution of RX by the pyranyl anion formed in the second cathodic wave (−1.5 to −1.6 V versus SCE). The rate constant for this reaction increases in the series RCl < RBr < RI.[917]

Other literature data on the solvation and reversibility during electro-reduction are available,[895,918] as well as comparisons with chemical reduction with chromous ions.[382]

Hünig and co-workers[311] investigated the polarography of 4,4'-bipy-rylium salts **44** (in a more comprehensive study involving also the *N*-methylpyridinium and the thiopyrylium congeners). The process is more complicated because it involves two one-electron processes, involving the dication **44**, the radical-cation **920**, and the neutral bipyranylidene **919** (cf. Section IV,A,3); in addition, **920** (R = Ph) dimerizes reversibly. In a demonstration of virtuosity, most compounds were obtained by unambiguous syntheses and studied separately to determine their properties so as to be able to obtain the formation constant of the "semi-quinone" $K = [Sem]^2/[Red][Ox]$ as a function of the heteroatom and the α-substituents. The two redox potentials $E_1 = [Sem]/[Red]$ and $E_2 = [Ox][Sem]$ in acetonitrile and dimethylformamide allow the determination of $\log K = (E_2 - E_1)/0.059$ (at 25°C, E in volts). The electrode processes

(919)　　　　　　(920)　　　　　　(44)

[Red]　　　　　　[Sem]　　　　　　[Ox]

Dimer

were shown to be reversible. The high K values (10^4 to 10^6) show that radicals **920** are fairly stable. The bipyrylium and bithiopyrylium salts have $E_1 = +0.03$ to $+0.25$ V (relative to Ag/AgCl) and $E_2 = +0.35$ to $+0.48$ V, while the N-methylpyridinium salts have negative values for the redox potentials. This explains difficulties encountered in the synthesis of **919** [Red] in the former case, of bithiopyranylidenes, and of **44** [Ox] in the latter case, as well as of bipyridylium dications. The influence of donor α-methyl groups is evident in the displacement of E_1 toward negative potentials, increasing thereby K, for the pyrylium system (but not for the pyridinium and thiopyrylium where both E_1 and E_2 are displaced for R = Me toward negative values leaving K unaffected).

Hünig and Ruider[506] also investigated the syntheses and polarographic behavior of phenylogous and diazavinylogous bipyrylium, bithiopyrylium, and bipyridinium salts and also their two-step reduction products.

The p-phenylogous system **932** ⇌ **933** ⇌ **671** has $E_1 = -0.15$ and $E_2 = -0.03$ V, $K = 10^3$ and the diazavinylogous system **934** ⇌ **935** ⇌ **936** only presents one half-wave reduction potential at $+0.76$ V, $K = 5$ (the bipyrylium system **919** ⇌ **920** ⇌ **44**, R = Ph had $E_1 = +0.15$, $E_2 = +0.47$ V, $K = 2.3 \times 10^5$). Thus the radical stability decreases in both cases by inserting a bridge, but K decreases much more with an —N=N— bridge than with a C_6H_4 bridge.

In a series of papers,[21,22,272,273] Evstifeev, Dorofeenko and co-workers studied the polarographic reduction of 5,6,7,8-tetrahydrobenzo[b]pyrylium salt derivatives formed by condensation with carbonyl compounds at the α-positions. Oscillography of **937** at various concentrations (2–20 N) of sulfuric acid indicated that at moderate acidities free radicals which dimerize are produced; at higher acidities these radicals are further reduced to 4H-pyrans. Adsorption phenomena on the dropping mercury electrode complicate the process.[22,272,273]

However, arylaminomethylenetetrahydrobenzo[b]pyrylium salts, which present two tautomeric forms **938 ⇌ 939**, present no such complicating adsorption phenomena.[22,272] In dimethylformamide the stability of the corresponding radicals formed by electroreduction increases relative to aqueous media; a Hammett correlation was found between the electro-reduction potential and the substituent constants of the N-aryl group.[22,272] Oscillographic studies helped in elucidating the mechanism of the hydrolysis of compounds **938 ⇌ 939** leading to $ArNH_2$, HCOOH and 2,4-diphenyl-5,6,7,8-tetrahydrobenzo[b]pyrylium.[21] Polarographic studies of flavylium salts revealed similar behavior.[919]

Oscillopolarographic studies in aprotic media (acetonitrile) revealed[895] that the presence of bulky (e.g., t-Bu) or conjugating groups (e.g., styryl) in α- and γ-positions leads to a reversibility coefficient close to 1, while unsubstituted γ-positions lead to low reversibility coefficients (0.3–0.5). Carbonyl and 2,6-di-*t*-butylphenoxy substituents of pyrylium salts were also investigated.[895] The reduction of bipyrylium salts proceeds in two stages leading first to a free radical, then to a bipyranylidene.[895] On reduction of the 2,4,6-tri-*t*-butylpyrylium cation the 2,4,6-tri-*t*-butylpyranyl radical is formed.[920]

On the applied side of electrochemical studies, the use of oscillopolarography for the determination of pyrylium salts was reported.[921–923] 2,4,6-Triphenylpyrylium tetrachloroferrate can be used in a liquid membrane electrode for the determination of iron(III): using di- or tetrachloroethane as solvents, the response is practically Nernstian in the range 10^{-4} to 10^{-1} M $FeCl_4^-$ (slope 58.5 mV). The electrode is highly selective in the presence of Zn^{2+}, Cu^{2+}, Al^{3+}, Ni^{2+}, Cd^{2+}, Mn^{2+}, NO_3^-, Br^-, SO_4^{2-}, and BF_4^-.

Cyclic voltametry of 2,4,6-triphenylpyrylium, 4-(*p*-diethylaminophenyl)-2,6-diphenylpyrylium fluoborates and the thiopyrylium analogs in acetonitrile or in dichloromethane as solvents afforded $E_{1/2}^{red}$ and $E_{1/2}^{ox}$ values which were interpreted in terms of HOMO and LUMO energies. Together with electronic absorption spectra, the data indicate the thiopyrylium moiety to be more electron withdrawing than pyrylium.[924] Previous studies by Degani *et al.*[338] on intermolecular hydride transfer reactions between 4*H*-pyran, 4*H*-thiopyran, tropylidene and the corresponding cations had indicated that thiopyrylium is more stable than pyrylium (cf. Section III,A,6,f).

Electrooxidations of 1,5-pentanediones, of 3,5-dien-1-ones or of 4,4′-bi-4*H*-pyrans were studied by Bratu and Balaban[340] using rotating platinum electrodes in acetonitrile. In all cases the products are pyrylium salts. The smaller the $E_{1/2}^{ox}$ value, the higher the yield of pyrylium salt when using hydride acceptors such as triphenylmethyl, in comparisons involving diones such as **940, 375, 160** (Scheme 19).

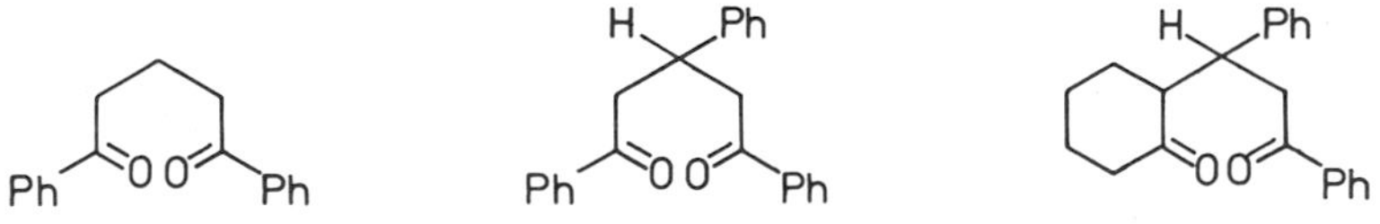

	(940)	(375)	(160)
$E_{1/2}^{ox}$	2.10	1.80	1.72

(Values are in volts vs. Ag/Ag⁺ 0.01 N in MeCN)

SCHEME 19

D. Chromatographic Separations

In preparative synthetic work, the analysis and separation of a mixture of pyrylium salts are often difficult problems. For analytical purposes, conversion by ammonia to pyridines, followed by gas-chromatography, is often useful[181,486] because the pyridines are formed in high yield and can be separated by preparative vapor phase chromatography and identified by NMR giving information about the original pyrylium salts. In the reaction of alkyl-substituted pyrylium salt mixtures with primary amines leading only to one crystalline pyridinium salt, or in similar reactions between excess 2,4,6-trimethylpyrylium and a precious amine, it is often useful to add ammonia at the end of the reaction, converting unreacted pyrylium salts to pyridines which are liquid and extractable by ether.

However, direct separation of pyrylium salts is very desirable. Thin layer chromatography on gypsum was studied[925] and R_f values were given for fifty pyrylium salts, using a mixture of benzene–chloroform (7 : 8 vol/vol) and UV fluorescent detection. As an application, in the synthesis of new pyrylium salts from acetals and two moles of ketone, the purity was checked by this TLC method.[159]

It can be safely assumed that the method of choice for the analysis and separation of mixtures of pyrylium salts will be high-performance high-pressure liquid chromatography. So far no report using this method has appeared in the literature.

E. Theoretical Calculations

A simple Hückel MO quantum-chemical calculation for the charge density in thiopyrylium salts[926] revealed the higher positive character of the α- than γ-positions, in agreement with the chemical reactivity toward most nucleophiles of thiopyrylium and pyrylium salts.

The first theoretical studies for pyrylium employed various semi-empirical methods and attempted rationalizations of electronic absorption spectra of pyrylium cations[927–930] or of charge-transfer spectra of pyrylium iodides[207] which had been studied experimentally shortly before that. Using the Hückel MO and the Goodman–Shull approximations,[931] satisfactory agreement was found for 2,4,6-triphenyl and/or methyl-substituted pyrylium salts[927,928]; for the effect of para substituents in phenyl groups of 2,6-diaryl-4-methylpyrylium and in 2,6-dimethyl-4-arylpyrylium salts (the para substituents are H, Me, MeO) the calculations[928–930] indicate a reversal of the two first absorption bands (x or ^1L_b and y or ^1L_a, cf. Section IV,A,1,a) in 2,6-dimethyl-4-arylpyrylium relative to other pyrylium salts in agreement with the assignment[362] based on the bathochromic effects of increased conjugation at the γ-position on the Y-band throughout the range of the pyrylium cations.

The charge-transfer band in the visible region of pyrylium iodide has energies[207] which may be satisfactorily correlated with experimental data (the energy of the y-band in the electronic absorption spectrum), but the correlation with the energy of lowest unoccupied MO (E_{LUMO}) is less satisfactory. However, the half-wave polarographic reduction potentials correlate well both with E_{LUMO} and with the energy of the x-band in the electronic absorption spectra.

Attempts to rationalize on the basis of HMO calculations charge densities for correlations with ^{1}H-NMR spectra yielded satisfactory agreement[874]; however, this simple method failed to explain the faster isotopic exchange of γ-methyls relative to α-methyls in pyrylium salts, predicting higher stability for α-methylenepyrans (pyrylium anhydrobases). Using a self-consistent version of the HMO method (Wheland's ω-technique) Boyd[932] obtained the correct order of energies for the anhydrobases, explaining satisfactorily the rates of side-chain deuteration. Calculations for α- and γ-methylenepyrans (charge densities) using the PPP method with Dewar's parametrization were used to explain the higher stability of γ- than α-methylenepyrans (the negative charge density has a lower absolute value on the exocyclic carbon for the γ- than for the α-methylenepyran).[933] The above ω-technique was applied by Boyd and Balaban[934] for correlating the chemical shifts of methyl protons in pyrylium salts with the calculated electron densities. PPP-Type calculations[369b] reproduced the results obtained by Boyd with the ω,β method for the kinetics of side-chain isotope exchange in pyrylium salts, and explained the rate differences for the two α-methyl groups in 2,3,4,6-tetramethylpyrylium: the 2-methyl is deuterated five times faster than the isolated 6-methyl. These calculations also confirmed that the ^{1}H-NMR peak of the 2-methyl group appears at lower field than that of the 6-methyl group, in agreement with the unambiguous synthesis of 2,3,4-trimethyl-6-[*methyl*-D₃]pyrylium by reaction of 3,4-dimethyl-3-penten-2-one with CD₃COCl and AlCl₃.[369a]

The simple LCAO–MO method was applied[935] to the calculation of the first transition energy (Δm) for 65 pyrylium salts whose experimental electronic absorption spectra had been reported by Wizinger *et al.*[222,487,794-796] and by Balaban *et al.*[207,362,797] A satisfactory agreement with one set of parameters was found with one regression line $\bar{\nu}$ (cm^{-1}) = 14700 Δm + 10500 (standard deviation 1100 cm^{-1}). An even better agreement is obtained when the noncoplanarity of aryl substituents with the pyrylium ring is taken into account. Boyd and Singer [935] pointed out an unexplained anomaly for the effects of methoxy and hydroxy substituents on the absorption of aryl-substituted pyrylium salts; this anomaly is probably due to solute–solvent interactions as shown by calculations using the ω, the ω,β, and the PPP LCAO–MO self-consistent techniques[936]; better results with fewer parameters were obtained with the ω,β technique than with the PPP method.

Dewar and Gleicher[937] calculated π-binding energies, resonance energies, bond lengths, and the heat of combustion of several oxygen and nitrogen heterocycles among which they included pyrylium using both the PPP and the SPO treatments.

Nucleophilic superdelocalizabilities, calculated after Fukui's method, were used[319,938] to interpret the α/γ-attack of borohydride or of Grignard reagents on 2,4,6-trisubstituted pyrylium salts. The same French group calculated[939] by the PPP method the electronic transition energies using the geometry assumed by Dewar and Gleicher[937] and obtained good agreement with experimental data including the fact that for 2,6-dimethyl-4-arylpyrylium the y-band appears at longer wavelength than the x-band, unlike other pyrylium salts whose longest wavelength absorption is caused by the x-band.

Karlsson and Mårtensson[940] performed iterative extended Hückel calculations for all valence electrons, using also Del Re's method for σ electrons and iterative PPP calculations in the variable electronegativity formalism (Nishimoto-Mataga's and Ohno's approximations converge toward the same values when these two methods are charge iterated) for the unsubstituted systems benzene, pyridine, and the pyrylium cation, and for fluorobenzene, obtaining charge distributions, orbital energies, and first excitation energies. The valence-electron distribution in the unsubstituted pyrylium ion[941,942] calculated by the CNDO/2 method (in a regular hexagon geometry) and by other all-valence methods indicates that there is a slightly negative charge at the oxygen atom (only the π-electron charge of the oxygen atom is positive, therefore the oxonium formula for pyrylium is misleading) and a higher positive charge for α- than for γ-carbon ring atoms. Ground and excited singlet states of methyl- and phenyl-substituted pyrylium rings were calculated using a CNDO/3 parametrization; calculated relative oscillator strengths agree well with electronic spectral data.[943]

Fabian, Mehlhorn, and Zahradnik[944] made extensive use of the PPP method using the variable β-approximation for calculating electronic absorption spectra of cyclopentapyrans and for explaining the sensitization by 2,4,6-triphenylpyrylium observed earlier for light-sensitive polymers (polyesters: vinyl cinnamate and vinyl cinnamylidene acetate) which become insoluble after development. Molecular diagrams for singlet and triplet states of 2,6-diphenyl- and 2,4,6-triphenylpyrylium are presented,[945] also with *para*-methyl or *para*-methoxy substituents; the absorption spectra are in agreement with calculations. The most powerful sensitizer for poly(vinyl cinnamate) is 2,4,6-triphenylpyrylium without any para substituents while 2,4,6-tri-*p*-anisylpyrylium is the best sensitizer for poly(vinyl cinnamylideneacetate) indicating that the triplet T_1 and T_2 states of pyrylium are involved in the sensitization.[946]

Japanese authors[938] calculated the electronic transition energies and the oscillator strengths for the unsubstituted pyrylium and thiopyrylium cations, as well as the electronic distributions and bond orders by the PPP method (after configuration interaction among all the singly excited states). Thiopyrylium was found to have more contribution (28.4%) of carbocationic resonance hybrid structures than pyrylium (14.6%).

Gheorghiu and Balaban,[822] using a similar PPP approach, investigated the electron densities, finding in agreement with ^{1}H- and ^{13}C-NMR data, higher positive charges in α- than in γ-positions (like most calculations, excepting those in Refs. 939 and 940), bond orders (finding that the first excited state of α- or γ-phenyl-substituted pyrylium salts is more planar than the ground state), singlet and triplet transition energies (calculated by the PPP method with or without configuration interactions), and the frontier orbitals of methyl- and/or phenyl-substituted pyrylium cations.

Ab initio nonempirical MO calculations for benzenium, pyridinium, pyrylium, and thiopyrylium cations were also reported.[947] The geometry assumed for the unsubstituted pyrylium cation showed unequal bond lengths and angles and was modeled after the pyridinium ion. As in the CNDO/2 calculations mentioned earlier, the oxygen heteroatom appears to be negatively charged (-0.44), while the α- ($+0.25$), β- ($+0.13$), and γ-carbon atoms ($+0.03$) appear positively charged.

Other theoretical calculations which were reported are: the ω-Hückel method for 2,4,6-triphenylpyrylium in order to compare the results with experimental X-ray data in terms of interionic charge-transfer interactions[853]; simple semiempirical (all valence electrons) MO calculations of the charge distribution and electronic spectrum[948] of 2,4,6-trimethylpyrylium[949]; calculated hydride ion affinities for correlating observed hydrogen transfers and disproportionations of 2*H*- and 4*H*-pyrans.[950]

V. Practical Applications of Pyrylium Salts

A. Introduction

Until recently, the procedures for obtaining pyrylium salts were aimed at convenience for laboratory purposes. Thus, the three 2,4,6-trimethylpyrylium salts included in *Organic Syntheses, Collective Volume 5*,[175,176,951,952] cannot be easily scaled up for the following reasons: the perchlorate[175,951] presents the danger of explosion; the trifluoromethanesulfonate[952] is very expensive; the fluoborate[176] gives a rather low yield relative to acetic anhydride because of the fairly low concentration in which fluorboric acid is available. However, the recently obtained[177] 2,4,6-trimethylpyrylium sulfoacetate has none of the above drawbacks and its production could easily be scaled up. 2,4,6-Triphenylpyrylium chloride is commercially available, thus confirming the prediction made in the last sentence of Part I[1]; in that part, Section III contained a few brief notes on practical applications.

B. Applications in the Photographic and Reprographic Industries

The largest number of patents involving pyrylium salts deals with applications in the photographic industry. The photographic technologies, especially those of color photography, have benefited from using pyrylium salts in the photosensitive layers of photographic paper and film. A summary of such uses includes:

1. Photosensitizers for positive emulsions, allowing the direct generation of positive images in a wide range of wavelengths.[953–957]

2. Photosensitizers for gelatin emulsions, allowing the light-induced cross-linking of gelatin under the action of radiations with lower energies (longer wavelengths) than in the absence of such pyrylium sensitizers.[958–960]

3. Photosensitizers in electrophotography, allowing the use of ordinary light sources which are less expensive for obtaining xerographic copies.[961–982]

4. Silver-free photographic films which, after exposure to light, give directly visible images either by coloring or by bleaching.[983–987]

5. Stabilizers for photographic emulsions which enable a longer conservation period without marked degradation.[988–990]

6. Internal labeling agents for photographic films, allowing their rapid identification through the fluorescence of certain pyrylium salts in various colors when exposed to ultraviolet light.[991]

7. Lithographic photosensitizers,[992] and additives for obtaining silver-free lithographic plates.[993–995]

8. Photosensitizers for photoconductive materials.[329c,996–1008]

Other data are quoted in Section IV,E.

C. APPLICATIONS AS ANTICORROSION AGENTS

The use of pyrylium salts as corrosion inhibitors becomes increasingly more widespread. The inhibiting effect increases on increasing the nucleophilicity of the substituents attached to the pyrylium ring.[1009–1013] An interesting linear correlation was observed between the longest wavelength absorption band and the corrosion-inhibiting coefficient of the pyrylium salt.[1014] Excellent results were obtained with pyrylium salts as corrosion inhibitors in acid or electrolytic polishing baths.[1015–1017] For the copper plating of steel, styrylpyrylium salts act as good surfactants inhibiting the anodic dissolution of iron in sulfuric acid and the anodic deposition of copper.[1018] For monitoring the quality of current-conducting coatings, 1,2-ethylenebipyrylium salts serve as color indicators in polymer films.[1019]

D. APPLICATIONS IN MACROMOLECULAR CHEMISTRY

Pyrylium salts have been successfully used as cationic polymerization initiators[1020] and as initiators for the stereospecific polymerization of 1,3-butadiene.[1021] They were also used as photosensitizers in cross-linking of polymers[1022] or in compounding light-sensitive polymers.[945] Other references are quoted in Part I,[1] using 2,4,6-trimethylpyrylium chloroferrate as polymerization initiator.

E. APPLICATIONS IN ORGANIC CHEMISTRY

In addition to the numerous applications discussed in preceding sections or in general reviews[2–17] which make pyrylium salts important key products in the synthesis of various carbocyclic or heterocyclic systems, two other applications deserve to be mentioned. On adding the pseudobase of 2,4,6-triphenylpyrylium, i.e., 1,3,5-triphenyl-1,5-pentenedione, to the reaction mixture of an alcohol and an acid chloride, the equilibrium is shifted toward ester formation because the hydrogen chloride is trapped as 2,4,6-triphenylpyrylium chloride, and the kinetics can be followed by the fluorescence of the pyrylium cation.[1023]

Since pyrylium perchlorates have characteristic melting points and are easily separated from organic reaction mixtures by precipitation with ether, the identification of olefins like isobutene can take place through their conversion to pyrylium salts by diacylation; thus the protodealkylation of 3,6-di-*t*-butylpyrocatechol was observed by trapping the isobutene as 2,4,6-trimethylpyrylium perchlorate.[1024]

F. Applications in Analytical Chemistry

Due to their low solubility, some 2,4,6-triarylpyrylium salts can be used for the quantitative gravimetric determination of anions (I^-, SCN^-, Cl_3CCOO^-, ClO_4^-, BF_4^-, MnO_4^-, $Cr_2O_7^{2-}$, $Fe(CN)_6^{4-}$) or of complex metallic anions (allowing the determination of these metals) obtained from Zn(II), Sn(II), Cd(II), Pt(II) or Au(III).[1025] Pyrylium salts have been used as constituents of specific membrane electrodes.[1026]

Some pyrylium salts like the pyrylocyanine **320** can be used as fluorescent acidimetric indicators for titrating weak organic bases in nonaqueous media with perchloric acid (cf. Section III,B,2,a).[156]

G. Applications in Electrochemistry

These are discussed in Section IV,C,3.

H. Applications as Fluorescent Dyestuffs
and in the Laser Technique

The fluorescent emission of pyrylium salts easily allows optical pumping, resulting in an inverse population of energy levels leading to light-activated stimulated emission of radiation (laser effect). The lasers thus obtained have the advantage of a varied range of wavelengths and of the convenient dissipation of thermal energy by continuously recirculating the pyrylium salt solution[838,1027,1028]; other references are cited in Part I[1] and in the present review in Section IV,A,1,b.

The fluorescence of pyrylium salts can be put to use in luminophors incorporated in plastics,[945,1029] in luminescent paints,[143,235,1006,1030] or in hydrology for tracing water courses.[1031,1032]

I. APPLICATIONS IN THE MANUFACTURE OF LABELED COMPOUNDS

The ready deuteration in D_2O or AcOD at benzylic positions of alkyl side chains bonded to 2-, 4-, or 6-positions of pyrylium salts allows the preparation of selectively deuterated pyrylium salts [because γ-positions are deuterated and dedeuterated more rapidly than α-positions (cf. Section III,A,7,b)], and hence of pyridines or other systems possessing deuterated side chains.[364,695] Such compounds [furans, phenols, anilines, benzene and naphthalene derivatives, etc. (cf. Section III,C)] cannot be obtained by alternative methods, or could only be prepared by laborious methods, not by direct hydrogen exchange.[372,438]

On using [^{14}C]acetic anhydride [$(CH_3—^{14}CO)_2O$], 2,4,6-[2,6-$^{14}C_2$]trimethylpyrylium perchlorate was obtained and from it a whole series of ring transformation products ^{14}C-labeled in ring positions (e.g., 2,4,6-[2,6-$^{14}C_2$]collidine, [1,3-$^{14}C_2$]mesitylene, [2,6-$^{14}C_2$]nitromesitylene, [2,6-$^{14}C_2$]mesitonitrile, 2,4,6-[2,6-$^{14}C_2$]trimethylacetophenone, 4,6,8-[4,8-$^{14}C_2$]trimethylazulene, 1,3-[1-^{14}C]dimethylnaphthalene.[437,477,1033] The ring transformation of pyrylium by $^{13}CH_3NO_2$ opened another way for introducing a labeled carbon atom into a definite ring position of benzene derivatives (cf. Section III,C,3,e).[708]

J. BIOLOGICAL EFFECTS

Because of the ease with which pyrylium salts react with amino acids and peptides (cf. Section III,C,3,c) it could be expected that they are not biologically indifferent compounds. Indeed, some pyrylium salts show remarkable activity as bactericides and/or fungicides.[47,1034–1036] Thus, e.g., 2,6-dimethyl-4-(2-benzothiazolyl)pyrylium perchlorate (423) proved to be effective against bacteria which cause dysentery, destroying *in vitro* bacterial cultures in a concentration of 0.01 mg/liter.[47] A recent patent[1037] describes the use of the 2,4,6-trimethylpyrylium cation in the manufacture of a new coccidiostatic, (941) which has an analogous structure with "Amprolium" (942).

(941)   (942)

A series of *in vitro* studies demonstrated that pyrylium salts may also possess genetic activity; in most cases chromosomial aberrations were observed.[1038–1043]

The study of pharmacological properties of 2,4,6-trimethylpyrylium perchlorate has shown that it has a pronounced sedative effect,[1044] and also neurotropic and analgetic activity with relatively low acute toxicity (in mice the LD_{50} is ~520 mg/kg).[47] In addition, this salt shows a certain antitumoral activity.[1045] Doses of 1–50 mg/kg of 2,4,6-trimethylpyrylium salts administrated intramuscularly in laboratory animals decrease the biopotential of thalamus, hypothalamus, and of the visual cortex.[1046]

Certain α-styrylpyrylium salts were reported to act as plant growth stimulants.[1047]

VI. Perspectives

Possessing the most electronegative heteroatom, pyrylium salts constitute an extreme case of monocyclic aromatic six-membered systems, namely a benzene ring with the strongest possible single perturbation, replacement of a CH group by an O^+ heteroatom. The consequences of this perturbation are manifold: characteristic physical properties varying monotonously in the series benzene, pyridine, pyridinium, pyrylium; acidification of α- or γ-oriented benzylic protons; facile ring opening under the action of nucleophiles, absence of electrophilic substitutions; easy formation from acyclic starting materials. These properties [which are due to, and benefit from, the reduced aromaticity (but not too drastically reduced)] make pyrylium salts an attractive intermediate step in the conversion of acyclic starting materials to a host of carbocyclic or heterocyclic rings, mostly aromatic, or acyclic conjugated systems.

Figure 12 presents once more the main syntheses of pyrylium salts, whereas Fig. 13 demonstrates again schematically important ring-interconversion reactions of the pyrylium cation.

In addition to their interest for organic chemical syntheses, pyrylium salts present both theoretical and practical interest: their physical properties can be nicely correlated with the perturbation introduced by the heteroatom into a benzene ring, while their practical uses reviewed in Section V have just begun to appear in the reprographic industries. The fact that the chemistry of pyrylium salts has been referred to more and more in internationally well-known review articles, and this in close connection with useful synthetic methods, demonstrates once more the broad and increasing chemical importance of this type of heterocycles.

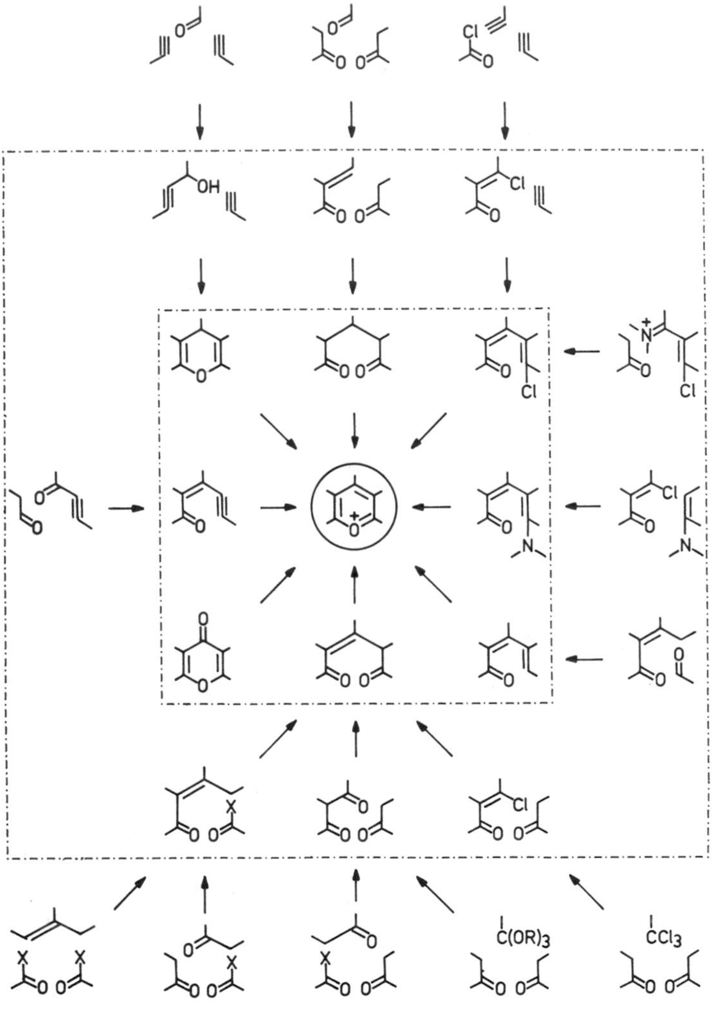

FIG. 12. Synthetic main routes to the pyrylium cation (the inner field contains one-component syntheses, followed by two- and three-component syntheses).

These heterocycles, indeed, behave for most synthetic purposes as special "masked" carbonyl compounds, especially as unsaturated 1,5-dicarbonyl species; and carbonyl compounds are, as stated elsewhere,[1048] the "backbone of organic synthesis". Pyrylium salts represent the most convincing example of the value of heterocycles as tools in organic syntheses (cf. also the remarks in the Introduction, Section I).

Possible future trends in pyrylium research is expected to profit from

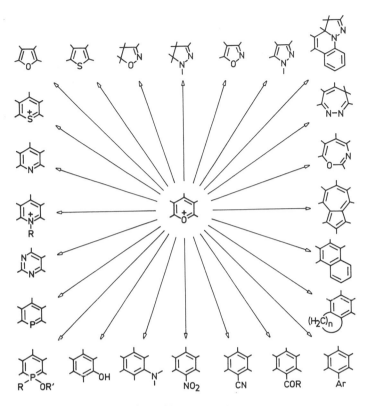

Fig. 13. Schematic representation of important ring transformation reactions of pyr-ylium salts.

HPLC separation of pyrylium salts when they occur in mixtures; till now, such separations were difficult. An attractive feature of preparing and handling pyrylium salts is that, since they are salts, they crystallize nicely and are easily purified from organic by-products by simple washing with either, or by recrystallization. In the experience of the authors, once one has started to explore the wide possibilities offered by the chemistry of the pyrylium ring, one can easily become an addict, being "hooked" by this fascinating field of research. The present review, together with Part I,[1] has tried to convey in a convenient form for the reader the main outlines of research and their results in an area of heterocyclic chemistry which till recently was mostly ignored or neglected. As Fig. 1 demonstrates, however, this period is now over, and the exponential increase of papers and patents is likely to continue in the foreseeable future.

Acknowledgment

Thanks are expressed to Dr. G. Nicolae for assistance in the literature search, and to Professor K. Dimroth for making available his Lecture.[4] Mr. T. Zimmermann dedicatedly assisted in typing the tables and proof reading.

Appendix: Tables XII–XXX

TABLE XII
STABLE α-ANHYDRO BASES OF PYRYLIUM SALTS: 2-METHYLENE-2H-PYRAN DERIVATIVES

Formula	X	R^1	R^2	M.P. (°C)	Reference
$C_{18}H_{14}O$	CH_2	Ph	Ph	90	1049
$C_{19}H_{14}O_2$	CHCHO	Ph	Ph	125	245, 250
$C_{21}H_{18}O_4$	CHCHO	$C_6H_4OMe(4)$	$C_6H_4OMe(4)$	132	250
$C_{24}H_{19}NO_3$	CH–CH=C(CN)(COOEt)	Ph	Ph	152	1050, 1051
$C_{24}H_{20}O_3$	CH–CH=C(COMe)(COMe)	Ph	Ph	101	1051
$C_{25}H_{22}O_4$	CH–CH=C(COMe)(COOEt)	Ph	Ph	151	1051
$C_{26}H_{16}O_2$	=⟨cyclohexadienone⟩=O	Ph	Ph	164	360
$C_{26}H_{16}O_3$	=⟨cyclohexadienone⟩=O	Ph	$C_6H_4OH(4)$	340	360
$C_{26}H_{24}O_5$	CH–CH=C(COOEt)(COOEt)	Ph	Ph	91	1051
$C_{27}H_{20}O_3$	CH–CH=C(Ph)(COOEt)	Ph	Ph	176	1051

TABLE XII (continued)

Formula	X	R^1	R^2	M.P. (°C)	Reference
C$_{27}$H$_{23}$NO$_3$	CH–CH=N–Ph	C$_6$H$_4$OMe(4)	C$_6$H$_4$OMe(4)	165	250
C$_{28}$H$_{26}$O$_5$	CH–CH=CH–CH=C⟨COOEt / COOEt	Ph	Ph	70	880
C$_{34}$H$_{24}$O$_3$	CH–CH=C⟨COPh / COPh	Ph	Ph	108	1051

TABLE XIII
STABLE α-ANHYDRO BASES OF PYRYLIUM SALTS: 5,6-DIHYDRO-7H-CHROMENE DERIVATIVES

Formula	R	M.P. ($^{\circ}$C)	Reference
$C_{16}H_{18}O_2$	CHO	153	243
$C_{21}H_{20}O$	CH=CH–CH=C(CN)(CN)	232	350
$C_{23}H_{22}O$	(CH=CH)$_2$–CH=C(CN)(CN)	209	350
$C_{25}H_{19}NO_2S_2$		277	350
$C_{26}H_{20}N_2O_4$		292	350
$C_{30}H_{22}N_2O_3$	CH=C(CN)(C$_6$H$_4$NO$_2$(4))	262	350
$C_{30}H_{22}O_3$		225	350

TABLE XIII (continued)

Formula	R	M.P. (°C)	Reference
$C_{31}H_{22}O_3$		231	350
$C_{31}H_{28}O_3$		224	350
$C_{32}H_{23}BF_2O_3$		380	1052
$C_{32}H_{24}N_2O_3$		244	350
$C_{32}H_{26}N_2O_2$		232	350
$C_{33}H_{24}O_3$		258	350
$C_{34}H_{25}BF_2O_3$		278	1052

TABLE XIII (continued)

Formula	R	M.P. (°C)	Reference
$C_{34}H_{26}N_2O_3$	$(CH=CH)_2-CH=C{<}^{CN}_{C_6H_4NO_2(4)}$	237	350
$C_{36}H_{30}N_2O_2$		227	350

TABLE XIV
STABLE α-ANHYDRO BASES OF PYRYLIUM SALTS: CYCLOPENTA[b]PYRAN DERIVATIVES

Formula	R^1	R^2	R^3	M.P. (°C)	Reference
$C_{24}H_{22}O$	H	Ph	t-Bu	137	204
$C_{26}H_{18}O$	H	Ph	Ph	242	204
$C_{32}H_{22}O$	Ph	Ph	Ph	184.5-185	343
$C_{33}H_{24}O_2$	Ph	Ph	$C_6H_4OMe(4)$	157-158	343
$C_{33}H_{24}O_2$	Ph	$C_6H_4OMe(4)$	Ph	206-208	343

TABLE XV
STABLE α-ANHYDRO BASES OF PYRYLIUM SALTS: INDENO[2,1-*b*]PYRAN DERIVATIVES

Formula	R^1	R^2	R^3	R^4	M.P. ($^\circ$C)	Reference
$C_{14}H_{12}O$	H	H	H	Et	119	347
$C_{14}H_{12}O$	H	Me	H	Me	89	346, 347
$C_{15}H_{14}O$	Me	Me	H	Me	124	347
$C_{15}H_{14}O$	H	Me	Me	Me	125	346, 347
$C_{15}H_{14}O$	H	H	H	Pr	193	347
$C_{15}H_{14}O$	H	H	H	i-Pr	137	347
$C_{16}H_{16}O$	H	Me	Et	Me	105	346, 347
$C_{17}H_{12}OS$	H	Me	H	2-thienyl	125	346, 347
$C_{17}H_{16}O$	H	H	$-(CH_2)_5-$		148	347
$C_{18}H_{11}BrO$	H	H	H	$C_6H_4Br(4)$	287	347
$C_{18}H_{11}ClO$	H	H	H	$C_6H_4Cl(4)$	243	347
$C_{18}H_{12}O$	H	H	H	Ph	215	128, 347
$C_{18}H_{14}OS$	Me	Me	H	2-thienyl	136	347

TABLE XV (continued)

Formula	R^1	R^2	R^3	R^4	M.P. (°C)	Reference
$C_{19}H_{14}O$	H	Ph	H	Me	118	67
$C_{19}H_{14}O$	H	Me	H	Ph	116	67, 68, 346, 347
$C_{19}H_{14}O$	H	H	H	$C_6H_4Me(4)$	219	347
$C_{19}H_{22}O$	H	Me	$i\text{-}C_6H_{11}$	Me	oil	347
$C_{20}H_{14}O$	H	H	$-(CH_2)_2C_6H_4(2)-$		277	347
$C_{20}H_{16}O$	H	Me	H	$C_6H_4Me(4)$	145	346, 347
$C_{20}H_{16}O$	Me	Me	H	Ph	155	67, 68, 347
$C_{20}H_{16}O$	Ph	Me	H	Me	148, 163	346, 347
$C_{20}H_{16}O$	H	Me	Ph	Me	90	346, 347
$C_{21}H_{16}O$	H	H	$-(CH_2)_3C_6H_4(2)-$		263	347
$C_{21}H_{18}O$	H	Ph	H	Pr	161	67
$C_{21}H_{18}O$	H	Ph	H	i-Pr	110	67
$C_{21}H_{18}O$	H	Me	CH_2Ph	Me	120	346, 347
$C_{21}H_{18}O$	Me	Me	H	$C_6H_4Me(4)$	136	346
$C_{22}H_{20}O$	H	Bu	H	Ph	32	67
$C_{23}H_{16}O$	H	Me	H	2-naphthyl	139	347
$C_{23}H_{16}OS$	Ph	Me	H	2-thienyl	201, 182	346, 347
$C_{23}H_{18}O$	2-indenyl	Me	H	Me	179	346, 347

TABLE XV (continued)

Formula	R^1	R^2	R^3	R^4	M.P. (°C)	Reference
$_4H_{16}O$	H	Ph	H	Ph	147	67, 68, 346, 347
$_4H_{18}O$	Me	Me	H	2-naphthyl	195	347
$_4H_{20}O$	2-indenyl	Me	Me	Me	215	346
$_5H_{18}O$	Me	Ph	H	Ph	151	67
$_5H_{18}O$	Ph	Me	H	Ph	215	346, 347
$_5H_{22}O$	2-indenyl	Me	Et	Me	195	346
$_6H_{18}OS$	2-indenyl	Me	H	2-thienyl	213	346
$_6H_{20}O$	Ph	Me	H	$C_6H_4Me(4)$	199	346, 347
$_7H_{20}O$	1-acenaphthylenyl	Me	Me	Me	219	67
$_8H_{20}O$	2-indenyl	Me	H	Ph	241	346, 347
$_8H_{22}O$	1-acenaphthylenyl	Me	Et	Me	95	67
$_8H_{24}O$	i-Bu	Ph	H	Ph	138	67
$_9H_{20}O$	Ph	Me	H	2-naphthyl	209	347
$_9H_{22}O$	2-indenyl	Me	Ph	Me	217	346
$_9H_{22}O$	2-indenyl	Me	H	$C_6H_4Me(4)$	224	346
$_9H_{22}O$	1-acenaphthylenyl	H	$-(CH_2)_5-$		178	67
$_0H_{20}O$	Ph	Ph	H	Ph	216	122, 347

TABLE XV (continued)

Formula	R^1	R^2	R^3	R^4	M.P. ($^\circ$C)	Reference
$C_{30}H_{24}O$	2-indenyl	Me	CH_2Ph	Me	234	346
$C_{31}H_{20}O$	1-acenaphthylenyl	Me	H	Ph	223	67
$C_{31}H_{20}O_2$	COPh	Ph	H	Ph	233	67
$C_{33}H_{22}O$	2-indenyl	Ph	H	Ph	216	346
$C_{33}H_{22}O$	1-acenaphthylenyl	H	$-(CH_2)_3C_6H_4(2)-$		186	67
$C_{33}H_{24}O$	1-acenaphthylenyl	Me	CH_2Ph	Me	181	67

TABLE XVI
Stable γ-Anhydro Bases of Pyrylium Salts

Formula	X	R^1	R^2	R^3	R^4	M.P. (°C)	Reference
$C_{10}H_8N_2O$	$C(CN)_2$	Me	H	H	Me	191	329a
$C_{10}H_9NO_2S_2$		Me	H	H	Me	260	329a
$C_{10}H_{10}N_2O_2$	$C(CN)CONH_2$	Me	H	H	Me	252	329a
$C_{10}H_{12}O_2$	CHCOMe	Me	H	H	Me	92	569
$C_{11}H_{10}N_2O_4$		H	H	H	H	214	1053
$C_{12}H_{12}O_6$	CHCOMe	COOMe	H	H	COOMe	186	328
$C_{12}H_{13}NO_2S_2$		Me	H	H	Me	202	330
$C_{12}H_{13}NO_3$	$C(CN)COOEt$	Me	H	H	Me	184	329a
$C_{14}H_{13}NO_3S$	$CHSC_6H_4NO_2(4)$	Me	H	H	Me	102	241
$C_{14}H_{14}O_7$	$C(COMe)_2$	COOMe	H	H	COOM	130	328
$C_{16}H_{13}NO_3$		Me	H	H	Me	187	39
$C_{17}H_{14}O_6$	CHCOPh	COOMe	H	H	COOMe	155	328
$C_{17}H_{16}O_7$		COOMe	H	H	COOMe	210	328

TABLE XVI (continued)

Formula	X	R^1	R^2	R^3	R^4	M.P. (°C)	Reference
$C_{18}H_{13}NO_3$	$CHNO_2$	Ph	H	H	Ph	170	228, 324
$C_{18}H_{14}O$	CH_2	Ph	H	H	Ph	155	901, 1049
$C_{18}H_{14}O_2$	CHOH	Ph	H	H	Ph	206	331
$C_{19}H_{12}O_2$	(anthraquinone structure)	H	H	H	H	129	34
$C_{19}H_{13}NO$	CHCN	Ph	H	H	Ph	147	38
$C_{20}H_{12}N_2O$	$C(CN)_2$	Ph	H	H	Ph	261	38
$C_{20}H_{13}NO_2S_2$	(thiazolidine structure)	Ph	H	H	Ph	319	285
$C_{20}H_{13}NO_3S$	(thiazolidine structure)	Ph	H	H	Ph	316	285
$C_{20}H_{14}N_2O_2$	$C(CN)CONH_2$	Ph	H	H	Ph	290	38
$C_{20}H_{16}N_2O_5S_2$	$C(SC_6H_4NO_2(4))(SC_6H_4NO_2(4))$	Me	H	H	Me	206	241
$C_{20}H_{16}O_2$	CHCOMe	Ph	H	H	Ph	110	224
$C_{20}H_{18}O$	CMe_2	Ph	H	H	Ph	145	324
$C_{20}H_{24}N_2O_5$	$CHC_6H_3(NO_2)_2(2,4)$	t-Bu	H	H	t-Bu	125	329c
$C_{21}H_{15}NO_3$	(oxazolone structure, Me)	Ph	H	H	Ph	178	39
$C_{21}H_{16}O_2$	(anthraquinone structure)	Me	H	H	Me	170	34

TABLE XVI (continued)

Formula	X	R^1	R^2	R^3	R^4	M.P. (°C)	Reference
$C_{22}H_{14}N_2O_2$	C(CN)COCH$_2$CN	Ph	H	H	Ph	261	38
$C_{22}H_{15}NO_3$	C(CN)COCOOMe	Ph	H	H	Ph	230	38
$C_{22}H_{17}NOS_3$	[thiazolidine ring: S=, NEt, =S]	Ph	H	H	Ph	320	285, 330
$C_{22}H_{17}NO_2S_2$	[thiazolidine ring: S=, NEt, =O]	Ph	H	H	Ph	210	330
$C_{22}H_{21}NO$	CHCH=NCHMe$_2$	Ph	H	H	Ph	114	231
$C_{23}H_{14}N_4O$	C(CN)C(NH$_2$)=C(CN)$_2$	Ph	H	H	Ph	311	38
$C_{23}H_{16}O_2$	[cyclohexadienone, o-]	Ph	H	H	Ph	131	1054
$C_{23}H_{16}O_2$	[cyclohexadienone, p-]	Ph	H	H	Ph	263	361
$C_{23}H_{16}O_3$	[cyclohexadienone, p-]	C$_6$H$_4$OH(4)	H	H	Ph	317	1055
$C_{23}H_{22}O$	[cyclohexylidene]	Ph	H	H	Ph	153	324
$C_{24}H_{15}Br_3O$	CHC$_6$H$_4$Br(4)	C$_6$H$_4$Br(4)	H	H	C$_6$H$_4$Br(4)	248	224
$C_{24}H_{15}N_3O_7$	CHC$_6$H$_2$(NO$_2$)$_3$(2,4,6)	Ph	H	H	Ph	110	329d
$C_{24}H_{16}Br_2O$	CHPh	C$_6$H$_4$Br(4)	H	H	C$_6$H$_4$Br(4)	204	224

TABLE XVI (continued)

Formula	X	R^1	R^2	R^3	R^4	M.P. ($^{\circ}$C)	Reference
$C_{24}H_{16}N_2O_5$	CHPh	$C_6H_4NO_2(4)$	H	H	$C_6H_4NO_2(4)$	302	224
$C_{24}H_{16}N_2O_5$	$CHC_6H_3(NO_2)_2(2,4)$	Ph	H	H	Ph	210	329c, 329d
$C_{24}H_{17}BrO$	$CHC_6H_4Br(4)$	Ph	H	H	Ph	195	224
$C_{24}H_{17}NO_3$	$CHC_6H_4NO_2(2)$	Ph	H	H	Ph	110	329c, 329d
$C_{24}H_{17}NO_3$	$CHC_6H_4NO_2(4)$	Ph	H	H	Ph	207	329d
$C_{24}H_{18}O$	CHPh	Ph	H	H	Ph	140	224, 228, 695
$C_{24}H_{18}O_7$	$C(COPh)_2$	COOMe	H	H	COOMe	190	328
$C_{24}H_{19}NO$	$CHC_6H_4NH_2(2)$	Ph	H	H	Ph	107	329c, 329d
$C_{24}H_{19}NO$	$CHC_6H_4NH_2(4)$	Ph	H	H	Ph	170	329d
$C_{24}H_{19}NO_3$	CHCH=C(CN)COOEt	Ph	H	H	Ph	162	1051
$C_{24}H_{20}N_2O$	$CHC_6H_3(NH_2)_2(2,4)$	Ph	H	H	Ph	153	329d
$C_{24}H_{20}O_3$	$CHCH=C(COMe)_2$	Ph	H	H	Ph	122	1051
$C_{25}H_{16}N_2O_3$	$C(CN)C_6H_4NO_2(2)$	Ph	H	H	Ph	204	329d, 1056
$C_{25}H_{16}N_2O_3$	$C(CN)C_6H_4NO_2(4)$	Ph	H	H	Ph	236	329d
$C_{25}H_{18}O_2$	CHCOPh	Ph	H	H	Ph	160	224
$C_{25}H_{22}O_4$	CHCH=C(COMe)COOEt	Ph	H	H	Ph	164	1051
$C_{25}H_{25}NO$	$CHCH=NC_6H_{11}$	Ph	H	H	Ph	175	231

TABLE XVI (continued)

Formula	X	R^1	R^2	R^3	R^4	M.P. (°C)	Reference
$C_{26}H_{16}BrNO_3$	oxazolone (Ph)	Ph	Br	H	Ph	196	39
$C_{26}H_{17}NO_3$	oxazolone (Ph)	Ph	H	H	Ph	246	39, 326, 427, 1057
$C_{26}H_{17}NO_3S$	thiazolidinedione (N–Ph)	Ph	H	H	Ph	301	285
$C_{26}H_{18}N_2O_2$	C(CN)CONHPh	Ph	H	H	Ph	240	38
$C_{26}H_{19}NO_4$	C(COOH)NHCOPh	Ph	H	H	Ph	204	326, 362
$C_{26}H_{21}N_3O_3$	C(NHCOPh)(CONHNH$_2$)	Ph	H	H	Ph	219	362
$C_{27}H_{21}NO_4$	C(COOMe)NHCOPh	Ph	H	H	Ph	249	326, 362
$C_{30}H_{22}O$	CH_2	Ph	Ph	Ph	Ph	230	392
$C_{30}H_{24}O$	triphenylcyclopentadiene (Ph, Ph, Ph)	Me	H	H	Me	261	676
$C_{31}H_{20}O_2$	anthraquinone	Ph	H	H	Ph	259	34
$C_{31}H_{32}O_2$	2,6-di-t-Bu cyclohexadienone	Ph	H	H	Ph	271	34
$C_{36}H_{28}O$	tetraphenylcyclopentadiene (Ph, Ph, Ph, Ph)	Me	H	H	Me	268	1058 1059
$C_{39}H_{24}N_2O_3$	C(CN)COC(CN)= (2,6-diphenylpyran)	Ph	H	H	Ph	345	38

TABLE XVII
STABLE PSEUDO BASES OF PYRYLIUM SALTS

Formula	R^1	R^2	R^3	R^4	R^5	M.P. (°C)	Reference
$C_8H_{12}O_2$	Me	H	Me	H	Me	oil	44, 390
$C_{14}H_{24}O_2$	t-Bu	H	Me	H	t-Bu	60	389
$C_{22}H_{23}NO_2$	Ph	H	piperidino	H	Ph	138	1060
$C_{23}H_{16}Br_2O_2$	$C_6H_4Br(4)$	H	Ph	H	$C_6H_4Br(4)$	125	404
$C_{23}H_{16}Cl_2O_2$	$C_6H_4Cl(4)$	H	Ph	H	$C_6H_4Cl(4)$	118	404
$C_{23}H_{17}ClO_2$	Ph	H	$C_6H_4Cl(4)$	H	Ph	112	404
$C_{23}H_{18}O_2$	Ph	H	Ph	H	Ph	120	32, 208, 404, 1025, 1061-1063

Formula						m.p.	Ref.
$C_{24}H_{17}O_2$	-C$_6$H$_4$CH$_2$(2)-		Ph	H	Ph	128–129	205
$C_{24}H_{20}O_2$	Ph	H	C$_6$H$_4$Me(4)	H	Ph	127	404
$C_{24}H_{20}O_2$	Ph	Me	Ph	H	Ph	103	393
$C_{24}H_{20}O_2$	Ph	H	Ph	Me	Ph	143	393
$C_{24}H_{20}O_3$	Ph	H	Ph	H	C$_6$H$_4$OMe(4)	90	388
$C_{24}H_{20}O_3$	Ph	H	C$_6$H$_4$OMe(4)	H	Ph	112	388, 394, 404
$C_{24}H_{20}O_3$	Ph	H	Ph	H	C$_6$H$_3$OH(2)Me(4)	124	1064
$C_{24}H_{20}O_4$	C$_6$H$_4$OH(2)	H	Ph	H	C$_6$H$_3$OH(2)Me(4)	152	1064
$C_{25}H_{20}O_4$	Ph	H	C$_6$H$_4$OAc(2)	H	Ph	113	1054
$C_{25}H_{22}O_2$	C$_6$H$_4$Me(4)	H	Ph	H	C$_6$H$_4$Me(4)	99	404

TABLE XVII (continued)

Formula	R^1	R^2	R^3	R^4	R^5	M.P. (°C)	Reference
$C_{25}H_{22}O_4$	Ph	H	$C_6H_4OMe(4)$	H	$C_6H_4OMe(4)$	98	1055
$C_{25}H_{22}O_4$	$C_6H_4OMe(4)$	H	Ph	H	$C_6H_3OH(2)Me(4)$	126	1064
$C_{25}H_{22}O_5$	$C_6H_4OMe(4)$	H	Ph	H	$C_6H_3OH(2)OMe(4)$	129	1064
$C_{25}H_{23}NO_2$	Ph	H	$C_6H_4NMe_2(4)$	H	Ph	137	1065
$C_{26}H_{24}O_2$	$C_6H_4Me(4)$	H	$C_6H_4Me(4)$	H	$C_6H_4Me(4)$	97	404
$C_{26}H_{24}O_3$	$C_6H_4Me(4)$	H	$C_6H_4OMe(4)$	H	$C_6H_4Me(4)$	108	404
$C_{26}H_{24}O_6$	$C_6H_4OMe(4)$	H	$C_6H_4OMe(4)$	H	$C_6H_3OH(2)OMe(4)$	146	1064
$C_{26}H_{24}O_7$	$C_6H_3CH(2)OMe(4)$	H	$C_6H_4OMe(4)$	H	$C_6H_3OH(2)OMe(4)$	128	1064
$C_{27}H_{29}NO_4$	t-Bu	H	(anthracene structure with NO_2)	H	t-Bu	95	1064
$C_{29}H_{22}O_2$	Ph	Ph	H	Ph	Ph	110	392

$C_{29}H_{22}O_2$	Ph	H	Ph	Ph	Ph	$113/147^a$	393
$C_{30}H_{22}O_2$	$-C_6H_4CH(Ph)(2)-$		Ph	H	Ph	153–154	205
$C_{30}H_{25}NO_2$	Ph	H	$C_6H_4N(Me)Ph(4)$	H	Ph	149	1060
$C_{31}H_{21}NO_4$	Ph	H	[structure, NO_2]	H	Ph	165	329c
$C_{31}H_{22}O_2$	Ph	H	[structure]	H	Ph	110	329c
$C_{35}H_{26}O_2$	Ph	Ph	Ph	Ph	Ph	151	1066
$C_{35}H_{27}NO_2$	Ph	H	$C_6H_4NPh_2(4)$	H	Ph	140	1060

[a] cis/trans.

TABLE XVIII
PYRIDINES OBTAINED FROM PYRYLIUM SALTS

Formula	R^1	R^2	R^3	R^4	R^5	M.P. (B.P.) (°C)	Reference
C_6H_7N	H	H	Me	H	H	163[a]	412
C_6H_7NO	H	H	OMe	H	H	171[a]	1067
C_7H_9N	Me	H	Me	H	H		1068
C_7H_9N	Me	H	H	H	Me	160[a] (143)	109, 412
$C_8H_{11}N$	Me	H	Me	H	Me	156–157[a] (168)	94, 172, 193, 438, 477, 512, 1068, 1069
$C_8H_{11}N$	$^{14}CH_3$	H	Me	H	$^{14}CH_3$		477

Formula					mp	Ref.
$C_8H_{11}N$	CD_3	H	CD_3	H		372
$C_8H_{11}N$	Me	Me	H	Me	146[a] (172–175)	109
$C_8H_{11}NO$	Me	H	OMe	Me	154[a]	29
$C_8H_{11}NS$	Me	H	SMe	Me	51[a] (88/2)	507, 1070
C_9H_7NS	2-thienyl	H	H	H		121
$C_9H_9NO_4$	COOMe	H	H	COOMe		69
$C_9H_{13}N$	Et	H	Me	Me	112–113[a] (181)	94, 172, 512
$C_9H_{13}N$	Me	H	Et	Me	119–122[a] (185–186)	172, 193, 486, 512, 1071

TABLE XVIII (continued)

Formula	R^1	R^2	R^3	R^4	R^5	M.P. (B.P.) (°C)	Reference
C$_9$H$_{13}$N	Me	Me	Me	H	Me	107[a] (203–204)	172, 193, 486, 512, 1071
C$_9$H$_{13}$N	Me	Et	H	H	Me	112[a] 160[a] (145–150)	172, 1072
C$_9$H$_{13}$N	Me	Me	H	Me	Me	76 177[a] (198)	172, 512
C$_9$H$_{13}$N	Et	H	H	H	Et	(85/35)	1073
C$_{10}$H$_{10}$N$_2$S	Me	H	2-thiazolyl	H	Me		99
C$_{10}$H$_{13}$NO	Me	H	CH$_2$Ac	H	Me		193
C$_{10}$H$_{15}$N	Et	H	Me	H	Et	140[a] (182–195)	94, 172, 512

						mp	Refs.
$C_{10}H_{15}N$	Me	H	Pr	H	Me		**193**
$C_{10}H_{15}N$	Me	H	i-Pr	H	Me	168[a] (196/751)	94, **193**, 512
$C_{10}H_{15}N$	Me	Pr	H	H	Me	111[a] (203–205)	1074
$C_{10}H_{15}N$	Me	Et	Me	H	Me		**193**
$C_{10}H_{15}N$	Me	Me	Et	H	Me		**193**
$C_{10}H_{15}N$	Me	Me	H	Et	Me	144–145[a] (203–210)	109
$C_{10}H_{15}N$	Me	Me	Me	Me	Me		**193**
$C_{11}H_9N$	H	H	H	H	Ph		121
$C_{11}H_{11}NS$	Me	H	2-thienyl	H	Me	240–241	99, 1025

TABLE XVIII (continued)

Formula	R^1	R^2	R^3	R^4	R^5	M.P. (B.P.) (°C)	Reference
$C_{11}H_{11}NS$	2-thienyl	H	Me	H	Me		109
$C_{11}H_{15}NO$	Me	H	CH(Me)Ac	H	Me		193
$C_{11}H_{15}NO$	Me	Me	CH_2Ac	H	Me	61	193
$C_{11}H_{16}N_2O$	Me	H	morpholino	H	Me	124	507, 1075
$C_{11}H_{17}N$	Et	H	Et	H	Et	136[a] (218)	94
$C_{11}H_{17}N$	Me	H	Bu	H	Me		1076
$C_{11}H_{17}N$	Et	Me	Me	H	Et	137[a]	172
$C_{11}H_{17}N$	Me	Pr	Me	H	Me		1076
$C_{11}H_{17}N$	Me	Bu	H	H	Me	103[a] (221-223)	1074
$C_{11}H_{17}N$	Me	Me	Et	Me	Me	20-21 (131/17)	512

$C_{12}H_{11}NO$	H	H	H	H	C$_6$H$_4$OMe(4)	121
$C_{12}H_{17}N$	Me	H	Me	-CH$_2$CH$_2$CH(Me)CH$_2$-		1076
$C_{12}H_{17}N$	Me	H	Me	-CH(Me)CH$_2$CH$_2$CH$_2$-		483, 1077
$C_{12}H_{17}NO$	Me	H	CH(Et)Ac	Me		193
$C_{12}H_{17}NO$	Me	H	C(Me)$_2$Ac	Me		193
$C_{12}H_{17}NO$	Me	Me	CH(Me)Ac	Me		193
$C_{12}H_{17}NO$	Me	Et	CH$_2$Ac	Me	66	193
$C_{12}H_{17}NO$	Me	Me	CH$_2$Ac	Me	93	193
$C_{12}H_{18}N_2$	Me	H	piperidino	Me	83	28
$C_{12}H_{19}N$	i-Pr	H	Me	i-Pr	171[a] (210)	94

TABLE XVIII (continued)

Formula	R^1	R^2	R^3	R^4	R^5	M.P. (B.P.) (°C)	Reference
$C_{12}H_{19}N$	Pr	H	Me	H	Pr	91[a] (225)	94
$C_{12}H_{19}N$	Me	Me	H	Bu	Me	133-135[a] (231-235)	109
$C_{12}H_{19}N$	Me	t-Bu	Me	H	Me		1068
$C_{13}H_9NS_2$	2-thienyl	H	H	H	2-thienyl		114, 153
$C_{13}H_{13}N$	Me	H	Ph	H	Me	59 223[a]	94, 172 512, 1076
$C_{13}H_{13}N$	Ph	H	Me	H	Me	135-137[a] (295)	94
$C_{13}H_{13}N$	Me	Ph	H	H	Me	168 (185/96)	121
$C_{13}H_{13}NO$	H	H	H	H	$C_6H_4OEt(4)$		121

Formula						mp	Ref.
$C_{13}H_{13}NO_2$	H	H	H	H	$C_6H_3(OMe)_2(3,4)$		121
$C_{13}H_{13}NO_2$	H	H	H	H	H	186	28
$C_{13}H_{21}N$	Me	C_5H_{11}	Me	H	Me		1076
$C_{13}H_{21}N$	Me	H	C_6H_{13}	H	Me		1076
$C_{13}H_{21}N$	Me	C_6H_{13}	H	H	Me	(247–253)	1074
$C_{14}H_{11}NS_2$	2-thienyl	H	2-thienyl	H	Me		503
$C_{14}H_{12}N_2S$	Me	-	2-benzthiazolyl	H	Me	105–107	47, 1025
$C_{14}H_{13}NO_2$	Me	H	$C_6H_4COOH(4)$	H	Me		1076
$C_{14}H_{13}N$	Me	H	$-C_6H_4(CH_2)_2(2)-$	H	H	107	1078

Ring structure (shown in table): 2,6-dimethyl-4-methoxy pyran ring, labels — OMe, Me, CH, Me.

TABLE XVIII (continued)

Formula	R¹	R²	R³	R⁴	R⁵	M.P. (B.P.) (°C)	Reference
$C_{14}H_{15}N$	Me	H	CH_2Ph	H	Me	143[a] (134/3)	179
$C_{14}H_{15}N$	Me	Ph	Me	H	Me	141[a] (119/3)	179
$C_{14}H_{15}NO$	Me	H	$C_6H_4OMe(4)$	H	Me		264
$C_{14}H_{15}NS$	Me	H	SCH_2Ph	H	Me	56 (166–172/5)	28, 507, 1070
$C_{14}H_{15}NS$	Me	Ph	SMe	H	Me	99–100	1079
$C_{14}H_{19}N$	Me	$-(CH_2)_4-$	H	$-CH_2CH(Me)CH_2CH_2-$			542
$C_{14}H_{20}N$	Me	$-(CH_2)_3$——CH_2 $-(CH_2)_3-$ H			Me	171[a]	184
$C_{14}H_{23}N$	Bu	H	Me	H	Bu	111[a] (261)	94
$C_{14}H_{23}N$	i-Bu	H	Me	H	i-Bu	70–71[a] (245)	94

Formula						mp	No.
$C_{14}H_{23}N$	t-Bu	H	Me	H	t-Bu	140[a] (226)	94
$C_{14}H_{23}N$	Me	C_7H_{15}	H	H	Me	96[a] (270)	1074
$C_{15}H_{13}NO$	[benzofuran structure]	H	Me	H	Me		124
$C_{15}H_{15}NO$	Me	Me	Me	H	Me	156[a] (243/96)	510
$C_{15}H_{15}NO_2$	Me	$C_6H_4Ac(4)$	H	H	$C_6H_4OMe(4)$	109	120
$C_{15}H_{16}N_2$	Me	H	[indoline structure]	H	Me	112	212
$C_{15}H_{17}N$	Et	H	Ph	H	Et	199[a]	94
$C_{15}H_{18}N_2$	Me	H	$C_6H_4NMe_2(4)$	H	Me		1076

TABLE XVIII (continued)

Formula	R^1	R^2	R^3	R^4	R^5	M.P. (B.P.) (°C)	Reference
C$_{16}$H$_{25}$N	Me	-(CH$_2$)$_4$——CH$_2$— / H		-(CH$_2$)$_4$-	Me		185
C$_{16}$H$_{25}$N	-(CH$_2$)$_5$-	H	Me	-(CH$_2$)$_5$- / H		104	511, 1080
C$_{17}$H$_{13}$N	Ph	H	Ph	H	H	70	98
C$_{17}$H$_{13}$N	Ph	H	H	H	Ph	81	66, 202, 266, 290, 1081
C$_{17}$H$_{13}$NO$_2$	C$_6$H$_4$OH(4)	H	H	H	C$_6$H$_4$OH(4)	220	114
C$_{17}$H$_{15}$N	1-naphthyl	H	Me	H	Me		109
C$_{17}$H$_{17}$NO$_2$		H	Me	H	Me		124
C$_{17}$H$_{29}$N	t-Bu	H	t-Bu	H	t-Bu	69	152

Formula							
$C_{18}H_{13}Br_2N$	Me	H	$C_6H_4Br(4)$	H	$C_6H_4Br(4)$	135	1082
$C_{18}H_{13}NO_2$	Ph	H	Ph	H	COOH	150	146
$C_{18}H_{13}NO_2$	Ph	H	COOH	H	Ph	278	202, 266
$C_{18}H_{13}N_3O_4$	$C_6H_4NO_2(3)$	H	$C_6H_4NO_2(3)$	H	Me		660
$C_{18}H_{13}N_3O_4$	Me	H	$C_6H_4NO_2(4)$	H	$C_6H_4NO_2(4)$		660, 1082
$C_{18}H_{15}N$	Ph	H	Me	H	Ph	182[a]	94
$C_{18}H_{15}N$	Ph	H	Ph	H	Me	75–76	415, 1083–1085
$C_{18}H_{15}NO_2$	$C_6H_4OH(4)$	H	$C_6H_4OH(4)$	H	Me	276	1086
$C_{18}H_{23}NO$	Pr	H	Pr	H	Pr		1087
$C_{18}H_{29}NO$	i-Pr	H	$C(Me)_2COCHMe_2$	H	i-Pr	150[a]	180

TABLE XVIII (continued)

Formula	R^1	R^2	R^3	R^4	R^5	M.P. (B.P.) (°C)	Reference
$C_{19}H_{13}NO_2$	2-furyl	H	Ph	H	2-furyl	128	502
$C_{19}H_{13}NS_2$	2-thienyl	H	Ph	H	2-thienyl	116	502
$C_{19}H_{14}BrN$	$C_6H_4Br(4)$	H	H	$-C_6H_4(CH_2)_2(2)-$		152	130
$C_{19}H_{14}ClN$	$C_6H_4Cl(4)$	H	H	$-C_6H_4(CH_2)_2(2)-$		148	130
$C_{19}H_{15}N$	Ph	H	H	$-C_6H_4(CH_2)_2(2)-$		130	130
$C_{19}H_{17}N$	Ph	H	Et	H	Ph	178-179[a]	94
$C_{19}H_{17}N$	Et	H	Ph	H	Ph		1082, 1084, 1085
$C_{19}H_{17}N$	5-acenaphthyl	H	Me	H	Me		109
$C_{19}H_{17}N$	Ph	Me	H	Me	Ph	136	202
$C_{19}H_{17}N$	Me	H	Ph	Ph	Me	172-173[a]	94

Formula					mp (°C)	Ref.
$C_{19}H_{17}NO_2$	$C_6H_4OMe(4)$	H	H	H	195	114
$C_{19}H_{17}NO_2$	$C_6H_4OMe(4)$	H	$C_6H_4OMe(4)$	H		796
$C_{19}H_{22}N_2$	Me	H	Me	Me	102	725
$C_{19}H_{29}N$	C_6H_{11}	H	Et	C_6H_{11}		1088
$C_{19}H_{29}N$	C_6H_{11}	Me	Me	C_6H_{11}	193–194[a]	1088
$C_{19}H_{29}N$	C_6H_{11}	Me	Me	C_6H_{11}	187–190[a]	1088
$C_{19}H_{29}N$	C_6H_{11}	H	Me	C_6H_{11}	210[a]	1088
$C_{20}H_{15}NO_2$	Ph	H	$CH(CHO)_2$	Ph	231	231
$C_{20}H_{17}N$	$C_6H_4Me(4)$	H	H	$-C_6H_4(CH_2)_2(2)-$	144	130
$C_{20}H_{17}NO_2$	Ph	H	CH_2COOMe	Ph	129	1082
$C_{20}H_{19}N$	Pr	H	Ph	Ph		1085

[structural formula of the ring system shown in the table header]

TABLE XVIII (continued)

Formula	R^1	R^2	R^3	R^4	R^5	M.P. (B.P.) (°C)	Reference
$C_{20}H_{19}N$	$C_6H_4Me(4)$	H	$C_6H_4Me(4)$	H	Me	96	411
$C_{20}H_{19}NO_2$	$C_6H_3OH(2)Me(3)$	H	$C_6H_3OH(2)Me(3)$	H	Me		1086
$C_{20}H_{19}NO_2$	$C_6H_3OH(2)Me(4)$	H	$C_6H_3OH(2)Me(4)$	H	Me		1086
$C_{20}H_{19}NO_2$	$C_6H_3OH(2)Me(5)$	H	$C_6H_3OH(2)Me(5)$	H	Me		1086
$C_{20}H_{19}NO_2$	$C_6H_4OMe(4)$	H	Me	H	$C_6H_4OMe(4)$		464
$C_{20}H_{19}NO_2$	Me	H	$C_6H_4OMe(4)$	H	$C_6H_4OMe(4)$	114 194[a]	503
$C_{21}H_{13}NO_2$		H	H	H			124
$C_{21}H_{14}INO$	Ph	H		H	Ph	128	1089
$C_{21}H_{15}NO$	2-furyl	H	Ph	H	Ph		502
$C_{21}H_{15}NS$	2-thienyl	H	Ph	H	Ph		502

Formula						
$C_{21}H_{17}N$	$-C_6H_4(CH_2)_2(2)-$	H	$-(CH_2)_2C_6H_4(2)-$	H	162	543,544
$C_{21}H_{17}N$	$-C_6H_4(CH_2)_2(2)-$	H	$-C_6H_4(CH_2)_2(2)-$	H	196	130
$C_{21}H_{19}N_3$	Ph	(1-methylpyrrol-2-yl)	(1-methylpyrrol-2-yl)	H		502
$C_{21}H_{20}NO_3$	CH=CHOH	$C_6H_4OMe(4)$	$C_6H_4OMe(4)$	H	121	250
$C_{21}H_{20}N_2O$	Ph	morpholino	Ph	H		1060
$C_{21}H_{21}N$	t-Bu	Ph	Ph	H		1085
$C_{21}H_{21}NO_2$	$C_6H_4OEt(4)$	H	$C_6H_4OEt(4)$	H	202	114
$C_{21}H_{21}NO_4$	$C_6H_3(OMe)_2(3,4)$	H	$C_6H_3(OMe)_2(3,4)$	H	152	114
$C_{22}H_{18}N_2$	Ph	(1-methylpyrrol-2-yl)	Ph	H	138	502
$C_{22}H_{19}N$	$-(CH_2)_2C_6H_4(2)-$	H	$-(CH_2)_3C_6H_4(2)-$	H	127	130

TABLE XVIII (continued)

Formula	R^1	R^2	R^3	R^4	R^5	M.P. (B.P.) (°C)	Reference
$C_{22}H_{19}NO_2$	Ph	H	$CH_2CH=CHCOOMe$	H	Ph	135	53
$C_{22}H_{19}NO_4$	Me	H	$C_6H_4COOMe(4)$	H	$C_6H_4COOMe(4)$		1086
$C_{22}H_{21}NO_2$	$C_6H_4OMe(4)$	H	Me	H	$CH=CHC_6H_4OMe(4)$	124–125	106
$C_{22}H_{22}N_2$	Ph	H	piperidino	H	Ph	197–208	228, 1060
$C_{22}H_{23}NO_2$	Me	H	$C_6H_4OEt(4)$	H	$C_6H_4OEt(4)$	120 / 152[a]	503
$C_{22}H_{23}NO_2$	Pr	H	$C_6H_4OMe(4)$	H	$C_6H_4OMe(4)$		1082
$C_{22}H_{23}NO_2$	$C_6H_2OH(2)(Me)_2(3,4)$	H	$C_6H_2OH(2)(Me)_2(3,4)$	H	Me		1086
$C_{22}H_{23}NO_4$	Me	H	$C_6H_3(OMe)_2(3,4)$	H	$C_6H_3(OMe)_2(3,4)$	113	1082
$C_{22}H_{23}NO_4$	Me	H	$C_6H_3(OMe)_2(2,5)$	H	$C_6H_3(OMe)_2(2,5)$	111	1082
$C_{23}H_{12}Cl_2N_4O_6$	$C_6H_3NO_2(3)Cl(4)$	H	$C_6H_4NO_2(4)$	H	$C_6H_3NO_2(3)Cl(4)$		1090

$C_{23}H_{14}Br_3N$	$C_6H_4Br(4)$	H	$C_6H_4Br(4)$	H	$C_6H_4Br(4)$	282	1091
$C_{23}H_{14}Cl_3N$	$C_6H_4Cl(4)$	H	$C_6H_4Cl(4)$	H	$C_6H_4Cl(4)$	212–214	404
$C_{23}H_{14}N_4O_6$	$C_6H_4NO_2(3)$	H	$C_6H_4NO_2(3)$	H	$C_6H_4NO_2(3)$		1082
$C_{23}H_{14}N_4O_6$	$C_6H_4NO_2(4)$	H	$C_6H_4NO_2(4)$	H	$C_6H_4NO_2(4)$	166–167	168
$C_{23}H_{15}Br_2N$	$C_6H_4Br(4)$	H	Ph	H	$C_6H_4Br(4)$	192–194	404
$C_{23}H_{15}Cl_2N$	$C_6H_4Cl(4)$	H	Ph	H	$C_6H_4Cl(4)$	189–190	404
$C_{23}H_{15}Cl_2N$	$C_6H_4Cl(4)$	H	$C_6H_4Cl(4)$	H	Ph	143	1092
$C_{23}H_{15}N_3O_4$	$C_6H_4NO_2(3)$	H	Ph	H	$C_6H_4NO_2(3)$		262
$C_{23}H_{16}BrN$	$C_6H_4Br(4)$	H	Ph	H	Ph	155	1081, 1093
$C_{23}H_{16}BrN$	Ph	H	$C_6H_4Br(4)$	H	Ph	131–132, 221–222[a]	168

TABLE XVIII (continued)

Formula	R^1	R^2	R^3	R^4	R^5	M.P. (B.P.) ($^{\circ}$C)	Reference
C$_{23}$H$_{16}$Br$_2$N$_2$	C$_6$H$_4$Br(4)	H	NHPh	H	C$_6$H$_4$Br(4)		1090
C$_{23}$H$_{16}$ClN	C$_6$H$_4$Cl(4)	H	Ph	H	Ph	138	1063, 1092
C$_{23}$H$_{16}$ClN	Ph	H	C$_6$H$_4$Cl(2)	H	Ph	112–113	404
C$_{23}$H$_{16}$ClN	Ph	H	C$_6$H$_4$Cl(4)	H	Ph	129–130 226[a]	168
C$_{23}$H$_{16}$FN	Ph	H	C$_6$H$_4$F(4)	H	Ph	137–138 230–231[a]	168
C$_{23}$H$_{16}$IN	Ph	H	C$_6$H$_4$I(4)	H	Ph	144–145 220–221[a]	168
C$_{23}$H$_{16}$N$_2$O$_2$	C$_6$H$_4$NO$_2$(3)	H	Ph	H	Ph		262, 1094
C$_{23}$H$_{16}$N$_2$O$_2$	Ph	H	C$_6$H$_4$NO$_2$(3)	H	Ph	152 185[a]	1094, 1095
C$_{23}$H$_{16}$N$_2$O$_2$	Ph	H	C$_6$H$_4$NO$_2$(4)	H	Ph	187 225[a]	262, 1096

$C_{23}H_{17}N$	Ph	H	Ph	H	Ph	108	1097
$C_{23}H_{17}NO$	Ph	H	Ph	H	H		1064
$C_{23}H_{17}NO$	Ph	H	Ph	H	$C_6H_4OH(2)$		1094
$C_{23}H_{17}NO$	Ph	H	Ph	H	$C_6H_4OH(3)$		1094
$C_{23}H_{17}NO$	Ph	H	$C_6H_4OH(2)$	H	$C_6H_4OH(4)$	178	1054
$C_{23}H_{17}NO$	Ph	H	$C_6H_4OH(4)$	H	Ph		1064
$C_{23}H_{17}NO_2$	$C_6H_4OH(4)$	H	Ph	H	$C_6H_4OH(4)$		360
$C_{23}H_{18}N_2$	Ph	H	$C_6H_4NH_2(4)$	H	Ph		1094, 1098
$C_{23}H_{18}N_2$	Ph	H	NHPh	H	Ph		1075
$C_{23}H_{17}N$	Ph	H	Ph	H	Ph	137-141	202, 404-406, 1062, 1063, 1081

TABLE XVIII (continued)

Formula	R¹	R²	R³	R⁴	R⁵	M.P. (B.P.) (°C)	Reference
$C_{23}H_{18}N_2O_2$	$C_6H_4OH(4)$	H	$C_6H_4NH_2(4)$	H	$C_6H_4OH(4)$		360
$C_{23}H_{19}NO_3$	Ph	H	$COCH=CHCH_2COOMe$	H	Ph	103	1089
$C_{23}H_{21}NO_2$	$-C_6H_3OMe(3)(CH_2)_2(6)-$		H	$-(CH_2)_2C_6H_3(6)OMe(3)-$		172	116
$C_{24}H_{16}N_2O$	Ph	H	(fused isoquinoline structure)	H	Ph	145–150	1056
$C_{24}H_{17}NO$	Ph	H	COPh	H	Ph	122	238
$C_{24}H_{17}NO_2$	Ph	H	COOPh	H	Ph	283	202
$C_{24}H_{19}N$	Ph	H	$C_6H_4Me(4)$	H	Ph	117–119	404, 1081
$C_{24}H_{19}N$	$C_6H_4Me(4)$	H	Ph	H	Ph	111	1063
$C_{24}H_{19}N$	Ph	H	CH_2Ph	H	Ph	122	238
$C_{24}H_{19}N$	Ph	Ph	Me	H	Ph	157[a]	179

$C_{24}H_{19}N$	Ph	H	Ph	Me	Ph	143	823, 1081
$C_{24}H_{19}NO$	Ph	H	$C_6H_4OMe(2)$	H	Ph	122	1054
$C_{24}H_{19}NO$	Ph	H	$C_6H_4OMe(4)$	H	Ph	99–100	404, 1062
$C_{24}H_{19}NO$	$C_6H_4OMe(2)$	H	Ph	H	Ph		1064
$C_{24}H_{19}NO$	$C_6H_4OMe(3)$	H	Ph	H	Ph		1094
$C_{24}H_{19}NO$	$C_6H_4OMe(4)$	H	Ph	H	Ph		1094, 1099
$C_{24}H_{19}NO$	$C_6H_3OH(2)Me(4)$	H	Ph	H	Ph		1064
$C_{24}H_{19}NO$	Ph	H	Ph	H	$C_6H_3OH(2)Me(4)$		1064
$C_{24}H_{19}NO$	$C_6H_3Me(3)OH(4)$	H	Ph	H	Ph		1097
$C_{24}H_{19}NO_2$	$C_6H_3OH(2)Me(4)$	H	Ph	H	$C_6H_4OH(2)$		360, 1064
$C_{24}H_{20}N_2$	Ph	H	N(Me)Ph	H	Ph	120	1060

TABLE XVIII (continued)

Formula	R^1	R^2	R^3	R^4	R^5	M.P. (B.P.) (°C)	Reference
C$_{24}$H$_{21}$NO$_2$	Ph	H	C$_6$H$_4$OMe(4)	H	C$_6$H$_3$OH(2)Me(4)		1064
C$_{24}$H$_{24}$N	Ph	H	Ph	–CH(Me)CH$_2$C(Me)$_2$CH$_2$–		132	823
C$_{24}$H$_{26}$ClN	C$_6$H$_4$Cl(4)	H	Ph	i-Pr	i-Bu	101	823
C$_{24}$H$_{27}$N	Ph	H	Ph	i-Pr	i-Bu	154^a	823
C$_{24}$H$_{27}$NO$_4$	Me	H	C$_6$H$_3$(OMe)$_2$(3,4)	H	C$_6$H$_3$(OEt)$_2$(3,4)		1082
C$_{25}$H$_{17}$NO$_2$	Ph	H	COCOPh	H	Ph	153	238
C$_{25}$H$_{18}$N$_4$O$_6$	C$_6$H$_3$NO$_2$(3)Me(4)	H	C$_6$H$_4$NO$_2$(4)	H	C$_6$H$_3$NO$_2$(3)Me(4)		1090
C$_{25}$H$_{19}$N	CH=CHPh	H	Ph	H	Ph	107	107, 824, 1081
C$_{25}$H$_{19}$N	Ph	H	CH=CHPh	H	Ph	123	107, 168, 824, 1081
C$_{25}$H$_{19}$NO	CH=CHC$_6$H$_4$OH(2)	H	Ph	H	Ph		107

$C_{25}H_{19}NO$	$CH=CHC_6H_4OH(4)$	H	Ph	H	Ph		107
$C_{25}H_{19}NO$	Ph	H	$CH=CHC_6H_4OH(2)$	H	Ph		107
$C_{25}H_{19}NO$	Ph	H	$CH=CHC_6H_4OH(4)$	H	Ph		107
$C_{25}H_{19}NO_2$	Ph	H	$CH(COOH)Ph$	H	Ph	138	53
$C_{25}H_{19}N_3$	Ph	H	[benzo-fused N–N–Me ring structure]	H	Ph		502
$C_{25}H_{19}N_3O_4$	$C_6H_3NO_2(2)Me(4)$	H	Ph	H	$C_6H_3NO_2(2)Me(4)$		1090
$C_{25}H_{20}ClN$	$C_6H_4Cl(4)$	H	$C_6H_4Me(4)$	H	$C_6H_4Me(4)$	188	1063
$C_{25}H_{20}N_2O$	Ph	H	$CH_2NHCOPh$	H	Ph		326, 331
$C_{25}H_{20}N_2O$	Ph	H	$C_6H_4NHCOMe(4)$	H	Ph		1098
$C_{25}H_{20}N_2O_4$	$C_6H_4OMe(4)$	H	$C_6H_4NO_2(3)$	H	$C_6H_4OMe(4)$	108 198a	1095, 1096

TABLE XVIII (continued)

Formula	R¹	R²	R³	R⁴	R⁵	M.P. (B.P.) (°C)	Reference
$C_{25}H_{20}N_2O_4$	$C_6H_4OMe(4)$	H	$C_6H_4NO_2(4)$	H	$C_6H_4OMe(4)$		1095, 1096
$C_{25}H_{21}N$	$C_6H_4Me(4)$	H	Ph	H	$C_6H_4Me(4)$	159-160	404
$C_{25}H_{21}N$	Ph	H	$C_6H_4Me(4)$	H	$C_6H_4Me(4)$	138	1063
$C_{25}H_{21}NO$	Ph	H	Ph	H	$C_6H_3OMe(2)Me(4)$		1064
$C_{25}H_{21}NO$	$C_6H_3Me(3)OMe(4)$	H	Ph	H	Ph		1091
$C_{25}H_{21}NO_2$	$C_6H_4OMe(4)$	H	Ph	H	$C_6H_4OMe(4)$		1099
$C_{25}H_{21}NO_2$	$C_6H_4OMe(4)$	H	$C_6H_4OMe(4)$	H	Ph		1099
$C_{25}H_{21}NO_2$	Ph	H	$C_6H_3(OMe)_2(3,4)$	H	Ph		168
$C_{25}H_{21}NO_2$	$C_6H_3OH(2)Me(4)$	H	Ph	H	$C_6H_4OMe(4)$		1064
$C_{25}H_{21}NO_3$	$C_6H_3OH(2)OMe(4)$	H	$C_6H_4OMe(4)$	H	Ph		1064
$C_{25}H_{21}NO_3$	$C_6H_4OMe(4)$	H	Ph	H	$C_6H_3OH(2)OMe(4)$		1064

Formula					mp	Ref
$C_{25}H_{21}NO_4$	(3-Me-5-OMe-benzofuran-2-yl)	H	H	H	(3-Me-5-OMe-benzofuran-2-yl)		124
$C_{25}H_{22}N_2$	$C_6H_4NMe_2(4)$	Ph	H	Ph	H	187 195[a]	795, 1100
$C_{25}H_{22}N_2$	Ph	$C_6H_4NMe_2(4)$	H	Ph	H		1065, 1095
$C_{25}H_{23}NO$	Ph	CH=CHPh	H	$-COCH_2C(Me)_2CH_2-$	124	824
$C_{25}H_{25}N$	Ph	Ph	H	(4-Me-cyclohexyl)	109	823
$C_{25}H_{25}NO_4$	$-C_6H_2(OMe)_2(2,3)(CH_2)_2(6)-$	H	$-(CH_2)_2C_6H_2(6)(OMe)_2(3,4)-$			116
$C_{26}H_{19}N$	Me	1-naphthyl	H	1-naphthyl	127	1082
$C_{26}H_{20}N_2$	Ph	(1-Me-indol-2-yl)	H	Ph	144	502

TABLE XVIII (continued)

Formula	R^1	R^2	R^3	R^4	R^5	M.P. (B.P.) (°C)	Reference
$C_{26}H_{21}N$	Ph	H	CH=CHPh	Me	Ph	146	824
$C_{26}H_{21}N$	CH=CHPh	H	Ph	Me	Ph	115	824
$C_{26}H_{21}N$	Ph	H	CH=CHC$_6$H$_4$OMe(4)	H	Ph		107
$C_{26}H_{21}N$	CH=CHC$_6$H$_4$OMe(4)	H	Ph	H	Ph		107
$C_{26}H_{21}NO_5$	C$_6$H$_3$OH(2)OMe(4)	H	C$_6$H$_4$OMe(4)	H	C$_6$H$_3$OH(2)OMe(4)		1064
$C_{26}H_{22}N_2O_2$	C$_6$H$_4$OMe(4)	H	C$_6$H$_4$NHCOMe(4)	H	Ph		1098
$C_{26}H_{23}N$	C$_6$H$_4$Me(4)	H	C$_6$H$_4$Me(4)	H	C$_6$H$_4$Me(4)	176–177	404, 1063
$C_{26}H_{23}NO$	C$_6$H$_4$Me(4)	H	C$_6$H$_4$OMe(4)	H	C$_6$H$_4$Me(4)	156–157	404
$C_{26}H_{23}NO_3$	C$_6$H$_4$OMe(4)	H	C$_6$H$_4$OMe(4)	H	C$_6$H$_4$OMe(4)	135	796, 1099
$C_{26}H_{23}NO_3$	Ph	H	C$_6$H$_4$OMe(4)	H	C$_6$H$_3$COOMe(2)Me(4)		1064
$C_{26}H_{23}NO_4$	C$_6$H$_3$OH(2)OMe(4)	H	C$_6$H$_4$OMe(4)	H	C$_6$H$_4$OMe(4)		1064

Formula	R		R'		R''		
$C_{26}H_{23}NO_8$	$C_6H_3(OAc)_2(2,4)$	H	$C_6H_3(OAc)_2(2,4)$	H	Me		1086
$C_{26}H_{24}N_2O$	$C_6H_4OMe(4)$	H	Ph	H	$C_6H_4NMe_2(4)$		796
$C_{26}H_{24}N_2O$	$C_6H_4OMe(4)$	H	$C_6H_4NMe_2(4)$	H	Ph		796
$C_{26}H_{31}NO_2$	$C_6H_2OH(2)Me(4)i\text{-}Pr(5)$	H	$C_6H_2OH(2)Me(4)i\text{-}Pr(5)$	H	Me		1086
$C_{27}H_{17}N$	Ph	H	Ph	H	[naphthalene structure]	168	1081
$C_{27}H_{19}N$	1-naphthyl	H	Ph	H	Ph	122	1081
$C_{27}H_{19}N$	2-naphthyl	H	Ph	H	Ph	124	1081
$C_{27}H_{19}N_3$	Ph	H	3-indolyl	H	3-indolyl		502
$C_{27}H_{21}N$	$(CH=CH)_2Ph$	H	Ph	H	Ph		107
$C_{27}H_{21}N$	CH=CHPh	H	Ph	H	CH=CHPh		794
$C_{27}H_{21}NO_2$	[benzofuran structure with OMe and Me substituents]	H	Ph	H	Ph		124

TABLE XVIII (continued)

Formula	R^1	R^2	R^3	R^4	R^5	M.P. (B.P.) (°C)	Reference
C$_{27}$H$_{24}$N$_2$	Ph	H	CH=CHC$_6$H$_4$NMe$_2$(4)	H	Ph		487
C$_{27}$H$_{24}$N$_2$	CH=CHC$_6$H$_4$NMe$_2$(4)	H	Ph	H	Ph		794
C$_{27}$H$_{24}$N$_2$O$_3$	C$_6$H$_4$OMe(4)	H	C$_6$H$_4$NHCOMe(4)	H	C$_6$H$_4$OMe(4)		1098
C$_{27}$H$_{25}$N	Ph	H	C$_6$H$_4$t-Bu(4)	H	Ph	178[a]	264
C$_{27}$H$_{26}$N$_2$	Ph	H	C$_6$H$_4$NEt$_2$(4)	H	Ph		404
C$_{27}$H$_{26}$N$_2$O$_2$	C$_6$H$_4$OMe(4)	H	C$_6$H$_4$NMe$_2$(4)	H	C$_6$H$_4$OMe(4)	185[a]	1095
C$_{27}$H$_{26}$N$_2$O$_2$	C$_6$H$_4$NMe$_2$(4)	H	C$_6$H$_4$OMe(4)	H	C$_6$H$_4$OMe(4)		796
C$_{27}$H$_{27}$N$_3$	C$_6$H$_4$NMe$_2$(4)	H	Ph	H	C$_6$H$_4$NMe$_2$(4)		796
C$_{27}$H$_{27}$N$_3$	C$_6$H$_4$NMe$_2$(4)	H	C$_6$H$_4$NMe$_2$(4)	H	Ph		796
C$_{27}$H$_{33}$N	C$_6$H$_4$i-Pr(4)	H	Ph	i-Pr	i-Bu	154[a]	823
C$_{28}$H$_{29}$N$_3$O	C$_6$H$_4$NMe$_2$(4)	H	C$_6$H$_4$OMe(4)	H	C$_6$H$_4$NMe$_2$(4)		796

Formula					m.p. (°C)	Ref.
$C_{28}H_{29}N_3O$	$C_6H_4NMe_2(4)$	H	H	$C_6H_4OMe(4)$		796
$C_{29}H_{21}N$	Ph	Ph	Ph	Ph	241	202
$C_{29}H_{21}N$	Ph	Ph	H	Ph	181–183	94, 1061, 1081
$C_{29}H_{21}N$	H	$C_6H_4Ph(4)$	H	Ph	123	824
$C_{29}H_{21}N$	$C_6H_4Ph(4)$	Ph	H	Ph	143	824
$C_{29}H_{21}N$	$C_6H_4Ph(4)$	H	H	$C_6H_4Ph(4)$		114
$C_{29}H_{21}NO_4$	CH=CH–[3,4-methylenedioxyphenyl]	Ph	H	CH=CH–[3,4-methylenedioxyphenyl]		794
$C_{29}H_{22}N_2$	Ph	NPh_2	H	Ph	205	1060
$C_{29}H_{25}NO_2$	$CH=CHC_6H_4OMe(4)$	Ph	H	$CH=CHC_6H_4OMe(4)$		794
$C_{29}H_{26}N_2$	Ph	[tetrahydrocarbazolyl, N-Me]	H	Ph	172	725

TABLE XVIII (continued)

Formula	R^1	R^2	R^3	R^4	R^5	M.P. (B.P.) (°C)	Reference
$C_{29}H_{28}N_2$	$CH=CHC_6H_4NEt_2(4)$	H	Ph	H	Ph		794
$C_{29}H_{28}N_2O_2$	$CH=CHC_6H_4NMe_2(4)$	H	$C_6H_4OMe(4)$	H	$C_6H_4OMe(4)$		487
$C_{29}H_{30}N_2O_2$	$C_6H_4OMe(4)$	H	$C_6H_4NEt_2(4)$	H	$C_6H_4OMe(4)$		796
$C_{29}H_{32}N_4$	$C_6H_4NMe_2(4)$	H	$C_6H_4NMe_2(4)$	H	$C_6H_4NMe_2(4)$		796
$C_{30}H_{21}NO$	Ph	H	Ph	COPh	Ph	174	824
$C_{30}H_{23}N$	Me	H	$C_6H_4Ph(4)$	H	$C_6H_4Ph(4)$	193	1082
$C_{30}H_{23}N$	2-naphthyl	H	CH=CHPh	Me	Ph	181	824
$C_{31}H_{17}N$			Ph			135	1081
$C_{31}H_{19}N$	1-naphthyl	H	Ph			121	1081
$C_{31}H_{20}N_2O_2$	1-naphthyl	H	$C_6H_4NO_2(4)$	H	1-naphthyl	258	1095

$C_{31}H_{21}N$	1-naphthyl	H	Ph	H	1-naphthyl	135	824
$C_{31}H_{21}N$	2-naphthyl	H	Ph	H	1-naphthyl		824
$C_{31}H_{21}N$	2-naphthyl	H	Ph	H	2-naphthyl	204	824
$C_{31}H_{23}N$	Ph	H	CH=CHPh	Ph	Ph	187	824
$C_{31}H_{25}N$	Me	CPh_3	H	H	Ph	109	106
$C_{31}H_{31}N_3$	CH=CHC$_6$H$_4$NMe$_2$(4)	H	Ph	H	CH=CHC$_6$H$_4$NMe$_2$(4)		487, 794
$C_{32}H_{23}NO$	Ph	H	CH=CHPh	COPh	Ph	176	824
$C_{32}H_{23}NO$	CH=CHPh	H	Ph	COPh	Ph	88	824
$C_{32}H_{25}NO$	CH=C(Ph)C$_6$H$_4$OMe(4)	H	Ph	H	Ph		795
$C_{33}H_{22}N$	2-naphthyl	H	Ph	H	C$_6$H$_4$Ph(4)	178	824

TABLE XVIII (continued)

Formula	R¹	R²	R³	R⁴	R⁵	M.P. (B.P.) (°C)	Reference
$C_{33}H_{25}N$	$C_6H_4Ph(4)$	H	Ph	H	2-naphthyl		824, 1096
$C_{33}H_{27}NO_2$	$CH=C\binom{C_6H_4OMe(4)}{C_6H_4OMe(4)}$	H	Ph	H	Ph		795
$C_{33}H_{28}N_2$	$CH=C\binom{Ph}{C_6H_4NMe_2(4)}$	H	Ph	H	Ph		795
$C_{35}H_{25}N$	Ph	Ph	Ph	Ph	Ph	238	202
$C_{35}H_{25}N$	$C_6H_4Ph(4)$	H	Ph	H	$C_6H_4Ph(4)$	191	824
$C_{35}H_{39}N_3$	$CH=CHC_6H_4NEt_2(4)$	H	Ph	H	$CH=CHC_6H_4NEt_2(4)$		794
$C_{36}H_{25}NS_3$	2-thienyl	H		H	2-thienyl		502
$C_{36}H_{26}N_2O_2$	Ph	H	H	CPh_3	$C_6H_4NO_2(4)$	141	106
$C_{38}H_{27}NO$	$C_6H_4Ph(4)$	H	CH=CHPh	COPh	Ph	156	824

$C_{40}H_{29}NS$	Ph	H	5-methylthiophen-2-yl–CPh_3	H	Ph	212	502
$C_{50}H_{35}NO_2$	Ph	H	2,6-diphenyl-4H-pyran-4-ylidene =$CC(Ph)=CHCOPh$	H	Ph	220	493
$C_{51}H_{35}N_3$	Ph	2,6-diphenylpyridin-4-yl	H	2,6-diphenylpyridin-4-yl	Ph	264	156
$C_{57}H_{41}NS_2$	Ph	H	5-methylthiophen-2-yl–CPh_3	H	5-methylthiophen-2-yl–CPh_3	502	

a melting point of the picrate derivative

TABLE XIX

BISPYRIDINES OBTAINED FROM PYRYLIUM SALTS

Formula	R	X	M.P. (°C)	Reference
$C_{32}H_{20}N_2S_4$	2-thienyl	$C_6H_4(4)$	decomp.	503, 504, 506
$C_{32}H_{20}N_2S_4$	1-thienyl	$C_6H_4(3)$		221
$C_{34}H_{24}N_2$	Ph	—	248	311
$C_{38}H_{26}N_2S$	Ph			221
$C_{40}H_{28}N_2$	Ph	$C_6H_4(4)$	274	221, 504, 506
$C_{40}H_{28}N_2$	Ph	$C_6H_4(3)$		221
$C_{48}H_{44}N_2O_8$	$C_6H_3(OMe)_2(3,4)$	$C_6H_4(4)$	decomp.	221, 504, 506
$C_{50}H_{34}N_4$	Ph		173	331

TABLE XIX (continued)

Formula	R	X	M.P. (°C)	Reference

Other Bispyridines obtained from pyrylium salts:

| $C_{32}H_{50}N_2$ | | | 103–105 | 371 |
| $C_{40}H_{28}N_2$ | | | 137 | 145 |

TABLE XX

BICYCLIC PYRIDINES OBTAINED FROM PYRYLIUM SALTS (*b*-FUSION)

Formula	R^1	R^2	R^3	n	M.P. (B.P.) (°C)	Reference
$C_{10}H_{13}N$	Me	H	H	4	156[a] (225)	534, 535
$C_{11}H_{15}N$	Me	Me	H	4		543, 544
$C_{11}H_{15}N$	Me	H	H	5		219
$C_{13}H_{13}NS$	2-thienyl	H	H	4	188[a]	534
$C_{14}H_{13}N$	Ph	H	H	3		128
$C_{14}H_{15}NS$	2-thienyl	H	H	5		219
$C_{15}H_{15}N$	Ph	H	H	4	163[a]	534, 535, 542
$C_{16}H_{17}N$	Ph	H	H	5		117, 128, 213, 534
$C_{16}H_{17}NO$	$C_6H_4OMe(4)$	H	H	4	145[a]	534
$C_{17}H_{17}N$	CH=CHPh	H	H	4		219
$C_{17}H_{19}NO$	$C_6H_4OEt(4)$	H	H	4	162[a]	534
$C_{17}H_{19}NO$	$C_6H_4OMe(4)$	Me	H	4		1087
$C_{17}H_{19}NO$	$C_6H_4OMe(4)$	H	H	5		219
$C_{17}H_{19}NO_2$	$C_6H_3(OMe)_2(3,4)$	H	H	4	175[a]	117, 534

TABLE XX (continued)

Formula	R^1	R^2	R^3	n	M.P. ($^{\circ}$C)	Reference
$C_{18}H_{19}N$	CH=CHPh	H	H	5		219
$C_{18}H_{19}NO$	CH=CHC$_6$H$_4$OMe(4)	H	H	4		219
$C_{18}H_{21}NO$	C$_6$H$_4$OEt(4)	Me	H	4	1087	
$C_{18}H_{21}NO$	C$_6$H$_3$(OMe)$_2$(3,4)	H	H	5		117
$C_{20}H_{17}N$	Ph	H	Ph	3		1081
$C_{21}H_{19}N$	Ph	H	Ph	4	107	264, 1077
$C_{25}H_{27}N$	Ph	H	C$_6$H$_4$t-Bu(4)	4	139	264
$C_{25}H_{27}N$	Ph	H	C$_6$H$_4$i-Bu(4)	4		44
$C_{29}H_{35}N$	Ph	H	C$_6$H$_3$(t-Bu)$_2$(3,5)	4	196	264

[a] melting point of the picrate derivative

TABLE XXI

Bicyclic Pyridines Obtained from Pyrylium Salts (*c*-Fusion)

Formula	R[1]	R[2]	n	R[3]	M.P. (B.P.) (°C)	Reference
$C_{10}H_{13}N$	Me	H	3	Me	121[a]	509
$C_{11}H_{15}N$	Me	Me	3	Me	147[a] (104)	190, 538
$C_{11}H_{15}N$	Me	H	4	Me	124[a] (248, 129/12)	492, 509, 537, 538
$C_{12}H_{17}N$	Et	H	3	Et	129[a] (171/100)	509
$C_{12}H_{17}N$	Me	H	4	Et	97[a]	538
$C_{12}H_{17}N$	Me	H	5	Me		1102
$C_{13}H_{13}NS$	2-thienyl	H	4	H		1103
$C_{13}H_{19}N$	Me	H	4	i-Pr	160–161[a]	538
$C_{13}H_{19}N$	Me	H	4	Pr	115–116[a]	538
$C_{13}H_{19}N$	Et	H	4	Et	(189/100)	509
$C_{14}H_{15}NS$	2-thienyl	H	4	Me	198[a]	538, 539
$C_{14}H_{21}N$	Me	H	4	t-Bu	139[a]	538
$C_{14}H_{21}N$	Me	H	4	i-Bu	139–140[a] (273–275)	538

TABLE XXI (continued)

Formula	R^1	R^2	n	R^3	M.P. (B.P.) (°C)	Reference
$C_{15}H_{14}ClN$	$C_6H_4Cl(4)$	H	4	H	97	1103
$C_{15}H_{15}N$	Ph	H	4	H	130	1103
$C_{15}H_{17}NO$	2-furfuryl	H	4	Me	182–183[a]	538, 539
$C_{15}H_{17}NS$	2-thienyl	H	4	Et	161[a]	538, 539
$C_{15}H_{23}N$	Me	H	4	C_5H_{11}	105[a] (301–303)	538
$C_{16}H_{17}N$	Ph	H	4	Me	76–77 175[a]	537, 538, 540
$C_{16}H_{17}N$	$C_6H_4Me(4)$	H	4	H	63	1103
$C_{16}H_{19}NO$	$C_6H_4OMe(4)$	H	4	H		1103
$C_{16}H_{19}NS$	2-thienyl	H	4	Pr	138[a]	538, 539
$C_{16}H_{19}NS$	2-thienyl	H	4	i-Pr	156[a]	538, 539
$C_{17}H_{19}N$	Ph	H	4	Et	153[a]	538, 539, 540
$C_{17}H_{19}N$	Ph	H	5	Me	142[a]	215
$C_{17}H_{19}NO$	$C_6H_4OMe(4)$	H	4	Me	89–90	206
$C_{17}H_{21}NO$	2-furfuryl	H	4	Pr	164–166	538, 539
$C_{17}H_{21}NS$	2-thienyl	H	4	i-Bu	142–144[a]	538, 539
$C_{18}H_{19}NO_2$	$C_6H_3(OMe)_2(3,4)$	H	4	H		1103
$C_{18}H_{21}N$	Ph	H	4	i-Pr	162[a]	538, 540

TABLE XXI (continued)

Formula	R^1	R^2	n	R^3	M.P. (B.P.) (°C)	Reference
$C_{18}H_{21}N$	Ph	H	4	Pr	139[a]	537, 538
$C_{18}H_{21}N$	Ph	H	5	Et	125[a]	215
$C_{18}H_{21}NO$	$C_6H_4OMe(4)$	H	4	Et		206
$C_{18}H_{21}NO_2$	$C_6H_3(OMe)_2(3,4)$	H	4	Me	65	206
$C_{18}H_{23}NS$	2-thienyl	H	4	C_5H_{11}	167[a]	538, 539
$C_{18}H_{23}NS$	2-thienyl	H	4	$CHEt_2$	146[a]	538, 539
$C_{19}H_{21}N$	Ph	H	4	CH=CHEt	70	213
$C_{19}H_{23}N$	Ph	H	4	Bu	141[a]	537, 538, 540
$C_{19}H_{23}N$	Ph	H	4	i-Bu	130–131[a]	538, 540
$C_{19}H_{23}N$	Ph	H	5	Pr	110	215
$C_{19}H_{23}N$	Ph	H	5	i-Pr	82	215
$C_{19}H_{23}N$	$C_6H_4OMe(4)$	H	4	Pr	60	206
$C_{20}H_{25}N$	Ph	H	4	CH=CHPr		213
$C_{20}H_{25}N$	Ph	H	4	C_5H_{11}		538, 540
$C_{20}H_{25}N$	Ph	H	4	$CHEt_2$	138[a]	538, 540
$C_{20}H_{25}N$	Ph	H	5	Bu		215
$C_{20}H_{25}N$	Ph	H	5	i-Bu		215
$C_{20}H_{25}NO_2$	$C_6H_3(OMe)_2(3,4)$	H	4	Pr	175[a]	206

TABLE XXI (continued)

Formula	R^1	R^2	n	R^3	M.P. (B.P.) (°C)	Reference
$C_{21}H_{18}BrNO$	Ph	H	4	CH=CH–(furanyl)–Br	73	213
$C_{21}H_{18}ClNO$	Ph	H	4	CH=CH–(furanyl)–Cl	71	213
$C_{21}H_{18}INO$	Ph	H	4	CH=CH–(furanyl)–I	54	213
$C_{21}H_{19}N$	Ph	H	4	Ph		206
$C_{21}H_{19}NO$	Ph	H	4	CH=CH–(furanyl)	66	213
$C_{21}H_{19}NS$	Ph	H	4	CH=CH–(thienyl)		213
$C_{21}H_{19}N_2$	Ph	H	4	CH=CH–(pyrrolyl)	99	213
$C_{22}H_{21}N$	Ph	H	4	CH_2Ph		206, 538, 540
$C_{22}H_{21}NO$	Ph	H	4	$C_6H_4OMe(3)$		206
$C_{22}H_{21}NO$	Ph	H	4	$C_6H_4OMe(4)$		206
$C_{22}H_{21}NO$	Ph	H	4	CH=CH–(furanyl)–Me	75	213
$C_{23}H_{20}N_2O_2$	Ph	H	4	$CH=CHC_6H_4NO_2(3)$	70 173[a]	214
$C_{23}H_{20}N_2O_2$	Ph	H	4	$CH=CHC_6H_4NO_2(4)$	220 228[a]	214

TABLE XXI (continued)

Formula	R^1	R^2	n	R^3	M.P. (B.P.) (°C)	Reference
$C_{23}H_{21}N$	Ph	H	4	CH=CHPh	234[a]	214
$C_{23}H_{21}NO$	Ph	H	4	CH=CHC$_6$H$_4$OH(2)	58	213
$C_{23}H_{23}NO_2$	C$_6$H$_4$OMe(4)	H	4	C$_6$H$_4$OMe(4)		206
$C_{24}H_{21}NO$	Ph	H	4	CH=CHCOPh	64	213
$C_{24}H_{23}NO$	Ph	H	4	CH=CHC$_6$H$_4$OMe(4)	75	213
$C_{24}H_{23}NO_2$	Ph	H	4	CH=CHC$_6$H$_3$OMe(3)OH(4)	48	213
$C_{25}H_{22}N_2$	Ph	H	4		73	213
$C_{25}H_{23}N$	Ph	H	4	(CH=CH)$_2$Ph	243[a]	214
$C_{25}H_{25}NO_2$	Ph	H	4	CH=CHC$_6$H$_3$(OMe)$_2$(3,4)	180 195[a]	214
$C_{25}H_{26}N_2$	Ph	H	4	CH=CHC$_6$H$_4$NMe$_2$(4)	142 198[a]	214
$C_{30}H_{29}N$	Ph	H	4			211
$C_{32}H_{33}N$	Ph	H	4			211

[a] melting point of the picrate derivative

TABLE XXII

TRICYCLIC PYRIDINES OBTAINED FROM PYRYLIUM SALTS (b,e-FUSION)

Formula	n	R	n'	M.P. ($^{\circ}$C)	Reference
$C_{12}H_{15}N$	3	H	4		1087
$C_{13}H_{17}N$	4	H	4	68	534, 535, 542
$C_{14}H_{19}N$	4	Me	4	155[a]	542
$C_{14}H_{19}N$	4	H	5		219
$C_{15}H_{21}N$	5	H	5	106–107	219

[a] melting point of the picrate derivative

TABLE XXIII

Tricyclic Pyridines Obtained from Pyrylium Salts (b,d-Fusion)

Formula	R	n	M.P. (B.P.) (°C)	Reference
$C_{12}H_{15}N$	Me	3	137 (277 145/5)	331, 541
$C_{13}H_{17}N$	Et	3		331
$C_{14}H_{19}N$	Pr	3	111[a] (294–296)	331, 541
$C_{14}H_{19}N$	Me	4		541
$C_{15}H_{21}N$	Et	4		541
$C_{16}H_{21}N$	Pr	4	203	331, 541
$C_{16}H_{23}N$	Me	5		541
$C_{17}H_{16}ClN$	$C_6H_4Cl(4)$	3	105	102
$C_{17}H_{17}N$	Ph	3	81	102
$C_{17}H_{25}N$	i-Bu	4		541
$C_{17}H_{25}N$	Et	5		541
$C_{18}H_{19}N$	$C_6H_4Me(4)$	3		102

TABLE XXIII (continued)

Formula	R	n	M.P. (B.P.) (°C)	Reference
$C_{18}H_{19}NO$	$C_6H_4OMe(4)$	3	97	102
$C_{18}H_{27}N$	Pr	5		541
$C_{19}H_{19}N$	CH=CHPh	3	88	541
$C_{19}H_{21}NO_2$	$C_6H_3(OMe)_2(3,4)$	3	78	541

[a] melting point of the picrate derivative

TABLE XXIV
Steroid Pyridines Obtained from Pyrylium Salts (I)

Formula	R^1	R^2	R^3	M.P. ($^\circ$C)	Reference
$C_{23}H_{31}N$	Me	H	H		546
$C_{24}H_{33}N$	Me	H	Me	120	546
$C_{25}H_{35}N$	Me	Me	Me	224 - 225	110
$C_{26}H_{31}NS$	1-thienyl	H	H		546
$C_{27}H_{37}N$	$-(CH_2)_4-$		Me	145 - 148	110
$C_{28}H_{33}N$	Ph	H	H	160	546
$C_{29}H_{35}N$	Ph	H	Me	160	546
$C_{29}H_{35}NO$	$C_6H_4OH(4)$	H	Me	175 - 177	110
$C_{29}H_{35}NO$	$C_6H_4OMe(4)$	H	H		546
$C_{30}H_{37}N$	$C_6H_4Me(4)$	H	Me	250 - 253	110
$C_{30}H_{37}NO$	$C_6H_4OEt(4)$	H	H	107	546
$C_{30}H_{37}NO$	$C_6H_4OMe(4)$	H	Me	111	546
$C_{30}H_{37}NO$	$C_6H_4OH(4)$	Me	Me	269 - 272	110
$C_{31}H_{37}NO_2$	$C_6H_4COOMe(4)$	H	Me	177	110
$C_{32}H_{39}NO_2$	$C_6H_4COOMe(4)$	Me	Me	272	110
$C_{32}H_{41}NO$	$C_6H_4OMe(4)$	Et	Me	234 - 236	110

TABLE XXV
STEROID PYRIDINES OBTAINED FROM PYRYLIUM SALTS (II)

Formula	R^1	R^2	R^3	M.P. ($^\circ$C)	Reference
$C_{26}H_{37}NO_2$	Me	Me	OAc	287–289	110
$C_{28}H_{39}NO_2$	$-(CH_2)_4-$		OAc	222–224	110
$C_{30}H_{36}BrNO_2$	$C_6H_4Br(4)$	H	OAc	218–220	110
$C_{30}H_{37}NO_3$	$C_6H_4OH(4)$	H	OAc	260–263	110
$C_{31}H_{39}NO_2$	$C_6H_4Me(4)$	H	OAc		110
$C_{31}H_{39}NO_3$	$C_6H_4OH(4)$	Me	OAc	258–259	110
$C_{32}H_{39}NO_4$	$C_6H_4COOMe(4)$	H	OAc	263	110
$C_{33}H_{41}NO_4$	$C_6H_4COOMe(4)$	Me	OAc	259	110
$C_{33}H_{43}NO_3$	$C_6H_4OMe(4)$	Et	OAc	322–326	110
$C_{36}H_{51}N$	Ph	H	$CH(Me)(CH_2)_2CHMe_2$	247–252	110
$C_{36}H_{51}NO_2$	$C_6H_4OH(4)$	H	$CH(Me)(CH_2)_2CHMe_2$	229–231	110
$C_{38}H_{53}N$	CH=CHPh	H	$CH(Me)(CH_2)_2CHMe_2$	156–159	110

TABLE XXVI
N-Substituted Pyridinium Salts Obtained from Pyrylium Salts[a]

Formula	R^1	R^2	R^3	R^4	R^5	R^6	Reference
$C_8H_{12}NO$	OH	Me	H	Me	H	Me	94, 463
$C_8H_{13}N_2$	NH$_2$	Me	H	Me	H	Me	465
$C_9H_{13}N_2O$	NHCHO	Me	H	Me	H	Me	1104
$C_9H_{14}N$	Me	Me	H	Me	H	Me	401
$C_9H_{14}NO$	Me	Me	H	OMe	H	Me	28, 507
$C_9H_{14}NO$	OH	Me	H	Et	H	Me	94
$C_9H_{14}NS$	Me	Me	H	SMe	H	Me	507
$C_9H_{14}N_3O$	NHCONH$_2$	Me	H	Me	H	Me	94

Formula							References
$C_9H_{14}N_3S$	$NHCSNH_2$	Me	H	Me	H	Me	1104
$C_9H_{15}N_2$	Me	Me	H	NHMe	H	Me	28, 507
$C_9H_{15}N_2O_2S$	$NHSO_2Me$	Me	H	Me	H	Me	661
$C_{10}H_{13}N_4$	1,2,4-triazol-4-yl	Me	H	Me	H	Me	585, 586, 587
$C_{10}H_{14}N$	$CH=CH_2$	Me	H	Me	H	Me	632, 1105, 1106
$C_{10}H_{14}NO_2$	CH_2COOH	Me	H	Me	H	Me	94, 573, 576
$C_{10}H_{15}BrN$	CH_2CH_2Br	Me	H	Me	H	Me	574, 1105, 1107
$C_{10}H_{15}ClN$	CH_2CH_2Cl	Me	H	Me	H	Me	632, 1105, 1108
$C_{10}H_{15}IN$	CH_2CH_2I	Me	H	Me	H	Me	574, 1105, 1107
$C_{10}H_{15}N_2O$	$NHCOCH_3$	Me	H	Me	H	Me	1104

TABLE XXVI (continued)

Formula	R^1	R^2	R^3	R^4	R^5	R^6	Reference
$C_{10}H_{15}N_2O_3$	$CH_2CH_2ONO_2$	Me	H	Me	H	Me	574, 1105, 1109
$C_{10}H_{16}N$	Et	Me	H	Me	H	Me	401
$C_{10}H_{16}NO$	CH_2CH_2OH	Me	H	Me	H	Me	632, 1105, 1109
$C_{10}H_{16}NO$	OH	Et	H	Me	H	Et	94
$C_{10}H_{16}N_3O$	$NHCONH_2$	Me	H	Et	H	Me	94
$C_{10}H_{17}N_2$	$CH_2CH_2NH_2$	Me	H	Me	H	Me	574, 1105
$C_{10}H_{17}N_2$	Me	Me	H	NMe_2	H	Me	507
$C_{11}H_{13}N_2S$	2-thiazolyl	Me	H	Me	H	Me	574, 587, 589, 1105
$C_{11}H_{14}N_3O$	$NHCOCH_2CN$	Me	H	Me	H	Me	1104
$C_{11}H_{15}N_2S$	2-thiazolinyl	Me	H	Me	H	Me	589, 1105, 1110

$C_{11}H_{16}N$	$CH_2CH=CH_2$	Me	H	Me	H	Me	632, 1105, 1111
$C_{11}H_{16}NO_2$	CH_2COOH	Me	H	Et	H	Me	94
$C_{11}H_{16}NO_2$	CH_2CH_2COOH	Me	H	Me	H	Me	573
$C_{11}H_{16}NO_2$	$CH(Me)COOH$	Me	H	Me	H	Me	577
$C_{11}H_{17}N_2O_2$	NHCOOEt	Me	H	Me	H	Me	1104
$C_{11}H_{18}N$	Pr	Me	H	Me	H	Me	401
$C_{11}H_{18}NO$	$CH(Me)CH_2OH$	Me	H	Me	H	Me	591, 1105
$C_{11}H_{18}NO$	OH	Et	H	Et	H	Et	94
$C_{12}H_{14}N_3O_2$	(heterocyclic structure)	Me	H	Me	H	Me	575
$C_{12}H_{15}F_3NO_2$	$CH_2CH_2OCOCF_3$	Me	H	Me	H	Me	632, 1105, 1109

TABLE XXVI (continued)

Formula	R¹	R²	R³	R⁴	R⁵	R⁶	Reference
$C_{12}H_{15}N_2O$	(3,5-dimethylisoxazol-4-yl), Me	Me	H	Me	H	Me	589, 1105, 1110
$C_{12}H_{15}N_2S$	(methylthiazolyl), Me	Me	H	Me	H	Me	584, 587
$C_{12}H_{15}N_4O_2$	(pyrimidine: NH_2, OH, HO)	Me	H	Me	H	Me	575
$C_{12}H_{16}N_5O$	(pyrimidine: NH_2, NH_2, HO)	Me	H	Me	H	Me	575
$C_{12}H_{17}N_2O_3$	$CH_2CONHCH_2COOH$	Me	H	Me	H	Me	577
$C_{12}H_{18}NO_2$	CH_2CH_2OAc	Me	H	Me	H	Me	632, 1105, 1110
$C_{12}H_{18}NO_2$	CH_2COOEt	Me	H	Me	H	Me	574, 576, 1105, 1112

$C_{12}H_{18}NO_2$	(CH$_2$)$_3$COOH	Me	H	Me	H	Me	574, 1105, 1112
$C_{12}H_{18}NO_2$	CH(Et)COOH	Me	H	Me	H	Me	577
$C_{13}H_{15}N_2$	3-pyridyl	Me	H	Me	H	Me	584
$C_{13}H_{15}N_2$	4-pyridyl	Me	H	Me	H	Me	574
$C_{12}H_{20}N$	Bu	Me	H	Me	H	Me	401
$C_{12}H_{20}NO$	Bu	Me	H	OMe	H	Me	1113
$C_{12}H_{20}NO$	CH(Et)CH$_2$OH	Me	H	Me	H	Me	1022, 1105, 1114
$C_{12}H_{20}NO$	OH	i-Pr	H	Me	H	i-Pr	94
$C_{13}H_{11}Cl_4N$	Me	Me	H	(tetrachlorophenyl)	H	Me	1115
$C_{13}H_{15}N_2$	2-pyridyl	Me	H	Me	H	Me	574, 589, 1105

TABLE XXVI (continued)

Formula	R¹	R²	R³	R⁴	R⁵	R⁶	Reference
$C_{13}H_{15}N_2O$	pyrid-2-on-1-yl	Me	H	Me	H	Me	585, 586
$C_{13}H_{16}NO$	2-furfuryl	Me	H	Me	H	Me	618
$C_{13}H_{16}N_3$	$NH-$ [pyridin-2-yl]	Me	H	Me	H	Me	587
$C_{13}H_{17}F_3NO_2$	$CH(Me)CH_2OCOCF_3$	Me	H	Me	H	Me	591, 1105
$C_{13}H_{19}N_3$	CH_2CH_2- [pyrazol-3-yl]	Me	H	Me	H	Me	574, 1105, 1116
$C_{13}H_{20}NO_2$	$CH(i\text{-}Pr)COOH$	Me	H	Me	H	Me	577
$C_{13}H_{20}NO_2$	$CH(Me)CH_2OCOCH_3$	Me	H	Me	H	Me	591
$C_{13}H_{21}N_2O_2$	$(CH_2)_3CH(NH_2)COOH$	Me	H	Me	H	Me	576
$C_{13}H_{22}N$	$i\text{-}C_5H_{11}$	Me	H	Me	H	Me	401
$C_{13}H_{22}N_3$	$-N$ [piperazin-1-yl] $N\text{-}Me$	Me	H	Me	H	Me	1117

Formula	Substituent							Ref.
$C_{14}H_{14}ClFN$	$C_6H_3Cl(3)F(4)$	Me	H	Me	H	H	Me	574
$C_{14}H_{14}N_3S$	NH_2	Me	H	2-benzthiazolyl	H	H	Me	47
$C_{14}H_{15}BrN$	$C_6H_4Br(2)$	Me	H	Me	H	H	Me	1118
$C_{14}H_{15}BrN$	$C_6H_4Br(4)$	Me	H	Me	H	H	Me	401
$C_{14}H_{15}ClN$	$C_6H_4Cl(4)$	Me	H	Me	H	H	Me	1113
$C_{14}H_{15}FN$	$C_6H_4F(2)$	Me	H	Me	H	H	Me	574
$C_{14}H_{15}FN$	$C_6H_4F(4)$	Me	H	Me	H	H	Me	574
$C_{14}H_{15}IN$	$C_6H_4I(4)$	Me	H	Me	H	H	Me	401
$C_{14}H_{15}N_2O_2$	$C_6H_4NO_2(4)$	Me	H	Me	H	H	Me	1119
$C_{14}H_{15}N_4O_4$	$NHC_6H_3(NO_2)_2(2,4)$	Me	H	Me	H	H	Me	1104
$C_{14}H_{16}N$	Ph	Me	H	Me	H	H	Me	401, 582, 875, 1118

TABLE XXVI (continued)

Formula	R^1	R^2	R^3	R^4	R^5	R^6	Reference
C$_{14}$H$_{16}$NO	C$_6$H$_4$OH(2)	Me	H	Me	H	Me	401
C$_{14}$H$_{15}$NO	C$_6$H$_4$OH(4)	Me	H	Me	H	Me	401
C$_{14}$H$_{16}$NS	Ph	Me	H	SMe	H	Me	507
C$_{14}$H$_{16}$N$_3$O	NHCO	Me	H	Me	H	Me	1104
C$_{14}$H$_{16}$N$_3$O$_2$	NHC$_6$H$_4$NO$_2$(4)	Me	H	Me	H	Me	1104
C$_{14}$H$_{16}$N$_3$O$_2$		Me	H	H	-(CH$_2$)$_4$-		575
C$_{14}$H$_{17}$N$_2$	C$_6$H$_4$NH$_2$(2)	Me	H	Me	H	Me	1118
C$_{14}$H$_{17}$N$_2$	C$_6$H$_4$NH$_2$(3)	Me	H	Me	H	Me	1118
C$_{14}$H$_{17}$N$_2$	C$_6$H$_4$NH$_2$(4)	Me	H	Me	H	Me	**401,** 1118
C$_{14}$H$_{17}$N$_2$	NHPh	Me	H	Me	H	Me	1120

$C_{14}H_{17}N_2$	Ph	Me		H	NHMe	H	Me	507
$C_{14}H_{17}N_2O_2S$	$NHSO_2Ph$	Me		H	Me	H	Me	1104
$C_{14}H_{17}N_2O_2S$	$C_6H_4SO_2NH_2$	Me		H	Me	H	Me	587
$C_{14}H_{17}N_2O_3S$	$NHC_6H_4SO_3H(4)$	Me		H	Me	H	Me	1104
$C_{14}H_{18}N_3O$		Me		H	Me	H	Me	574, 1037, 1105
$C_{14}H_{19}N_2$		Me		H	Me	H	Me	586
$C_{14}H_{19}N_4$		Me		H	Me	H	Me	574, 1037, 1105
$C_{14}H_{20}N$	Me		$-(CH_2)_4-$	H	H	$-(CH_2)_4-$		545
$C_{14}H_{20}NO_2$	CH_2COOEt	Me		H	H	H	$-(CH_2)_4-$	575
$C_{14}H_{21}N_2O_3$	$CH_2CONHCH_2COOEt$	Me		H	Me	H	Me	575
$C_{14}H_{22}N$	C_6H_{11}	Me		H	Me	H	Me	401

TABLE XXVI (continued)

Formula	R^1	R^2	R^3	R^4	R^5	R^6	Reference
$C_{14}H_{22}NO_2$	$CH(Et)CH_2OCOCH_3$	Me	H	Me	H	Me	667
$C_{14}H_{22}NO_2$	$(CH_2)_5COOH$	Me	H	Me	H	Me	574, 1105, 1112
$C_{14}H_{23}N_2O_2$	$(CH_2)_4CH(NH_2)COOH$	Me	H	Me	H	Me	576, 580
$C_{14}H_{23}N_2O_2$	$CH(COOH)(CH_2)_4NH_2$	Me	H	Me	H	Me	580
$C_{14}H_{24}N_3$	NMe_2	Me	H	piperidino	H	Me	85a
$C_{14}H_{25}N_2$	$(CH_2)_6NH_2$	Me	H	Me	H	Me	574, 1105
$C_{14}H_{25}N_2$	$CH_2CH_2NEt_2$	Me	H	Me	H	Me	574, 1105
$C_{15}H_{14}BrN_2S$	(structure)	Me	H	Me	H	Me	584
$C_{15}H_{15}F_3N$	$C_6H_4CF_3(2)$	Me	H	Me	H	Me	574
$C_{15}H_{15}F_3N$	$C_6H_4CF_3(3)$	Me	H	Me	H	Me	574
$C_{15}H_{16}NO_2$	$C_6H_4COOH(2)$	Me	H	Me	H	Me	573
$C_{15}H_{16}NO_2$	$C_6H_4COOH(3)$	Me	H	Me	H	Me	573
$C_{15}H_{16}NO_2$	$C_6H_4COOH(4)$	Me	H	Me	H	Me	573

$C_{15}H_{16}N_3$	1-benzimidazolyl	Me	H	Me	H	Me	586
$C_{15}H_{16}N_3$	2-benzimidazolyl	Me	H	Me	H	Me	574, 1105
$C_{15}H_{16}N_3O_3$	$NHCOC_6H_4NO_2(3)$	Me	H	Me	H	Me	466
$C_{15}H_{16}N_3O_3$	$NHCOC_6H_4NO_2(4)$	Me	H	Me	H	Me	466
$\mathbf{C_{15}H_{17}N_2O}$	NHCOPh	Me	H	Me	H	Me	465, 466, 1104
$C_{15}H_{18}N$	$C_6H_4Me(2)$	Me	H	Me	H	Me	574, 589, 1105
$C_{15}H_{18}N$	$C_6H_4Me(3)$	Me	H	Me	H	Me	574, 589, 1105
$C_{15}H_{18}N$	$C_6H_4Me(4)$	Me	H	Me	H	Me	44, 401, 582, 595
$C_{15}H_{18}N$	CH_2Ph	Me	H	Me	H	Me	401
$\mathbf{C_{15}H_{18}N}$	**Me**	**Me**	**Ph**	**Me**	**H**	**Me**	**179**

TABLE XXVI (continued)

Formula	R^1	R^2	R^3	R^4	R^5	R^6	Reference
C$_{15}$H$_{18}$NO	C$_6$H$_4$OMe(4)	Me	H	Me	H	Me	340, 589, 1105
C$_{15}$H$_{18}$NO	OCH$_2$Ph	Me	H	Me	H	Me	389
C$_{15}$H$_{18}$N$_3$O	NHCONHPh	Me	H	Me	H	Me	1104
C$_{15}$H$_{18}$N$_3$S	NHCSNHPh	Me	H	Me	H	Me	1104
C$_{15}$H$_{19}$N$_2$	Ph	Me	H	NMe$_2$	H	Me	507
C$_{15}$H$_{19}$N$_2$	N(Me)Ph	Me	H	Me	H	Me	1104
C$_{15}$H$_{19}$N$_2$O$_2$S	NHSO$_2$C$_6$H$_4$Me(4)	Me	H	Me	H	Me	1104
C$_{15}$H$_{19}$N$_2$O$_2$S	CH$_2$C$_6$H$_4$SO$_2$NH$_2$(4)	Me	H	Me	H	Me	574, 1105, 1121
C$_{15}$H$_{19}$N$_4$	C$_6$H$_4$N=C$\langle^{NH_2}_{NH_2}$ (4)	Me	H	Me	H	Me	587
C$_{15}$H$_{19}$N$_4$O$_2$S	C$_6$H$_4$SO$_2$N=C$\langle^{NH_2}_{NH_2}$ (4)	Me	H	Me	H	Me	587

Formula	R¹	R²	R³	R⁴	R⁵	Ref
$C_{15}H_{22}N$	Et		$-(CH_2)_4-$	H	$-(CH_2)_4-$	545
$C_{15}H_{22}NO$	CH_2CH_2OH		$-(CH_2)_4-$	H	$-(CH_2)_4-$	1122
$C_{15}H_{26}N$	C_7H_{15}	Me	Me	H	Me	401
$C_{16}H_{17}NF_3$	$C_6H_4CF_3(2)$	Et	Me	H	Me	1123
$C_{16}H_{17}N_2OS$	(2-methyl-benzothiazolyl, OMe structure)	Me	Me	H	Me	584
$C_{16}H_{18}NO$	Me	Me	$CH=CHC_6H_4OH(4)$	H	Me	1124
$C_{16}H_{18}NO_2$	$C_6H_4COOMe(2)$	Me	Me	H	Me	574, 589, 1105
$C_{16}H_{18}NO_2$	$CH(Ph)COOH$	Me	Me	H	Me	577
$C_{16}H_{18}N_3$	(benzimidazoline CH_2, N–H structure)	Me		H	Me	584
$C_{16}H_{18}N_3$	(benzimidazole N–Me structure)	Me		H	Me	1125

TABLE XXVI (continued)

Formula	R¹	R²	R³	R⁴	R⁵	R⁶	Reference
$C_{16}H_{18}N_3O_2$		$-(CH_2)_3-$	H		$-(CH_2)_4-$		575
$C_{16}H_{19}NCl$	$C_6H_4Cl(2)$	i-Pr	H	Me	H	Me	1123
$C_{16}H_{19}NF$	$C_6H_4F(2)$	i-Pr	H	Me	H	Me	1123
$C_{16}H_{19}N_2$	$N=C\binom{Me}{Ph}$	Me	H	Me	H	Me	1104
$C_{16}H_{20}N$	$CH(Me)Ph$	Me	H	Me	H	Me	1123
$C_{16}H_{20}N$	CH_2CH_2Ph	Me	H	Me	H	Me	574, 1105, 1116
$C_{16}H_{20}N$	$C_6H_3Me_2(2,3)$	Me	H	Me	H	Me	574, 589, 1105
$C_{16}H_{20}N$	$C_6H_3Me_2(3,4)$	Me	H	Me	H	Me	574, 589, 1105
$C_{16}H_{20}N$	$C_6H_3Me_2(2,6)$	Me	H	Me	H	Me	574, 589, 1105

Formula	R1	R2	R3	R4	R5	R6	Ref.
$C_{16}H_{20}N$	Ph	i-Pr	H	Me	H	Me	876
$C_{16}H_{20}N$	$C_6H_4Me(2)$	Et	H	Me	H	Me	876
$C_{16}H_{20}N$	$C_6H_4Me(3)$	Et	H	Me	H	Me	1123
$C_{16}H_{20}N$	$C_6H_4Me(4)$	Et	H	Me	H	Me	876
$C_{16}H_{20}NO$	$CH(Ph)CH_2OH$	Me	H	Me	H	Me	591
$C_{16}H_{20}NO$	$C_6H_4OMe(3)$	Et	H	Me	H	Me	1123
$C_{16}H_{21}N_2$	$C_6H_4NMe_2(4)$	Me	H	Me	H	Me	574, 1105
$C_{16}H_{21}N_2O_3S$	[penicillanic-acid nucleus: S, gem-Me₂ (Me, Me), COOH, β-lactam (N, O)]	Me	$-(CH_2)_4-$		H	Me	1105, 1126
$C_{16}H_{22}N$	$CH_2CH=CH_2$	H	H	Me	$-(CH_2)_4-$		545
$C_{16}H_{23}N_4O_5$	$(CH_2CONH)_3CH_2COOH$	Me	H	Me	H	Me	577

TABLE XXVI (continued)

Formula	R¹	R²	R³	R⁴	R⁵	R⁶	Reference
$C_{16}H_{24}N$	Pr	$-(CH_2)_4-$		H	$-(CH_2)_4-$		545
$C_{16}H_{27}N_2$	NH_2	Me	$-(CH_2)_4$———CH_2———$(CH_2)_4-$ H		H	Me	186
$C_{17}H_{17}N_2O$		Me	H	Me	H	Me	586
$C_{17}H_{18}N_3O$		Me	H	Me	H	Me	586
$C_{17}H_{18}N_3O_2S_2$	$C_6H_4SO_2NH$—(4)	Me	H	Me	H	Me	587
$C_{17}H_{19}NF_3$	$C_6H_4CF_3(2)$	i-Pr	H	Me	H	Me	1123
$C_{17}H_{19}NF_3$	$C_6H_4CF_3(3)$	i-Pr	H	Me	H	Me	1123
$C_{17}H_{19}N_2O$	Me	Me	H		H	Me	1127

$C_{17}H_{19}N_2O_3$	$C_6H_4CONHCH_2COOH(4)$	Me	H	Me	H	Me	1105, 1110
$C_{17}H_{19}N_2S$	Me	Me	H	Me–CH (benzothiazole)	H	Me	1127
$C_{17}H_{20}NO_2$	$C_6H_4COOEt(4)$	Me	H	Me	H	Me	**589**, 1105, 1110
$C_{17}H_{20}NO_2$	$CH(COOH)CH_2Ph$	Me	H	Me	H	Me	**577**
$C_{17}H_{20}NO_2$	Me	Me	H	CH=CH (aryl OMe/OH)	H	Me	1124
$C_{17}H_{20}N_3$	CH_2CH_2 (benzimidazole, H)	Me	H	Me	H	Me	1125
$C_{17}H_{20}N_3$	$CH(Me)$– (benzimidazole, H)	Me	H	Me	H	Me	1125
$C_{17}H_{20}N_3O_2$	(pyrimidine, HO / OH)	$-(CH_2)_4-$	H	H	$-(CH_2)_4-$	575

TABLE XXVI (continued)

Formula	R^1	R^2	R^3	R^4	R^5	R^6	Reference
C$_{17}$H$_{21}$N$_2$O	Ph	Me	H	morpholino	H	Me	507
C$_{17}$H$_{21}$N$_2$O$_3$	NHCOC$_6$H$_3$(OMe)$_2$(3,5)	Me	H	Me	H	Me	1104
C$_{17}$H$_{22}$N	C$_6$H$_2$Me$_3$(2,4,6)	Me	H	Me	H	Me	574
C$_{17}$H$_{22}$N	C$_6$H$_4$Me(2)	i-Pr	H	Me	H	Me	876
C$_{17}$H$_{22}$N	C$_6$H$_4$Me(3)	i-Pr	H	Me	H	Me	876
C$_{17}$H$_{22}$N	C$_6$H$_4$Me(4)	i-Pr	H	Me	H	Me	876
C$_{17}$H$_{22}$NO	C$_6$H$_4$OMe(2)	i-Pr	H	Me	H	Me	1123
C$_{17}$H$_{22}$NO	C$_6$H$_4$OMe(4)	i-Pr	H	Me	H	Me	1123
C$_{17}$H$_{22}$NO$_3$S	CH$_2$CH$_2$OTs	Me	H	Me	H	Me	632, 1105, 1109
C$_{17}$H$_{24}$NO$_2$	CH$_2$COOEt	-(CH$_2$)$_4$-		H	-(CH$_2$)$_4$-		575
C$_{17}$H$_{26}$N	Bu	-(CH$_2$)$_4$-		H	-(CH$_2$)$_4$-		545

Formula							Ref.
$C_{17}H_{28}N$	Me	Me	H	$-(CH_2)_4\!\!-\!\!CH_2(H)\!\!-\!\!(CH_2)_4-$	H	Me	186
$C_{18}H_{16}NO$	OH	Ph	H	Me	H	Ph	94
$C_{18}H_{18}N$	1-naphthyl	Me	H	Me	H	Me	574, 589, 1105
$C_{18}H_{18}N$	2-naphthyl	Me	H	Me	H	Me	574, 589, 1105
$C_{18}H_{18}NO$	Ph	Me	H	furfuryl	H	Me	48
$C_{18}H_{18}NO$	Ph	Me	H	[5-methyl-2-furyl structure, Me]	H	Me	48
$C_{18}H_{18}NS$	[4-phenyl-2-methylthiazolyl structure]	Me	H	Me	H	Me	584
$C_{18}H_{19}N_2$	[8-amino-1-naphthyl structure, NH₂]	Me	H	Me	H	Me	574, 1105

TABLE XXVI (continued)

Formula	R^1	R^2	R^3	R^4	R^5	R^6	Reference
$C_{18}H_{19}N_4O_2S$	$C_6H_4SO_2NH$-(pyrimidin-2-yl)(4)	Me	H	Me	H	Me	587
$C_{18}H_{21}N_2$	CH_2CH_2-(indol-3-yl)	Me	H	Me	H	Me	574, 1105, 1127a
$C_{18}H_{21}N_2S$	Bu	Me	H	2-benzthiazolyl	H	Me	47
$C_{18}H_{22}N$	Me	Et	H	CH=CHPh	H	Et	212
$C_{18}H_{22}NO_2$	Me	Me	H	CH=CH-(2-OH-5-OEt-phenyl)	H	Me	1124
$C_{18}H_{22}NO_2$	$CH(Ph)CH_2OCOCH_3$	Me	H	Me	H	Me	591
$C_{18}H_{22}NO_2$	$CH(Me)CH_2OCOPh$	Me	H	Me	H	Me	591
$C_{18}H_{23}N_2$	Me	Me	H	$CH=CHC_6H_4NMe_2(4)$	H	Me	1124
$C_{18}H_{23}N_2O$	CH_2Ph	Me	H	morpholino	H	Me	507

Formula							Ref
$C_{19}H_{15}Cl_4N$	CH_2Ph	Me	H	[tetrachlorocyclopentadiene ring, Cl_4]	H	Me	1115
$C_{19}H_{18}N$	Me	Me	H	Ph	H	Ph	1128
$C_{19}H_{18}N$	Ph	Ph	H	Me	H	Ph	1129
$C_{19}H_{18}N$	Me	H	H	$CHPh_2$	H	Ph	1130
$C_{19}H_{18}N_3$	NH_2	Me	H	9-carbazolyl	H	Me	725
$C_{19}H_{19}N_2$	Ph	Me	H	NHPh	H	Me	28
$C_{19}H_{19}N_2$	NHMe	Me	H	Ph	H	Ph	467
$C_{19}H_{20}N_3O_2S$	$C_6H_4SO_2NH$(pyridyl)(4)	Me	H	Me	H	Me	587
$C_{19}H_{21}BrN$	$C_6H_4Br(4)$	$-(CH_2)_4-$		H	$-(CH_2)_4-$		1131
$C_{19}H_{21}ClN$	$C_6H_4Cl(4)$	$-(CH_2)_4-$		H	$-(CH_2)_4-$		1131
$C_{19}H_{21}FN$	$C_6H_4F(4)$	$-(CH_2)_4-$		H	$-(CH_2)_4-$		1131

TABLE XXVI (continued)

Formula	R^1	R^2	R^3 R^4	R^5 R^6	Reference
$C_{19}H_{21}IN$	$C_6H_4I(4)$	$-(CH_2)_4-$	H	$-(CH_2)_4-$	1131
$C_{19}H_{21}N_2$	Me	Me	H	H Me	1127
$C_{19}H_{21}N_2$	Me	Me	H	H Me	1127
$C_{19}H_{21}N_2O_2$	CH(COOH)CH$_2$-	Me	H Me	H Me	577
$C_{19}H_{21}N_2O_2$	$C_6H_4NO_2(3)$	$-(CH_2)_4-$	H	$-(CH_2)_4-$	1122
$C_{19}H_{22}N$	Ph	$-(CH_2)_4-$	H	$-(CH_2)_4-$	1131 , 1132
$C_{19}H_{22}NO$	$C_6H_4OH(2)$	$-(CH_2)_4-$	H	$-(CH_2)_4-$	1122
$C_{19}H_{22}NO$	$C_6H_4OH(3)$	$-(CH_2)_4-$	H	$-(CH_2)_4-$	1122
$C_{19}H_{22}NO$	$C_6H_4OH(4)$	$-(CH_2)_4-$	H	$-(CH_2)_4-$	1122

$C_{19}H_{22}N_3O$	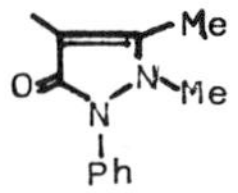	Me	H	Me	H	Me	587, 589, 1105, 1110
$C_{19}H_{24}NO_2$	$CH(Et)CH_2OCOPh$	Me	H	Me	H	Me	591
$C_{19}H_{25}N_2$	CH_2Ph	Me	H	piperidino	H	Me	507
$C_{20}H_{17}N_2S$	Ph	Me	H	2-benzthiazolyl	H	Me	47
$C_{20}H_{17}N_4$	**1,2,4-triazol-4-yl**	Me	H	Ph	H	Ph	585, 586
$C_{20}H_{18}N$	$CH=CH_2$	Me	H	Ph	H	Ph	632, 1105, 1106
$C_{20}H_{18}NO$	Me	CH=CH-OH	H	Ph	H	Ph	250
$C_{20}H_{18}N_3S$	NHPh	Me	H	2-benzthiazolyl	H	Me	47
$C_{20}H_{19}ClN$	CH_2CH_2Cl	Me	H	Ph	H	Ph	632, 1105, 1108
$C_{20}H_{19}N_2$	9-carbazolyl	Me	H	Me	H	Me	585, 586

TABLE XXVI (continued)

Formula	R^1	R^2	R^3	R^4	R^5	R^6	Reference
$C_{20}H_{20}N$	$C_6H_4Me(4)$	Me	H	Ph	H	Ph	44
$C_{20}H_{20}N$	Ph	Me	Ph	Me	H	Me	179
$C_{20}H_{20}NO$	CH_2CH_2OH	Me	H	Ph	H	Me	632, 1105 1109
$C_{20}H_{21}N_2$	NPh_2	Me	H	Me	H	Me	389
$C_{20}H_{23}N_2O$	NHCOPh	$-(CH_2)_4-$	H		$-(CH_2)_4-$		446
$C_{20}H_{24}N$	$C_6H_4Me(2)$	$-(CH_2)_4-$	H		$-(CH_2)_4-$		1131
$C_{20}H_{24}N$	$C_6H_4Me(3)$	$-(CH_2)_4-$	H		$-(CH_2)_4-$		1131
$C_{20}H_{24}N$	$C_6H_4Me(4)$	$-(CH_2)_4-$	H		$-(CH_2)_4-$		1131
$C_{20}H_{24}NO$	$C_6H_4OMe(2)$	$-(CH_2)_4-$	H		$-(CH_2)_4-$		1122
$C_{20}H_{24}NO$	$C_6H_4OMe(3)$	$-(CH_2)_4-$	H		$-(CH_2)_4-$		1122
$C_{20}H_{24}NO$	$C_6H_4OMe(4)$	$-(CH_2)_4-$	H		$-(CH_2)_4-$		1122

$C_{20}H_{26}N$	Me	i-Pr	H	CH=CH-Ph	H	i-Pr	212
$C_{20}H_{26}NO_5$	CH_2CH_2OCO- 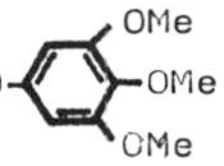	Me	H	Me	H	Me	574, 632, 1105
$C_{20}H_{27}N_2O_4$	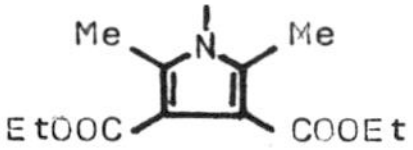	Me	H	Me	H	Me	586
$C_{21}H_{20}Br_2N$	$CH_2CH(Br)CH_2Br$	Me	H	Ph	H	Ph	574, 1105
$C_{21}H_{20}NO_2$	CH_2CH_2COOH	Me	H	Ph	H	Ph	574, 1105 1112
$C_{21}H_{20}N$	$CH_2CH=CH_2$	Me	H	Ph	H	Ph	632, 1105, 1111
$C_{21}H_{21}N_4O$	$NHCOC_6H_4N=NPh(4)$	Me	H	Me	H	Me	1104
$C_{21}H_{22}N$	CH_2Ph	$C_6H_4Me(2)$	H	Me	H	Me	1123

TABLE XXVI (continued)

Formula	R^1	R^2	R^3	R^4	R^5	R^6	Reference
$C_{21}H_{22}N$	$C_6H_4Me(2)$	CH_2Ph	H	Me	H	Me	1123
$C_{21}H_{22}NO$	$CH(Me)CH_2OH$	Me	H	Ph	H	Ph	1105
$C_{21}H_{22}NO$	$C_6H_4OMe(2)$	CH_2Ph	H	Me	H	Me	1123
$C_{21}H_{22}NS$	CH_2Ph	Me	H	SCH_2Ph	H	Me	507
$C_{21}H_{23}NO_8$	Me	Me	H	(cyclopentadiene: MeOOC, COOMe, MeOOC, COOMe)	H	Me	1115
$C_{21}H_{26}NO_2$	Me	i-Pr	H	$CH=CH_2$ (benzodioxole)	H	i-Pr	212
$C_{21}H_{28}NO$	Me	i-Pr	H	$CH=CHC_6H_4OMe(4)$	H	i-Pr	212
$C_{22}H_{19}F_3NO_2$	$CH_2CH_2OCOCF_3$	Me	H	Ph	H	Ph	632, 1105, 1109
$C_{22}H_{19}N_2O$	(isoxazole, Me)	Me	H	Ph	H	Ph	1105, 1110

Formula						
$C_{22}H_{20}N$	1-anthryl	Me	H	Me	H	1133
$C_{22}H_{20}N$	2-anthryl	Me	H	Me	H	1133
$C_{22}H_{20}N$	9-anthryl	Me	H	Me	H	1133
$C_{22}H_{21}N_2S$	Ph	(structure)	H	Me	H	1127
$C_{22}H_{22}N$	Me	Ph	H	Ph	$-(CH_2)_4-$	243
$C_{22}H_{22}NO_2$	CH_2CH_2OAc	Ph	H	Ph	H	632, 1105, 1109
$C_{22}H_{22}NO_2$	CH_2COOEt	Ph	H	Ph	H	574, 1105, 1112
$C_{22}H_{22}NO_2$	$(CH_2)_3COOH$	Ph	H	Ph	H	574, 1105, 1112
$C_{22}H_{23}N_3$	Bu	Me	H	Me	H	1134

TABLE XXVI (continued)

Formula	R^1	R^2	R^3	R^4	R^5	R^6	Reference
$C_{22}H_{23}N_4$	N=C(Ph)C(Ph)=NNH$_2$	Me	H	Me	H	Me	1104
$C_{22}H_{24}N$	Bu	Me	H	Ph	H	Ph	1135
$C_{22}H_{24}NO$	CH(Et)CH$_2$OH	Me	H	Ph	H	Ph	1105, 1114
$C_{22}H_{24}NO$	CH$_2$C$_6$H$_4$OMe(4)	C$_6$H$_4$Me(2)	H	Me	H	Me	1123
$C_{22}H_{25}N_4$	C$_6$H$_4$N=NC$_6$H$_4$NMe$_2$(4,4')	Me	H	Me	H	Me	1123
$C_{22}H_{31}N_2$	Me	i-Pr	H	CH=CHC$_6$H$_4$NMe$_2$(4)	H	i-Pr	212
$C_{23}H_{16}Br_2NO$	OH	C$_6$H$_4$Br(4)	H	Ph	H	C$_6$H$_4$Br(4)	463
$C_{23}H_{17}BrNO$	OH	Ph	H	C$_6$H$_4$Br(4)	H	Ph	463
$C_{23}H_{18}NO$	OH	Ph	H	Ph	H	Ph	463, 500
$C_{23}H_{19}N_2$	NH$_2$	Ph	H	Ph	H	Ph	471
$C_{23}H_{23}N_3$	CH$_2$CH$_2$	Me	H	Ph	H	Ph	574, 1105, 1116

Molecular formula						
$C_{23}H_{24}N$	1-naphthyl	Me	H	-(CH$_2$)$_4$-	-(CH$_2$)$_4$-	1131
$C_{23}H_{24}N$	2-naphthyl	Ph	H	-(CH$_2$)$_4$-	-(CH$_2$)$_4$-	1131
$C_{23}H_{24}NO_2$	CH(Ph)CH$_2$OCOPh	Me	Me	H	Me	591
$C_{23}H_{25}N_2$	Ph	piperidino	H	H	Ph	228
$C_{23}H_{26}N_3$	[ring: –N⌒N–Me]	Me	Ph	H	Ph	1117
$C_{24}H_{17}N_2O_4$	$C_6H_4NO_2(4)$	COOH	Ph	H	Ph	268
$C_{24}H_{18}Br_3N_2$	NHC$_6$H$_4$Br(4)	Me	C$_6$H$_4$Br(4)	H	C$_6$H$_4$Br(4)	418
$C_{24}H_{18}Cl_2N$	C$_6$H$_4$Cl(4)	C$_6$H$_4$Cl(4)	Ph	H	Ph	406
$C_{24}H_{18}N_5$	5-tetrazolyl	Ph	Ph	H	Ph	1119
$C_{24}H_{19}Br_2N_2$	NHPh	Me	C$_6$H$_4$Br(3)	H	C$_6$H$_4$Br(3)	418

TABLE XXVI (continued)

Formula	R^1	R^2	R^3	R^4	R^5	R^6	Reference
$C_{24}H_{19}Br_2N_2$	NHPh	Me	H	$C_6H_4Br(4)$	H	$C_6H_4Br(4)$	418
$C_{24}H_{19}ClN$	$C_6H_4Cl(2)$	Me	H	Ph	H	Ph	1105 , 1110
$C_{24}H_{19}Cl_2N_2$	NHPh	Me	H	$C_6H_4Cl(3)$	H	$C_6H_4Cl(3)$	418
$C_{24}H_{19}Cl_2N_2$	NHPh	Me	H	$C_6H_4Cl(4)$	H	$C_6H_4Cl(4)$	418
$C_{24}H_{19}N_4O_4$	NHPh	Me	H	$C_6H_4NO_2(3)$	H	$C_6H_4NO_2(3)$	660
$C_{24}H_{19}N_4O_4$	NHPh	Me	H	$C_6H_4NO_2(4)$	H	$C_6H_4NO_2(4)$	660
$C_{24}H_{19}N_4O_4$	$NHC_6H_4(NO_2)_2(2,4)$	Me	H	Ph	H	Ph	660
$C_{24}H_{19}N_4O_6S$	$NHSO_2Ph$	Me	H	$C_6H_4NO_2(3)$	H	$C_6H_4NO_2(3)$	470
$C_{24}H_{19}N_4O_6S$	$NHSO_2Ph$	Me	H	$C_6H_4NO_2(4)$	H	$C_6H_4NO_2(4)$	470
$C_{24}H_{20}N$	Me	Ph	H	Ph	H	Ph	406, 408, 614, 615, 621, 622, 669, 796, 1135, 1136

Formula							Ref.
$C_{24}H_{20}N$	Ph	Me	H	Ph	H	Ph	464, 582
$C_{24}H_{20}N$	Ph	Ph	H	Me	H	Ph	464
$C_{24}H_{2c}N$	Ph	H	H	$CHPh_2$	H	H	1130
$C_{24}H_{20}NO$	$C_6H_4OH(4)$	Me	H	Ph	H	Ph	597
$C_{24}H_{20}NO$	$C_6H_4OH(4)$	Ph	H	Me	H	Ph	597
$C_{24}H_{20}NO$	OH	Pn	Me	Ph	H	Ph	463
$C_{24}H_{20}N_3O_2$	$NHC_6H_4NO_2(4)$	Me	H	Ph	H	Ph	660
$C_{24}H_{21}N_2$	$C_6H_4NH_2(2)$	Me	H	Ph	H	Ph	1118
$C_{24}H_{21}N_2$	NHPh	Me	H	Ph	H	Ph	416, 660
$C_{24}H_{21}N_2$	NHPh	Ph	H	Me	H	Ph	464
$C_{24}H_{21}N_2O_2S$	$NHSO_2Ph$	Me	H	Ph	H	Ph	166

TABLE XXVI (continued)

Formula	R^1	R^2	R^3	R^4	R^5	R^6	Reference
$C_{24}H_{26}NO_2$	$(CH_2)_5COOH$	Me	H	Ph	H	Ph	574, 1105, 1112
$C_{24}H_{28}N$	C_6H_{13}	Me	H	Ph	H	Ph	1135
$C_{24}H_{29}N_2$	$CH_2CH_2NEt_2$	Me	H	Ph	H	Ph	574, 1105
$C_{24}H_{29}N_4$	$C_6H_4N=NC_6H_4NEt_2(4,4')$	Me	H	Me	H	Me	1105
$C_{24}H_{35}N_2$	Me	i-Pr	H	$CH=CHC_6H_4NEt_2(4)$	H	i-Pr	212
$C_{24}H_{35}N_2O_4$	$(CH_2)_6NHCO$-(trimethoxyphenyl)	Me	H	Me	H	Me	574, 1105
$C_{25}H_{19}N_3$	Me	$CH_2N=NPh$	H	Ph	H	Ph	239
$C_{25}H_{19}N_4$	1,2,4-triazol-4-yl	Ph	H	Ph	H	Ph	585, 586
$C_{25}H_{19}N_4$	1,2,4-triazol-3-yl	Ph	H	Ph	H	Ph	1119
$C_{25}H_{20}Cl_2N$	Et	$C_6H_4Cl(4)$	H	Ph	H	$C_6H_4Cl(4)$	406

$C_{25}H_{20}N$	CH=CH$_2$	Ph	Ph	H	Ph	H	632, 1105, 1106
$C_{25}H_{20}NO_2$	CH$_2$COOH	Ph	Ph	H	Ph	H	576
$C_{25}H_{20}N3$	CH$_2$—[benzimidazol-2-yl]	Ph	H	H	Ph	H	1125
$C_{25}H_{21}Br_2N_2$	NHC$_6$H$_4$Me(4)	Me	C$_6$H$_4$Br(4)	H	C$_6$H$_4$Br(4)	H	418
$C_{25}H_{21}ClN$	CH$_2$CH$_2$Cl	Ph	Ph	H	Ph	H	632, 1105, 1108
$C_{25}H_{21}ClN$	CH$_2$C$_6$H$_4$Cl(2)	Me	Ph	H	Ph	H	1135
$C_{25}H_{21}Cl_2N_2$	NHC$_6$H$_4$Me(3)	Me	C$_6$H$_4$Cl(4)	H	C$_6$H$_4$Cl(4)	H	418
$C_{25}H_{21}Cl_2N_2$	NHC$_6$H$_4$Me(4)	Me	C$_6$H$_4$Cl(3)	H	C$_6$H$_4$Cl(3)	H	418
$C_{25}H_{21}Cl_2N_2$	NHC$_6$H$_4$Me(4)	Me	C$_6$H$_4$Cl(4)	H	C$_6$H$_4$Cl(4)	H	418

TABLE XXVI (continued)

Formula	R^1	R^2	R^3	R^4	R^5	R^6	Reference
$C_{25}H_{21}N_2O$	NHCOMe	Ph	H	Ph	H	Ph	471, 667
$C_{25}H_{21}N_2O_2$	$C_6H_4NH_2(2)$	COOMe	H	Ph	H	Ph	268
$C_{25}H_{22}N$	CH_2Ph	Me	H	Ph	H	Ph	464, 620, 1135
$C_{25}H_{22}N$	$C_6H_4Me(3)$	Me	H	Ph	H	Ph	582
$C_{25}H_{22}N$	$C_6H_4Me(4)$	Me	H	Ph	H	Ph	582
$C_{25}H_{22}N$	Et	Ph	H	Ph	H	Ph	13, 406, 608, 614, 615, 621, 622
$C_{25}H_{22}NO$	CH_2CH_2OH	Ph	H	Ph	H	Ph	406, 574, 622, 632, 1105
$C_{25}H_{22}BrN_2$	$NHC_6H_4Br(4)$	Et	H	Ph	H	Ph	418
$C_{25}H_{23}N_2$	NHPh	Et	H	Ph	H	Ph	418

Formula							Reference
$C_{25}H_{23}N_2$	N(Me)Ph	Me	H	Ph	H	Ph	464
$C_{25}H_{23}N_2$	$CH_2CH_2NH_2$	Ph	H	Ph	H	Ph	632, 1105
$C_{25}H_{23}N_2O_2S$	$CH_2C_6H_4SO_2NH_2(4)$	Me	H	Ph	H	Ph	574, 1105, 1121
$C_{25}H_{42}N$	$C_{12}H_{25}$	$-(CH_2)_4-$	H		$-(CH_2)_4-$		545
$C_{26}H_{19}N_2S$	2-thiazolyl	Ph	H	Ph	H	Ph	608, 609
$C_{26}H_{21}N_2S$	2-thiazolinyl	Ph	H	Ph	H	Ph	1105, 1110
$C_{26}H_{22}N$	$CH_2CH=CH_2$	Ph	H	Ph	H	Ph	608, 612, 629, 632, 1105, 1111, 1136-1138
$C_{26}H_{22}N$	Me	CH=CHPh	H	Ph	H	Ph	
$C_{26}H_{23}N_2O_2$	NHCOOEt	Ph	H	Ph	H	Ph	471
$C_{26}H_{24}BrN_2$	$NHC_6H_4Br(4)$	Me	H	$C_6H_4Me(4)$	H	$C_6H_4Me(4)$	418

TABLE XXVI (continued)

Formula	R^1	R^2	R^3	R^4	R^5	R^6	Reference
$C_{26}H_{24}BrN_2O_2$	$NHC_6H_4Br(4)$	Me	H	$C_6H_4OMe(4)$	H	$C_6H_4OMe(4)$	418
$C_{26}H_{24}N$	Pr	Ph	H	Ph	H	Ph	621, 622
$C_{26}H_{24}N$	i-Pr	Ph	H	Ph	H	Ph	13, 608, 1136, 1143
$C_{26}H_{24}N$	CH_2CH_2Ph	Me	H	Ph	H	Ph	574, 1105, 1116
$C_{26}H_{24}N$	$CH_2C_6H_4Me(4)$	Me	H	Ph	H	Ph	1135
$C_{26}H_{24}N$	Ph	Me	H	$C_6H_4Me(4)$	H	$C_6H_4Me(4)$	411
$C_{26}H_{24}NO$	$CH_2CH_2CH_2OH$	Ph	H	Ph	H	Ph	608, 609, 621, 622, 632
$C_{26}H_{24}NO$	$CH_2CH(Me)OH$	Ph	H	Ph	H	Ph	632, 1105
$C_{26}H_{24}NO$	Me	Me	H		H	Me	1127

							Ref.
$C_{26}H_{24}NO$	Me	Me	H	(see structure)	H	Me	1127
$C_{26}H_{24}N_3O_4$	$NHC_6H_4NO_2(4)$	Me	H	$C_6H_4OMe(4)$	H	$C_6H_4OMe(4)$	660
$C_{26}H_{25}N_2$	$CH_2CH_2CH_2NH_2$	Ph	H	Ph	H	Ph	632, 1105
$C_{26}H_{25}N_2$	NHPh	Me	H	$C_6H_4Me(3)$	H	$C_6H_4Me(3)$	418
$C_{26}H_{25}N_2$	NHPh	Me	H	$C_6H_4Me(4)$	H	$C_6H_4Me(4)$	411, 418
$C_{26}H_{25}N_2$	NHPh	i-Pr	H	Ph	H	Ph	560
$C_{26}H_{25}N_2$	$NHC_6H_4Me(4)$	Et	H	Ph	H	Ph	418
$C_{26}H_{25}N_2$	Me	Ph	H	$C_6H_4NMe_2(4)$	H	Ph	464
$C_{26}H_{25}N_2O_2$	NHPh	Me	H	$C_6H_4OMe(4)$	H	$C_6H_4OMe(4)$	418
$C_{26}H_{32}N$	C_8H_{17}	Me	H	Ph	H	Ph	1135
$C_{26}H_{48}N$	$C_{18}H_{37}$	Me	H	Me	H	Me	401

Structure (heading): ring bearing O with two Ph groups and a CH.

TABLE XXVI (continued)

Formula	R^1	R^2	R^3	R^4	R^5	R^6	Reference
$C_{27}H_{21}F_3NO_2$	$CH_2CH_2OCOCF_3$	Ph	H	Ph	H	Ph	632, 1105, 1109
$C_{27}H_{21}N_2O$		Ph	H	Ph	H	Ph	1105, 1110
$C_{27}H_{21}N_4O_2$		Ph	H	Ph	H	Ph	1119
$C_{27}H_{24}NO_2$	CH_2COOEt	Ph	H	Ph	H	Ph	574, 576, 1105
$C_{27}H_{24}N$	CH_2Ph	$-(CH_2)_3-$		Ph	H	Ph	620
$C_{27}H_{24}N$	Ph	$-(CH_2)_4-$		Ph	H	Ph	243
$C_{27}H_{24}NO_2$	CH_2CH_2OAc	Ph	H	Ph	H	Ph	632, 1105, 1109
$C_{27}H_{24}NO_2$	$(CH_2)_3COOH$	Ph	H	Ph	H	Ph	574, 578, 1105

Formula							References
$C_{27}H_{24}NO_2$	$C_6H_4COOEt(4)$	Me	H	Ph	H	Ph	1105, 1110
$C_{27}H_{25}N_2O$	$NHCOPr$	Ph	H	Ph	H	Ph	471, 667
$C_{27}H_{26}N$	Bu	Ph	H	Ph	H	Ph	13, 607, 608, 609, 611, 614, 615, 621, 622, 1137
$C_{27}H_{26}N$	i-Bu	Ph	H	Ph	H	Ph	621, 622
$C_{27}H_{26}N$	sec-Bu	Ph	H	Ph	H	Ph	608, 1136
$C_{27}H_{26}N$	Me	$C_6H_4Me(4)$	H	$C_6H_4Me(4)$	H	$C_6H_4Me(4)$	406
$C_{27}H_{26}N$	CH_2Ph	Me	H	$C_6H_4Me(4)$	H	$C_6H_4Me(4)$	411
$C_{27}H_{26}NO$	$CH(Et)CH_2OH$	Ph	H	Ph	H	Ph	1105, 1114
$C_{27}H_{26}NO$	$CH_2C(Me_2)OH$	Ph	H	Ph	H	Ph	632, 1105
$C_{27}H_{26}NO$	$C_6H_4OH(4)$	t-Bu	H	Ph	H	Ph	597

TABLE XXVI (continued)

Formula	R¹	R²	R³	R⁴	R⁵	R⁶	Reference
$C_{27}H_{26}NO_3$	$C_6H_4OMe(4)$	Me	H	$C_6H_4OMe(4)$	H	$C_6H_4OMe(4)$	1138a
$C_{27}H_{26}NO_3S$	CH_2CH_2OTs	Me	H	Ph	H	Ph	632, 1105, 1109
$C_{27}H_{27}N_2$	NHPh	Et	H	$C_6H_4Me(4)$	H	$C_6H_4Me(4)$	418
$C_{27}H_{27}N_2$	NHPh	t-Bu	H	Ph	H	Ph	660
$C_{27}H_{27}N_2$	$NHC_6H_4Me(3)$	Me	H	$C_6H_4Me(4)$	H	$C_6H_4Me(4)$	418
$C_{27}H_{27}N_2$	$NHC_6H_4Me(4)$	Me	H	$C_6H_4Me(4)$	H	$C_6H_4Me(4)$	418
$C_{27}H_{27}N_2$	N(Me)Ph	Me	H	$C_6H_4Me(4)$	H	$C_6H_4Me(4)$	411
$C_{27}H_{27}N_2O_2$	$NHC_6H_4Me(4)$	Me	H	$C_6H_4OMe(4)$	H	$C_6H_4OMe(4)$	418
$C_{28}H_{20}N_5$	6-purinyl	Ph	H	Ph	H	Ph	1119
$C_{28}H_{21}N_2$	2-pyridyl	Ph	H	Ph	H	Ph	500, 607, 608, 609, 611, 612, 1119, 1139

Formula	Substituent						Ref.
$C_{28}H_{21}N_2$	4-pyridyl	Ph	H	Ph	H	Ph	500, 1119
$C_{28}H_{21}N_2O$	(N-pyridone)	Ph	H	Ph	H	Ph	586
$C_{28}H_{22}N$	1-naphthyl	Me	H	Ph	H	Ph	1118
$C_{28}H_{22}NO$	2-furuyl	Ph	H	Ph	H	Ph	429, 618, 667
$C_{28}H_{23}N_2$	H_2N-(naphthyl)	Me	H	Ph	H	Ph	1118
$C_{28}H_{24}N$	Me	$(CH=CH)_2Ph$	H	Ph	H	Ph	1138
$C_{28}H_{25}N_2$	CH_2CH_2-(indol-3-yl, N–H)	Me	H	Ph	H	Ph	574, 1105, 1127a
$C_{28}H_{25}N_3$	CH_2CH_2-(imidazolyl, N–NH)	Ph	H	Ph	H	Ph	574, 1105, 1116
$C_{28}H_{26}N$	CH_2Ph	$-(CH_2)_4-$		Ph		H	620

TABLE XXVI (continued)

Formula	R^1	R^2	R^3	R^4	R^5	R^6	Reference
$C_{28}H_{27}N_2O$	NHCOPh	t-Bu	H	Ph	H	Ph	471
$C_{28}H_{28}N$	CH_2Ph	t-Bu	H	Ph	H	Ph	620
$C_{28}H_{28}N$	C_5H_{11}	Ph	H	Ph	H	Ph	13, 621, 622
$C_{28}H_{28}N$	Et	$C_6H_4Me(4)$	H	$C_6H_4Me(4)$	H	$C_6H_4Me(4)$	406
$C_{28}H_{30}N_3$	Me	Ph	H	$C_6H_4NMe_2(4)$	H	$C_6H_4NMe_2(4)$	796
$C_{28}H_{30}N_3$	Me	$C_6H_4NMe_2(4)$	H	Ph	H	$C_6H_4NMe_2(4)$	794, 796
$C_{29}H_{20}Cl_2NO$	$C_6H_4OH(4)$	$C_6H_4Cl(4)$	H	Ph	H	$C_6H_4Cl(4)$	597
$C_{29}H_{20}N_3O_5$	$C_6H_4OH(4)$	$C_6H_4NO_2(4)$	H	Ph	H	$C_6H_4NO_2(4)$	597
$C_{29}H_{21}BrN$	$C_6H_4Br(2)$	Ph	H	Ph	H	Ph	1118, 1139
$C_{29}H_{21}BrN$	$C_6H_4Br(4)$	Ph	H	Ph	H	Ph	1139
$C_{29}H_{21}ClN$	$C_6H_4Cl(4)$	Ph	H	Ph	H	Ph	183, 601, 608

$C_{29}H_{21}ClNO$	$C_6H_4OH(4)$	$C_6H_4Cl(4)$	H	Ph	597
$C_{29}H_{21}ClNO$	$C_6H_4OH(4)$	Ph	H	Ph	597
$C_{29}H_{21}N_2O_2$	$C_6H_4NO_2(2)$	Ph	H	Ph	1119
$C_{29}H_{21}N_2O_2$	$C_6H_4NO_2(4)$	Ph	H	Ph	500, 1119
$C_{29}H_{21}N_2O_3$	$C_6H_4NO_2(3)$	$C_6H_4OH(4)$	H	Ph	594
$C_{29}H_{21}N_2O_3$	$C_6H_4OH(4)$	Ph	H	Ph	597
$C_{29}H_{21}N_2O_3$	$C_6H_4OH(4)$	$C_6H_4NO_2(4)$	H	Ph	597
$C_{29}H_{22}N$	Ph	Ph	H	Ph	13, 500, 581, 582, 595, 601, 608, 611, 612, 622, 1118, 1140
$C_{29}H_{22}NO$	$C_6H_4OH(2)$	Ph	H	Ph	598
$C_{29}H_{22}NO$	$C_6H_4OH(3)$	Ph	H	Ph	1140

TABLE XXVI (continued)

Formula	R^1	R^2	R^3	R^4	R^5	R^6	Reference
C$_{29}$H$_{22}$NO	C$_6$H$_4$OH(4)	Ph	H	Ph	H	Ph	597, 601, 1140
C$_{29}$H$_{22}$NO	Ph	Ph	H	C$_6$H$_4$OH(4)	H	Ph	592
C$_{29}$H$_{22}$NO	OH	Ph	Ph	H	Ph	Ph	463
C$_{29}$H$_{22}$NO	OH	Ph	Ph	Ph	H	Ph	463
C$_{29}$H$_{22}$NO$_3$	Ph	C$_6$H$_4$OH(4)	H	C$_6$H$_4$OH(4)	H	C$_6$H$_4$OH(4)	594
C$_{29}$H$_{22}$NS	C$_6$H$_4$SH(4)	Ph	H	Ph	H	Ph	598, 601
C$_{29}$H$_{23}$N$_2$	NHPh	Ph	H	Ph	H	Ph	406, 464, 500, 1066
C$_{29}$H$_{23}$N$_2$	C$_6$H$_4$NH$_2$(2)	Ph	H	Ph	H	Ph	1118
C$_{29}$H$_{23}$N$_2$	C$_6$H$_4$NH$_2$(3)	Ph	H	Ph	H	Ph	1118
C$_{29}$H$_{23}$N$_2$	C$_6$H$_4$NH$_2$(4)	Ph	H	Ph	H	Ph	1118
C$_{29}$H$_{23}$N$_2$	2-picolyl	Ph	H	Ph	H	Ph	614, 615, 617, 629, 1137, 1139

C$_{29}$H$_{23}$N$_2$	3-picolyl	Ph	H	Ph	H	Ph	614, 615, 617, 626, 1137
C$_{29}$H$_{23}$N$_2$	4-picolyl	Ph	H	Ph	H	Ph	617, 621, 622, 626, 629, 667, 1137
C$_{29}$H$_{23}$N$_2$	3-Me-2-picolyl	Ph	H	Ph	H	Ph	607
C$_{29}$H$_{23}$N$_2$	4-Me-2-picolyl	Ph	H	Ph	H	Ph	607
C$_{29}$H$_{23}$N$_2$	5-Me-2-pyridyl	Ph	H	Ph	H	Ph	608
C$_{29}$H$_{23}$N$_2$	6-Me-2-pyridyl	Ph	H	Ph	H	Ph	608
C$_{29}$H$_{23}$N$_2$O	C$_6$H$_4$NH$_2$(3)	Ph	H	Ph	H	Ph	594
C$_{29}$H$_{23}$N$_2$O	C$_6$H$_4$NH$_2$(4)	C$_6$H$_4$OH(4)	H	Ph	H	Ph	594
C$_{29}$H$_{23}$N$_2$O	C$_6$H$_4$NH$_2$(4)	Ph	H	C$_6$H$_4$OH(4)	H	Ph	594

TABLE XXVI (continued)

Formula	R^1	R^2	R^3	R^4	R^5	R^6	Reference
C$_{29}$H$_{26}$N$_3$O		Me	H	Ph	H	Ph	1105, 1110
C$_{29}$H$_{28}$N	C$_6$H$_{11}$	Ph	H	Ph	H	Ph	406, 608
C$_{29}$H$_{28}$NO$_2$	(CH$_2$)$_5$COOH	Ph	H	Ph	H	Ph	13, 574, 1105
C$_{29}$H$_{31}$N$_2$	CH$_2$CH$_2$NEt$_2$	Ph	H	Ph	H	Ph	574, 1105
C$_{30}$H$_{20}$BrN$_2$S		Ph	H	Ph	H	Ph	1119
C$_{30}$H$_{21}$N$_2$S	2-benzthiazolyl	Ph	H	Ph	H	Ph	608
C$_{30}$H$_{22}$ClN$_2$O	NHCOC$_6$H$_4$Cl(4)	Ph	H	Ph	H	Ph	471, 667
C$_{30}$H$_{22}$ClN$_2$S	NHCSC$_6$H$_4$Cl(4)	Ph	H	Ph	H	Ph	471
C$_{30}$H$_{22}$Cl$_2$N	CH$_2$C$_6$H$_3$Cl$_2$(2,4)	Ph	H	Ph	H	Ph	609
C$_{30}$H$_{22}$NO$_2$	C$_6$H$_4$COOH(2)	Ph	H	Ph	H	Ph	1139
C$_{30}$H$_{22}$NO$_2$	C$_6$H$_4$COOH(4)	Ph	H	Ph	H	Ph	1139

$C_{30}H_{22}N_3$	2-benzimidazolyl	Ph	H	Ph	H	Ph	32, 500, 1119
$C_{30}H_{22}N_3O_2$	$NHCOC_6H_4NO_2(4)$	Ph	H	Ph	H	Ph	471
$C_{30}H_{22}N_3O_4$	$CH_2C_6H_3(NO_2)_2(2,4)$	Ph	H	Ph	H	Ph	660
$C_{30}H_{23}ClN$	$CH_2C_6H_4Cl(2)$	Ph	H	Ph	H	Ph	609, 611, 1137
$C_{30}H_{23}ClN$	$CH_2C_6H_4Cl(4)$	Ph	H	Ph	H	Ph	607, 608, 609, 611, 617, 621
$C_{30}H_{23}N_2O$	NHCOPh	Ph	H	Ph	H	Ph	471, 667
$C_{30}H_{23}N_2O_2$	$CH_2C_6H_4NO_2(4)$	Ph	H	Ph	H	Ph	338, 574, 1121
$C_{30}H_{23}N_2S$	NHCSPh	Ph	H	Ph	H	Ph	471

TABLE XXVI (continued)

Formula	R^1	R^2	R^3	R^4	R^5	R^6	Reference
$C_{30}H_{24}N$	CH_2Ph	Ph	H	Ph	H	Ph	13, 574, 607, 608, 609, 611, 614, 615, 617, 620, 621, 622, 629, 667, 1105, 1121, 1136, 1137, 1140
$C_{30}H_{24}N$	$C_6H_4Me(2)$	Ph	H	Ph	H	Ph	500, 582, 1140
$C_{30}H_{24}N$	$C_6H_4Me(4)$	Ph	H	Ph	H	Ph	500, 582, 595, 601, 608
$C_{30}H_{24}NO$	$C_6H_4OMe(2)$	Ph	H	Ph	H	Ph	582, 595
$C_{30}H_{24}NO$	$C_6H_4OMe(4)$	Ph	H	Ph	H	Ph	500, 581, 582, 595, 601

$C_{30}H_{24}NO$	$C_6H_4OH(4)$	$C_6H_4Me(4)$	H	Ph	H	Ph	597
$C_{30}H_{24}NO$	$C_6H_4OH(4)$	Ph	H	$C_6H_4Me(4)$	H	Ph	597
$C_{30}H_{24}NOS$	$C_6H_4SH(4)$	Ph	H	$C_6H_4OMe(4)$	H	Ph	598
$C_{30}H_{24}NO_2$	$C_6H_4OH(2)$	Ph	H	$C_6H_4OMe(4)$	H	Ph	598
$C_{30}H_{24}NO_2$	$C_6H_4OH(4)$	$C_6H_4OMe(4)$	H	Ph	H	Ph	597
$C_{30}H_{24}NO_2$	$C_6H_4OH(4)$	Ph	H	$C_6H_4OMe(4)$	H	Ph	597
$C_{30}H_{25}N_2$	NPh_2	Me	H	Ph	H	Ph	464
$C_{30}H_{25}N_2O_2S$	$CH_2C_6H_4SO_2NH_2(4)$	Ph	H	Ph	H	Ph	574, 1105, 1121
$C_{30}H_{26}N$	Me	$(CH=CH)_3Ph$	H	Ph	H	Ph	1138
$C_{30}H_{30}N$	CH_2Ph	t-Bu	H	Ph	$-(CH_2)_2C_6H_4(2)-$		620
$C_{30}H_{32}N$	C_7H_{15}	Ph	H	Ph	H	Ph	611, 612

TABLE XXVI (continued)

Formula	R^1	R^2	R^3	R^4	R^5	R^6	Reference
$C_{30}H_{35}N_4$	Me	$C_6H_4NMe_2(4)$	H	$C_6H_4NMe_2(4)$	H	$C_6H_4NMe_2(4)$	796
$C_{31}H_{23}ClN$	$C_6H_4Cl(3)$	Ph	H	Ph	$-(CH_2)_2C_6H_4(2)-$		623
$C_{31}H_{23}ClN$	$C_6H_4Cl(4)$	Ph	H	Ph	$-(CH_2)_2C_6H_4(2)-$		623
$C_{31}H_{24}N$	Ph	Ph	H	Ph	$-(CH_2)_2C_6H_4(2)-$		623
$C_{31}H_{24}N$	Ph	CH=CHPh	H	Ph	H	Ph	500, 581
$C_{31}H_{24}N_3$		Ph	H	Ph	H	Ph	32, 500, 1119
$C_{31}H_{24}N$		Ph	H	Ph	H	Ph	1119, 1125
$C_{31}H_{24}N_3$		Ph	H	Ph	H	Ph	1125, 1141
$C_{31}H_{25}N_2O$	$NHCOCH_2Ph$	Ph	H	Ph	H	Ph	471, 667

Formula							Data
$C_{31}H_{25}N_2O$	$NHCOC_6H_4Me(4)$	Ph	H	Ph	H	Ph	471, 667
$C_{31}H_{25}N_2OS$	$NHCSC_6H_4OMe(4)$	Ph	H	Ph	H	Ph	471
$C_{31}H_{25}N_2O_2$	$NHCOC_6H_4OMe(4)$	Ph	H	Ph	H	Ph	471, 667
$C_{31}H_{25}N_2S$	$NHCSC_6H_4Me(4)$	Ph	H	Ph	H	Ph	471
$C_{31}H_{26}N$	CH_2Ph	Ph	Me	Ph	Me	Ph	620
$C_{31}H_{26}N$	CH_2CH_2Ph	Ph	H	Ph	H	Ph	574, 609, 614, 615, 621, 622, 1105, 1116, 1137
$C_{31}H_{26}N$	$CH_2C_6H_4Me(2)$	Ph	H	Ph	H	Ph	614, 615
$C_{31}H_{26}N$	$CH_2C_6H_4Me(4)$	Ph	H	Ph	H	Ph	574, 607, 608, 609, 611, 617, 621, 622, 1105, 1121, 1136, 1137, 1140

TABLE XXVI (continued)

Formula	R^1	R^2	R^3	R^4	R^5	R^6	Reference
$C_{31}H_{26}NO$	$CH_2C_6H_4OMe(4)$	Ph	H	Ph	H	Ph	574, 609, 621, 1105, 1121, 1136
$C_{31}H_{26}NO$	$C_6H_4OEt(4)$	Ph	H	Ph	H	Ph	595
$C_{31}H_{26}NO$	$C_6H_4OH(4)$	$C_6H_4Me(4)$	H	Ph	H	$C_6H_4Me(4)$	597
$C_{31}H_{26}NO$	$C_6H_2Me_2(2,6)OH(4)$	Ph	H	Ph	H	Ph	597
$C_{31}H_{26}NO$	$C_6H_2Me_2(3,5)OH(4)$	Ph	H	Ph	H	Ph	597
$C_{31}H_{26}NO_2$	Ph	$C_6H_3OH(2)Me(4)$	H	Ph	H	$C_6H_4OMe(4)$	1064
$C_{31}H_{26}NO_3$	$C_6H_4OH(4)$	$C_6H_4OMe(4)$	H	$C_6H_4OMe(4)$	H	Ph	597
$C_{31}H_{26}NO_3$	$C_6H_4OH(4)$	$C_6H_4OMe(4)$	H	Ph	H	$C_6H_4OMe(4)$	597
$C_{31}H_{26}NO_3$	Ph	$C_6H_3OH(2)OMe(4)$	H	$C_6H_4OMe(4)$	H	Ph	1064
$C_{31}H_{26}NO_3$	Ph	$C_6H_3OH(2)OMe(4)$	H	Ph	H	$C_6H_4OMe(4)$	1064
$C_{31}H_{26}NO_4$	Ph	$C_6H_3OH(2)OMe(4)$	H	$C_6H_4OMe(4)$	H	$C_6H_4OH(4)$	1064

Formula							Ref.
$C_{31}H_{26}N_3O_2$	$C_6H_4NMe_2(4)$	Ph	H	$C_6H_4NO_2(4)$	H	Ph	361
$C_{31}H_{27}N_2$	$C_6H_4NMe_2(4)$	Ph	H	Ph	H	Ph	**597, 601**
$C_{31}H_{27}N_2O_3S$	[penicillin-type structure: Me, Me, COOH, S, N, O]	Ph	H	Ph	H	Ph	1105, 1126
$C_{31}H_{34}N$	C_8H_{17}	Ph	H	Ph	H	Ph	**611, 612,** 1137
$C_{32}H_{23}N_2S$	[thiazole structure with Ph, Me, N, S]	Ph	H	Ph	H	Ph	608
$C_{32}H_{24}N_3O$	[quinazolinone structure with N, Me]	Ph	H	Ph	H	Ph	586
$C_{32}H_{25}ClN$	$C_6H_3Cl(3)Me(4)$	Ph	H	Ph	$-(CH_2)_2C_6H_4(2)-$		623
$C_{32}H_{25}N_2O$	$NHCOCH=CHPh$	Ph	H	Ph	H	Ph	**471, 667**
$C_{32}H_{26}N$	$C_6H_4Me(4)$	CH=CHPh	H	Ph	H	Ph	500
$C_{32}H_{26}N$	CH_2Ph	Ph	H	Ph	$-(CH_2)_2C_6H_4(2)-$		620
$C_{32}H_{26}N$	$C_6H_4Me(3)$	Ph	H	Ph	$-(CH_2)_2C_6H_4(2)-$		623

TABLE XXVI (continued)

Formula	R^1	R^2	R^3	R^4	R^5	R^6	Reference
$C_{32}H_{26}N$	$C_6H_4Me(4)$	Ph	H	Ph	$-(CH_2)_2C_6H_4(2)-$		623
$C_{32}H_{26}NO_2$	$C_6H_4COOEt(4)$	Ph	H	Ph	H	Ph	1105, 1110
$C_{32}H_{26}N_3$	CH_2CH_2 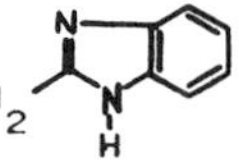	Ph	H	Ph	H	Ph	1119, 1125
$C_{32}H_{26}N_3$	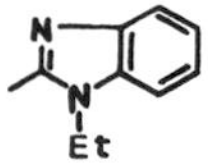 Et	Ph	H	Ph	H	Ph	32, 500, 1119
$C_{32}H_{28}N$	Me	$(CH=CH)_4Ph$	H	Ph	H	Ph	1138
$C_{32}H_{28}NO$	$C_6H_4OH(2)$	$C_6H_4Me(4)$	H	$C_6H_4Me(4)$	H	$C_6H_4Me(4)$	1140
$C_{32}H_{28}NO_3S$	CH_2CH_2OTs	Ph	H	Ph	H	Ph	632, 1105, 1109
$C_{32}H_{28}NO_4$	$C_6H_4OH(4)$	$C_6H_4OMe(4)$	H	$C_6H_4OMe(4)$	H	$C_6H_4OMe(4)$	597
$C_{33}H_{24}N$	1-naphthyl	Ph	H	Ph	H	Ph	1118
$C_{33}H_{24}NO$	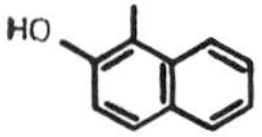	Ph	H	Ph	H	Ph	1140

Formula	Structure						Ref.
$C_{33}H_{24}NO$		Ph	H	Ph	H	Ph	597
$C_{33}H_{24}NO$		Ph	H	Ph	H	Ph	597
$C_{33}H_{34}N_3$	2-pyrimidinyl	Ph	Ph	H	Ph	Ph	583
$C_{33}H_{25}N_2$		Ph	H	Ph	H	Ph	1118
$C_{33}H_{26}N$	Ph	CH=CHPh	H	Ph	H	CH=CHPh	794
$C_{33}H_{26}N$	CH_2Ph	$-C_6H_4CH_2(2)-$	Ph	$-(CH_2)_2C_6H_4(2)-$			620
$C_{33}H_{27}N_2$		Ph	H	Ph	H	Ph	574, 1105, 1127a
$C_{33}H_{28}N_3$		Ph	H	Ph	H	Ph	500
$C_{33}H_{29}N_2$	Ph	$CH=CHC_6H_4NMe_2(4)$	H	Ph	H	Ph	794

TABLE XXVI (continued)

Formula	R¹	R²	R³	R⁴	R⁵	R⁶	Reference
$C_{33}H_{30}NO$	$C_6H_2Et_2(3,5)OH(4)$	Ph	H	Ph	H	Ph	597
$C_{33}H_{30}NO_3$	CH_2–(3,4,5-triOMe-C_6H_2)	Ph	H	Ph	H	Ph	574, 1105, 1121
$C_{33}H_{36}N_3$	Et	$CH=CHC_6H_4NMe_2(4)$	H	Ph	H	$CH=CHC_6H_4NMe_2(4)$	794
$C_{34}H_{25}N_2$	2-pyridyl	Ph	Ph	H	Ph	Ph	583
$C_{34}H_{27}N_2S$	CH_2–(thiazolyl)–S–$C_6H_4Me(4)$	Ph	H	Ph	H	Ph	574, 1105
$C_{34}H_{27}N_4O_2$	benzimidazolyl (N-Et)	$CH=CHC_6H_4NO_2(4)$	H	Ph	H	Ph	500
$C_{34}H_{28}N$	CH_2Ph	$-C_6H_4(CH_2)_2(2)-$		Ph	$-(CH_2)_2C_6H_4(2)-$		620
$C_{34}H_{28}N_3O$	pyrazol-5-one (Me, Me, N-Ph, O)	Ph	H	Ph	H	Ph	1105, 1110
$C_{34}H_{30}N$	Me	$(CH=CH)_5Ph$	H	Ph	H	Ph	1138
$C_{34}H_{32}N$	$CH_2C_6H_4Me(4)$	$C_6H_4Me(4)$	H	$C_6H_4Me(4)$	H	$C_6H_4Me(4)$	1140

Formula						
$C_{34}H_{32}NO$	$C_6H_4OH(4)$	$C_6H_4Et(4)$	$C_6H_4Me(4)$	H	$C_6H_4Et(4)$	597
$C_{34}H_{40}N$	$C_{11}H_{23}$	Ph	Ph	H	Ph	612
$C_{35}H_{25}ClN$	$C_6H_4Cl(4)$	Ph	H	Ph	Ph	583
$C_{35}H_{26}N$	Ph	Ph	H	Ph	Ph	583
$C_{35}H_{26}NO$	OH	Ph	Ph	Ph	Ph	463
$C_{35}H_{26}NO$	$C_6H_4OPh(4)$	Ph	Ph	H	Ph	601
$C_{35}H_{26}NO$	$C_6H_4C_6H_4OH(4,4')$	Ph	H	H	Ph	597
$C_{35}H_{26}N_3$	$C_6H_4N=NPh(4)$	Ph	H	H	Ph	1142
$C_{35}H_{27}N_4$	[structure: N=C–NHPh; 4-pyridyl; benzimidazole with Me, Me, Me]	Ph	H	H	Ph	670
$C_{35}H_{29}N_4O_2$	$CH=CHC_6H_4NO_2(4)$	Ph	H	H	Ph	**500**
$C_{35}H_{34}N$	C_6H_{13}	Ph	H	Ph	Ph	583
$C_{35}H_{34}NO$	$C_6H_2(i\text{-}Pr)_2(3,5)OH(4)$	Ph	Ph	H	Ph	597

TABLE XXVI (continued)

Formula	R^1	R^2	R^3	R^4	R^5	R^6	Reference
$C_{36}H_{27}ClN$	$CH_2C_6H_4Cl(4)$	Ph	Ph	H	Ph	Ph	583
$C_{36}H_{27}ClN_3$	$N=C(Ph)NHC_6H_4Cl(2)$	Ph	H	Ph	H	Ph	669, 670
$C_{36}H_{27}ClN_3$	$N=C(Ph)NHC_6H_4Cl(4)$	Ph	H	Ph	H	Ph	669, 670
$C_{36}H_{28}N$	CH_2Ph	Ph	Ph	H	Ph	Ph	583
$C_{36}H_{28}NO$	Me	Ph	H	[pyran ring with Ph substituents, CH]	H	Ph	1143
$C_{36}H_{28}N_3$	$N=C(Ph)NHPh$	Ph	H	Ph	H	Ph	669, 670
$C_{36}H_{30}NO$	[hydroxynaphthyl, HO]	$C_6H_4Me(4)$	H	$C_6H_4Me(4)$	H	$C_6H_4Me(4)$	1140
$C_{36}H_{32}N$	Me	$(CH=CH)_6Ph$	H	Ph	H	Ph	1138
$C_{36}H_{34}NO$	[phenol with $(CH_2)_7$ bridge, OH]	Ph	H	Ph	H	Ph	597

Formula							Ref.
$C_{36}H_{36}N$	C_7H_{15}	Ph	Ph	H	Ph	Ph	583
$C_{37}H_{30}N$	Et	Ph	Ph	Ph	Ph	Ph	609
$C_{37}H_{30}N_3$	$N=C(Ph)NHC_6H_4Me(2)$	Ph	H	Ph	H	Ph	669, 670
$C_{37}H_{30}N_3$	$N=C(Ph)NHC_6H_4Me(3)$	Ph	H	Ph	H	Ph	669, 670
$C_{37}H_{30}N_3$	$N=C(Ph)NHC_6H_4Me(4)$	Ph	H	Ph	H	Ph	670
$C_{37}H_{30}N_3$	$N=CNHPh \atop \quad C_6H_4Me(4)$	Ph	H	Ph	H	Ph	670
$C_{37}H_{30}N_3O$	$N=C(Ph)NHC_6H_4OMe(4)$	Ph	H	Ph	H	Ph	670
$C_{37}H_{31}N_2$	Me	Ph	CH= [ring: Ph, N–Me, Ph]	H	Ph	H	1143
$C_{37}H_{36}N_3$	Ph	$CH=CHC_6H_4NMe_2(4)$	H	Ph	H	$CH=CHC_6H_4NMe_2(4)$	794
$C_{37}H_{38}N$	C_8H_{17}	Ph	Ph	Ph	Ph	Ph	583

TABLE XXVI (continued)

Formula	R^1	R^2	R^3	R^4	R^5	R^6	Reference
$C_{37}H_{38}NO$	$C_6H_2(t\text{-}Bu)_2(3,5)OH(4)$	Ph	H	Ph	H	Ph	597
$C_{38}H_{30}N$	$CH_2CH=CH_2$	Ph	Ph	Ph	Ph	Ph	609
$C_{38}H_{30}NO$	Me	Ph	H	CH=CHCH (2,6-diphenylpyran-4-yl)	H	Ph	1143
$C_{38}H_{38}NO$	(2-hydroxyphenyl bridged by $(CH_2)_9$)	Ph	H	Ph	H	Ph	597
$C_{39}H_{28}N_3$	(2-methyl-4,6-diphenylpyrimidin-5-yl)	Ph	H	Ph	H	Ph	673
$C_{39}H_{33}N_2$	Me	Ph	H	CH=CHCH (N-Me-2,6-diphenyl)	H	Ph	1143
$C_{39}H_{34}N$	Bu	Ph	H	Ph	H	Ph	609

Formula	R	R'	R''	R'''	R''''	R'''''	Ref.
$C_{39}H_{34}N_3$	$N=C(Ph)NHC_6H_2Me_3(2,4,6)$	Ph	H	Ph	H	Ph	670
$C_{39}H_{35}N_4$	$N=NC_6H_4C_6H_4NEt_2(4,4')$	Ph	H	Ph	H	Ph	1105
$C_{39}H_{40}N_3$	[benzimidazole structure, C_9H_{19}]	Ph	H	Ph	H	Ph	500
$C_{39}H_{42}NO$	$C_6H_4OH(4)$	$C_6H_4(i\text{-}C_5H_{11})(4)$	H	Ph	H	$C_6H_4(i\text{-}C_5H_{11})(4)$	597
$C_{4C}H_{29}N_2$	2-pyridyl	Ph	Ph	Ph	Ph	Ph	609
$C_{4C}H_{32}NO$	Me	Ph	H	$(CH=CH)_2CH$ [pyranyl: Ph, Ph, O, Ph]	H	Ph	1143
$C_{40}H_{44}NO$	$C_6H_2(t\text{-}Bu)_2(3,5)OH(4)$	$C_6H_4Me(4)$	H	$C_6H_4Me(4)$	H	$C_6H_4Me(4)$	601
$C_{41}H_{29}BrNO$	$C_6H_2Ph_2(3,5)OH(4)$	Ph	H	$C_6H_4Br(4)$	H	Ph	601
$C_{41}H_{30}NO$	$C_6H_2Ph_2(3,5)OH(4)$	Ph	H	Ph	H	Ph	579, 1144
$C_{41}H_{35}N_2$	Me	Ph	H	$(CH=CH)_2CH$ [pyridinium: Ph, Ph, N–Me, Ph]	H	Ph	1143

TABLE XXVI (continued)

Formula	R¹	R²	R³	R⁴	R⁵	R⁶	Reference
$C_{41}H_{44}NO$	(structure: phenol with –OH and –(CH₂)₁₂– bridge)	Ph	H	Ph	H	Ph	579
$C_{42}H_{30}N_3$	(structure: benzimidazole with Ph, Ph)	Ph	H	Ph	H	Ph	1119
$C_{42}H_{32}N$	CH_2Ph	Ph	Ph	Ph	Ph	Ph	609
$C_{42}H_{34}N_2$	NHPh	Me	H	(cyclopentadienyl, Ph/Ph/Ph/Ph)	H	Me	1059
$C_{43}H_{35}N$	CH_2Ph	Me	H	(cyclopentadienyl, Ph/Ph/Ph/Ph)	H	Me	1059
$C_{43}H_{35}N$	$C_6H_4Me(4)$	Me	H	(cyclopentadienyl, Ph/Ph/Ph/Ph)	H	Me	1059
$C_{43}H_{36}N_2$	N(Me)Ph	Me	H	(cyclopentadienyl, Ph/Ph/Ph/Ph)	H	Me	1059
$C_{44}H_{36}NO$	$C_6H_2Ph_2(3,5)OH(4)$	$C_6H_4Me(4)$	H	$C_6H_4Me(4)$	H	$C_6H_4Me(4)$	601

a To save space the anion, and hence the melting point, are not listed.

TABLE XXVII

N,N'-Linked Bispyridinium Salts Obtained from Pyrylium Salts

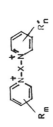

Formula	R_m	X	R'_n	Reference
$C_{17}H_{24}N_4O$	$2,4,6-Me_3$	$-NHCONH-$	$2,4,6-Me_3$	1104
$C_{18}H_{24}N_4O_2$	$2,4,6-Me_3$	$-NHCOCONH-$	$2,4,6-Me_3$	1104
$C_{18}H_{26}N_2$	$2,4,6-Me_3$	$-(CH_2)_2-$	$2,4,6-Me_3$	574, 1105
$C_{20}H_{30}N_2$	$2,4,6-Me_3$	$-(CH_2)_4-$	$2,4,6-Me_3$	574, 1105
$C_{20}H_{30}N_4$	$2,4,6-Me_3$	$-N\diagdown N-$	$2,4,6-Me_3$	1117
$C_{22}H_{26}N_2$	$2,4,6-Me_3$	$-C_6H_4(4)-$	$2,4,6-Me_3$	587, 1145
$C_{22}H_{26}N_2$	$2,4,6-Me_3$	$-C_6H_4(3)-$	$2,4,6-Me_3$	1145
$C_{22}H_{34}N_2$	$2,4,6-Me_3$	$-(CH_2)_6-$	$2,4,6-Me_3$	574, 1105

TABLE XXVII (continued)

Formula	R_m	X	R'_n	Reference
$C_{24}H_{38}N_2$	2,4,6-Me$_3$	$-(CH_2)_8-$	2,4,6-Me$_3$	574, 1105
$C_{28}H_{28}Cl_2N_2$	2,4,6-Me$_3$		2,4,6-Me$_3$	1145a
$C_{28}H_{30}N_2$	2,4,6-Me$_3$	$-C_6H_4(4)-C_6H_4(4)-$	2,4,6-Me$_3$	587
$C_{28}H_{30}N_2$	2,4,6-Me$_3$	$-C_6H_4(4)-C_6H_4(2)-$	2,4,6-Me$_3$	1145a
$C_{28}H_{30}N_2O_2S$	2,4,6-Me$_3$	$-C_6H_4(4)-SO_2-C_6H_4(4)-$	2,4,6-Me$_3$	587
$C_{28}H_{30}N_2S_2$	2,4,6-Me$_3$	$-C_6H_4(4)-SS-C_6H_4(4)-$	2,4,6-Me$_3$	587
$C_{28}H_{30}N_4$	2,4,6-Me$_3$	$-C_6H_4(4)-N=N-C_6H_4(4)-$	2,4,6-Me$_3$	587
$C_{29}H_{32}N_2$	2,4,6-Me$_3$	$-C_6H_4(4)-CH_2-C_6H_4(4)-$	2,4,6-Me$_3$	587
$C_{30}H_{34}N_2$	2,4,6-Me$_3$	$-C_6H_4(4)-(CH_2)_2-C_6H_4(4)-$	2,4,6-Me$_3$	1145a
$C_{30}H_{34}N_2O_2$	2,4,6-Me$_3$		2,4,6-Me$_3$	1145a

Formula		Linking group		Ref
$C_{32}H_{30}N_2$	2-Me-4,6-Ph$_2$	-C$_6$H$_4$(4)-	2,4,6-Me$_3$	1145
$C_{32}H_{30}N_2$	2-Me-4,6-Ph$_2$	-C$_6$H$_4$(3)-	2,4,6-Me$_3$	1145
$C_{35}H_{36}N_2$	2,4,6-Me$_3$	-C$_6$H$_4$(4)-CH(Ph)-C$_6$H$_4$(4)-	2,4,6-Me$_3$	1145a
$C_{37}H_{32}N_2$	2,4,6-Ph$_3$	-C$_6$H$_4$(3)-	2,4,6-Me$_3$	1145
$C_{37}H_{41}N_3$	2,4,6-Me$_3$	-C$_6$H$_4$(4)-CH-C$_6$H$_4$(4)- / C$_6$H$_4$NMe$_2$(4)	2,4,6-Me$_3$	1145a
$C_{40}H_{38}N_4$	2-Me-4,6-Ph$_2$	-N⟨piperazine⟩N-	2-Me-4,6-Ph$_2$	1117
$C_{43}H_{46}N_3$	2,4,6-Me$_3$	-C$_6$H$_4$(4)-CH-C$_6$H$_4$(4)- (mesityl pyridinium cation)		1145a
$C_{49}H_{40}N_2$	2,4,6-Ph$_3$	-(CH$_2$)$_3$-	2,4,6-Me$_3$	609
$C_{50}H_{42}N_2$	2,4,6-Ph$_3$	-(CH$_2$)$_4$-	2,4,6-Ph$_3$	607, 608, 612, 622, 627

TABLE XXVII (continued)

Formula	R_m	X	R'_n	Reference
$C_{50}H_{42}N_4$	$2,4,6-Ph_3$	$-N\!\!\diagup\!\!\diagdown\!\!N-$ (piperazine)	$2,4,6-Ph_3$	1117
$C_{51}H_{44}N_2$	$2,4,6-Ph_3$	$-(CH_2)_5-$	$2,4,6-Ph_3$	608, 609,
$C_{52}H_{38}N_2$	$2,4,6-Ph_3$	$-C_6H_4(4)-$	$2,4,6-Ph_3$	1139, 1145
$C_{52}H_{38}N_2$	$2,4,6-Ph_3$	$-C_6H_4(3)-$	$2,4,6-Ph_3$	608, 1139
$C_{52}H_{46}N_2$	$2,4,6-Ph_3$	$-(CH_2)_6-$	$2,4,6-Ph_3$	612
$C_{58}H_{42}N_2$	$2,4,6-Ph_3$	$-C_6H_4(4)-C_6H_4(4)-$	$2,4,6-Ph_3$	587
$C_{58}H_{42}N_2O_2S$	$2,4,6-Ph_3$	$-C_6H_4(4)-SO_2-C_6H_4(4)-$	$2,4,6-Ph_3$	587
$C_{58}H_{42}N_2S_2$	$2,4,6-Ph_3$	$-C_6H_4(4)-SS-C_6H_4(4)-$	$2,4,6-Ph_3$	587, 598
$C_{58}H_{42}N_4$	$2,4,6-Ph_3$	$-C_6H_4(4)-N=N-C_6H_4(4)-$	$2,4,6-Ph_3$	587
$C_{58}H_{58}N_2$	$2,4,6-Ph_3$	$-(CH_2)_{12}-$	$2,4,6-Ph_3$	609
$C_{59}H_{44}N_2$	$2,4,6-Ph_3$	$-C_6H_4(4)-CH_2-C_6H_4(4)-$	$2,4,6-Ph_3$	587
$C_{60}H_{46}N_2O_2S_2$	$2,6-Ph_2-4-C_6H_4OMe(4)$	$-C_6H_4(4)-SS-C_6H_4(4)-$	$2,6-Ph_2-4-C_6H_4OMe(4)$	598

TABLE XXVIII

PHOSPHABENZENES OBTAINED FROM PYRYLIUM SALTS

Formula	R^1	R^2	R^3	R^4	R^5	M.P. (°C)	Reference
$C_8H_{11}P$	Me	H	Me	H	Me	oil	1146
$C_{13}H_{13}P$	Me	H	Ph	H	Me	62–63	1147
$C_{17}H_{29}P$	t-Bu	H	t-Bu	H	t-Bu	88	152
$C_{18}H_{15}P$	Me	H	Ph	H	Ph	79–81	1148
$C_{18}H_{15}P$	Ph	H	Me	H	Ph	118–120	1148
$C_{18}H_{23}P$	t-Bu	H	i-Pr	H	Ph	88	681
$C_{19}H_{17}P$	Ph	H	Et	H	Ph	65–66	1149
$C_{19}H_{25}OP$	t-Bu	H	$C_6H_4OH(4)$	H	t-Bu	141	1150

TABLE XXVIII (continued)

Formula	R^1	R^2	R^3	R^4	R^5	M.P. (°C)	Reference
$C_{19}H_{25}P$	t-Bu	H	Ph	H	t-Bu	104-105	681
$C_{20}H_{19}P$	Ph	H	i-Pr	H	Ph		1149
$C_{20}H_{27}OP$	t-Bu	H	$C_6H_4OMe(2)$	H	t-Bu		1150
$C_{20}H_{27}OP$	t-Bu	H	$C_6H_4OMe(4)$	H	t-Bu	116	681
$C_{21}H_{27}O_2P$	t-Bu	H	$C_6H_4OCOMe(4)$	H	t-Bu	127	1150
$C_{23}H_{14}Cl_3P$	$C_6H_4Cl(4)$	H	$C_6H_4Cl(4)$	H	$C_6H_4Cl(4)$	181-182	681
$C_{23}H_{16}BrP$	Ph	H	$C_6H_4Br(4)$	H	Ph	148-149	1151
$C_{23}H_{16}ClP$	Ph	H	Ph	H	$C_6H_4Cl(4)$	166-167	681
$C_{23}H_{17}P$	Ph	H	Ph	H	Ph	171-172	683, 1148 1152
$C_{23}H_{22}ClP$	t-Bu	H	$C_6H_4Cl(4)$	$-(CH_2)_2C_6H_4(2)-$		150-152	1153
$C_{23}H_{23}P$	t-Bu	H	Ph	$-(CH_2)_2C_6H_4(2)-$		135-136	523, 1153
$C_{24}H_{19}OP$	Ph	H	$C_6H_4OMe(4)$	H	Ph	106-110	681, 682 683, 1154

$C_{24}H_{19}OP$	Ph	H	Ph	H	$C_6H_4OMe(4)$	161–163	681
$C_{24}H_{19}P$	Ph	H	Ph	H	$C_6H_4Me(4)$	155–157	681
$C_{24}H_{19}P$	Ph	H	CH_2Ph	H	Ph	97	1149
$C_{24}H_{25}OP$	t-Bu	H	$C_6H_4OMe(4)$	H	$-(CH_2)_2C_6H_4(2)-$	160–163	1153
$C_{24}H_{25}P$	t-Bu	H	$C_6H_4Me(4)$	H	$-(CH_2)_2C_6H_4(2)-$	99–101	1153
$C_{25}H_{21}O_2P$	Ph	H	$C_6H_4OMe(4)$	H	$C_6H_4OMe(4)$	134–136	681
$C_{25}H_{21}O_2P$	$C_6H_4OMe(4)$	H	Ph	H	$C_6H_4OMe(4)$	136–137	681, 683
$C_{25}H_{21}P$	$C_6H_4Me(4)$	H	Ph	H	$C_6H_4Me(4)$	133–134	683
$C_{25}H_{22}NP$	Ph	H	$C_6H_4NMe_2(4)$	H	Ph	116–117	682, 1154
$C_{26}H_{23}O_3P$	$C_6H_4OMe(4)$	H	$C_6H_4OMe(4)$	H	$C_6H_4OMe(4)$	105–106	683
$C_{26}H_{23}P$	$C_6H_4Me(4)$	H	$C_6H_4Me(4)$	H	$C_6H_4Me(4)$	167–170	1153
$C_{27}H_{19}P$	Ph	H	Ph	H	1-naphthyl	163–164	681

TABLE XXVIII (continued)

Formula	R¹	R²	R³	R⁴	R⁵	M.P. (°C)	Reference
$C_{27}H_{20}ClP$	$-C_6H_4(CH_2)_2(2)-$		$C_6H_4Cl(4)$	$-(CH_2)_2C_6H_4(2)-$		194–199	1153
$C_{27}H_{21}P$	$-C_6H_4(CH_2)_2(2)-$		Ph	$-(CH_2)_2C_6H_4(2)-$		193–197	1153
$C_{28}H_{23}OP$	$-C_6H_4(CH_2)_2(2)-$		$C_6H_4OMe(4)$	$-(CH_2)_2C_6H_4(2)-$		204–209	1153
$C_{28}H_{23}P$	$-C_6H_4(CH_2)_2(2)-$		$C_6H_4Me(4)$	$-(CH_2)_2C_6H_4(2)-$		178–181	1153
$C_{29}H_{21}P$	Ph	Ph	Ph	H	Ph	209–210	152, 683, 1148, 1155
$C_{30}H_{23}OP$	Ph	H	$C_6H_4Ph(4)$	H	$C_6H_4OMe(4)$	148–150	681
$C_{35}H_{25}P$	Ph	Ph	Ph	Ph	Ph	253–254	152, 683, 1155
$C_{39}H_{49}P$	$C_6H_3(t-Bu)_2(2,4)$	H	Ph	H	$C_6H_3(t-Bu)_2(2,4)$	220	15
$C_{40}H_{28}P_2$	Ph	H	$C_6H_4(4)$	H	Ph	218	685

TABLE XXIX
NITROBENZENES OBTAINED FROM PYRYLIUM SALTS

Formula	R^1	R^2	R^3	R^4	R^5	M.P. (°C)	Reference
$C_9H_{11}NO_2$	Me	H	Me	H	Me	41-42	700-702
$C_9H_{11}NO_3$	Me	H	OMe	H	Me	50	1156
$C_9H_{11}NO_2S$	Me	H	SMe	H	Me	62	329a
$C_{14}H_{13}NO_2$	Me	H	Ph	H	Me	49	1156
$C_{19}H_{15}NO_2$	Me	H	Ph	H	Ph	96-97	700-702
$C_{21}H_{19}NO_2$	i-Pr	H	Ph	H	Ph	91-92	1157
$C_{22}H_{19}NO_2$	Ph	H	Ph	$-(CH_2)_4-$		165	1141
$C_{22}H_{21}NO_2$	t-Bu	H	Ph	H	Ph	96-97	700-702
$C_{22}H_{21}NO_2$	Ph	H	Bu	H	Ph	190-191	321
$C_{22}H_{21}NO_2$	Ph	H	sec-Bu	H	Ph	126-127	1141
$C_{24}H_{14}Cl_3NO_2$	$C_6H_4Cl(4)$	H	$C_6H_4Cl(4)$	H	$C_6H_4Cl(4)$	205-207	1158
$C_{24}H_{15}Cl_2NO_2$	$C_6H_4Cl(4)$	H	Ph	H	$C_6H_4Cl(4)$	179	1159
$C_{24}H_{16}BrNO_2$	Ph	H	Ph	H	$C_6H_4Br(2)$	110-112	1160
$C_{24}H_{16}BrNO_2$	Ph	H	Ph	H	$C_6H_4Br(3)$	136-137	1160
$C_{24}H_{16}BrNO_2$	Ph	H	Ph	H	$C_6H_4Br(4)$	157-158	700-702

TABLE XXIX (continued)

Formula	R^1	R^2	R^3	R^4	R^5	M.P. (°C)	Reference
$C_{24}H_{16}BrNO_2$	Ph	H	$C_6H_4Br(2)$	H	Ph	243–244	1160
$C_{24}H_{16}BrNO_2$	Ph	H	$C_6H_4Br(3)$	H	Ph	155–156	1160
$C_{24}H_{16}BrNO_2$	Ph	H	$C_6H_4Br(4)$	H	Ph	142	1161
$C_{24}H_{16}ClNO_2$	Ph	H	Ph	H	$C_6H_4Cl(4)$	164–165	700-702
$C_{24}H_{16}N_2O_4$	Ph	H	Ph	H	$C_6H_4NO_2(4)$	166–167	1157
$C_{24}H_{17}NO_2$	Ph	H	Ph	H	Ph	144–145	2-4, 699, 700-702 1162
$C_{24}H_{23}NO_2$	Ph	H	C_6H_{11}	H	Ph	202–204	1157
$C_{25}H_{19}NO_2$	Ph	H	Ph	H	$C_6H_4Me(4)$	126–127	700-702
$C_{25}H_{19}NO_3$	Ph	H	Ph	H	$C_6H_4OMe(4)$	119–120	324
$C_{25}H_{19}NO_3$	Ph	H	$C_6H_4OMe(4)$	H	Ph	120–122	324
$C_{26}H_{21}NO_2$	$C_6H_4Me(4)$	H	Ph	H	$C_6H_4Me(4)$	140–142	700-702
$C_{26}H_{21}NO_4$	$C_6H_4OMe(4)$	H	Ph	H	$C_6H_4OMe(4)$	150	700-702
$C_{26}H_{21}NO_4$	Ph	H	$C_6H_4OMe(4)$	H	$C_6H_4OMe(4)$	115–116	700-702
$C_{26}H_{22}N_2O_2$	Ph	H	$C_6H_4NMe_2(4)$	H	Ph	160–161	1163
$C_{27}H_{23}NO_2$	$C_6H_4Me(4)$	H	$C_6H_4Me(4)$	H	$C_6H_4Me(4)$	136–138	1158
$C_{27}H_{23}NO_5$	$C_6H_4OMe(4)$	H	$C_6H_4OMe(4)$	H	$C_6H_4OMe(4)$	124–126	700-702

TABLE XXIX (continued)

Formula	R^1	R^2	R^3	R^4	R^5	M.P. (°C)	Reference
$C_{42}H_{29}NO_2$	$C_6H_4Ph(4)$	H	$C_6H_4Ph(4)$	H	$C_6H_4Ph(4)$	138–140	1164
$C_{42}H_{29}NO_3$	Ph	H	PhC= (2,6-diphenyl-4H-pyran)	H	Ph	202–225	1165
$C_{28}H_{19}NO_2$	Ph	H	Ph	H	1-naphthyl	164–165	1164
$C_{30}H_{21}NO_2$	Ph	H	Ph	H	$C_6H_4Ph(4)$	189–190	1164
$C_{30}H_{21}NO_2$	Ph	Ph	Ph	H	Ph	221	1156
$C_{36}H_{24}N_2O_4$	Ph	H	2,6-diphenyl-4-nitrophenyl	H	Ph	342–344	1164
$C_{36}H_{25}NO_2$	Ph	Ph	Ph	Ph	Ph	292	1156
$C_{36}H_{25}NO_2$	$C_6H_4Ph(4)$	H	Ph	H	$C_6H_4Ph(4)$	264–265	1164

TABLE XXX
Azulenes Obtained from Pyrylium Salts

Formula	R^1	R^2	R^3	R^4	M.P. (°C)	Reference
$C_{11}H_{10}$	H	H	OMe	H	82–83	413
$C_{13}H_{14}$	H	Me	Me	Me	81–82	729, 730
$C_{14}H_{16}$	Me	Me	Me	Me	100–101	729
$C_{14}H_{16}$	H	Me	Et	Me	36–38	1166
$C_{14}H_{16}O$	H	Me	OEt	Me	88–89	729
$C_{16}H_{14}S$	H	Me	2-thienyl	Me		371
$C_{16}H_{20}$	H	Me	n-Bu	Me	oil	1166
$C_{16}H_{20}$	H	Me	t-Bu	Me	33–34	729
$C_{18}H_{16}$	H	Me	Ph	Me	100–101	729
$C_{19}H_{15}NS$	H	Me	2-benzthiazolyl	Me		371
$C_{20}H_{18}$	H	Me	Me	CH=CHPh	119–120	1167
$C_{20}H_{20}$	H	Me	CH_2CH_2Ph	Me	83–84	1093
$C_{22}H_{20}$	H	Me	Me	$(CH=CH)_2Ph$	138–139	1167
$C_{23}H_{18}$	H	Me	Ph	Ph		729

TABLE XXX (continued)

Formula	R^1	R^2	R^3	R^4	M.P. (°C)	Reference
C$_{27}$H$_{26}$	H	Me	Me		226–227	1167
C$_{28}$H$_{22}$	H	Me	Me		204–205	1167

Note Added In Proof

Synthesis of 2- and 4-carboxypyrylium salts from α-ketoacids and chalcones followed by hydride abstraction using Ph$_3$C$^+$CO$_4^-$ was described.[1168] In handling pyrylium perchlorates special care should be taken to avoid explosions.[1170] Acylation of β-benzoylpropionic acid or ester (or of other related γ-ketoesters) affords pyrylium salts with a condensed lactonic ring: 2-oxo-3H-furo[3,2–c]pyrylium.[1170a] The mercuration of pyrylium salts by mercuric trifluoroacetate was reported.[1171] If β-chlorovinylketones are reacted with 2,6-di-t-butyl-4-methylpyrylium (149) under the conditions described on p. 37, pentamethine pyrylocyanines, vinylogs of 150,[1171a] are obtained.

The addition of the methoxide anion to 2,6-diphenylpyrylium yields a 4H-pyran as the kinetically favored product; and therefrom the acyclic diphenyl-2-pentadien-1,5-dione as the thermodynamically favored product, while 4-methoxy-2,6-diphenylpyrylium gives both 4- and 2-adducts;[1172] 2,4,6-triphenylpyrylium was shown by ¹H-NMR to afford with methoxide a 2H-pyran.[1173] Katritzky has continued to investigate the pyridinium ring as a leaving group,[1174–1180] especially when sterically constrained,[1181] as in 5,6,8,9-tetrahydro-7-phenyl-bisbenzo[a,b]acridinium salts. A primary amine, after reaction with triphenylpyrylium or other pyrylium salts, can be converted to a variety of functional groups in addition to those in Table III (pp. 122–123); alkenes through a mild alternative to the Hofmann degradation,[1182] various sulfur functionalities,[1183] and the hydroxyl group as an alternative to nitrous acid deamination can be mentioned.[1184] Alkyl nitrites cause α-demethylation of 1,2,4,6-tetramethylpyridinium salts yielding a 1,4,6-trimethyl-2H-pyridone.[1185] Pyridinium salts were obtained from aminopyridines,[1186] N-aminoheterocycles,[1187] urea, thiourea, and isothiourea derivatives (in these three last cases, pyrimidines were also formed).[1188] Guanidine also converts pyrylium salts to pyrimidines.[1189] The reaction of pyrylium salts with hydrazine, methylhydrazine, other monosubstituted hydrazines and 1,1-disubstituted hydrazines has been studied in detail.[1190–1193] It was found that 2,4,6-trialkylpyrylium salts having tert- or isopropyl groups in α-positions afford, with hydrazine, exclusively 1,2-diazepines, whereas α-ethyl or α-methyl groups suppress this reaction completely, leading to other products; small yields of 1,2-diazepines can, however, be obtained from hydrazine and the pseudo-bases of α-ethyl- or α-methylpyrylium salts in ethyl ether, a fact which proved to be useful for mass spectral assignments (cf. p. 155).[1193] With pyrylium salts, aminoacetaldehyde diethylacetal affords interesting pyridinium cations which can be converted into the corresponding aldehydes, useful starting materials for e.g., cyclizations to indolizines.[1194] From reactions of type 2,6-[C_5 + C_2] (cf. Table II, p. 88 and Table XXX, p. 366), the formation of an azulene from the pyrylophanium perchlorate (**108**) and cyclopentadiene was reported.[1195] The study of phototranspositions in the pyrylium ring using "ring permutations" to rationalize the results was continued.[1196]

Charge-transfer complexes of 2,4,6-tri-, 2,3,4,6-tetra-, 2,3,5,6-tetra-, and 2,3,4,5,6-pentaphenylpyrylium salts with tetracyanoquinodimethane (TCNQ) were obtained by a new method (TCNQ and a pyrylium pseudobase were refluxed in acetonitrile) and their spectra investigated.[1197] Voltammetric investigations of pyrylium salts were reported.[1198] NMR studies allowed the determination of coupling constants in the unsubstituted pyrylium cation, namely both $^{13}C,^1H$ and $^{13}C,^{13}C$ type J values; $^1J(\text{C-3, C-4}) = 50.4$ Hz is one of the lowest values for ^1J in aromatic systems, proving again the special situation of pyrylium among other six-membered aromatics with one heteroatom.[1199] 2,4,6-Triphenylpyrylium halides evidence in nonpolar aromatic solvents ESR spectra indicative of charge transfer.[1200] Cation-radicals of heterocyclics were reviewed, including those derived from pyrylium salts.[1201] Extensive CNDO/S studies on'the electronic structure of substituted pyrylium salts taking into account the effects of the anion and of the solvent gave good agreement with electronic absorption spectra,[1202] electrochemical properties,[1202] and photoelectron spectra; such XPS spectra of 2,6-diphenyl-4-(*p*-diethylaminophenyl)pyrylium tetrafluoroborate were determined in solid state for comparison with theoretical results.[1203] Among newer applications of pyrylium salts, heptamethinepyrylocyanines were prepared, and their uses as ultrafast saturable absorbers for Nd : lasers were described.[1204]

References

1. A. T. Balaban, W. Schroth, and G. Fischer, *Adv. Heterocycl. Chem.* **10**, 241 (1969).
2. K. Dimroth, *Angew. Chem.* **72**, 331 (1960); K. Dimroth, *in* "Neuere Methoden der präparativen organischen Chemie" (W. Foerst, ed.), Vol. 3, p. 239. Verlag Chemie, Weinheim, 1960.
3. K. Dimroth and K. H. Wolf, *in* "Newer Methods of Preparative Organic Chemistry" (W. Foerst, ed.), Vol. 3, p. 357. Academic Press, New York, 1964.
4. K. Dimroth, Lecture presented at the refreshment course 20/78 of the Gesellschaft Deutscher Chemiker (1978).
5. G. N. Dorofeenko, Yu. A. Zhdanov, V. I. Dulenko, and S. V. Krivun, "Perchloric Acid and Its Compounds in Organic Synthesis" (in Russian). Izd. Rostov. Univ., Rostov-on-Don, 1965; see also G. N. Dorofeenko, S. V. Krivun, V. I. Dulenko, and Yu. A. Zhdanov, *Usp. Khim.* **34**, 219 (1965) [*CA* **62**, 12993 (1965)].
6. G. N. Dorofeenko, E. I. Sadekova, and E. V. Kuznetsov, "Preparative Chemistry of Pyrylium Salts" (in Russian). Izd. Rostov. Univ., Rostov-on-Don, 1972 [*CA* **78**, 43238 (1973)].
7. I. F. Bel'skii, G. N. Dorofeenko, N. S. Prostakov, V. P. Sherstyuk, and Yu. I. Chumakov, "Heterocycles in Organic Syntheses" (in Russian), p. 108. Izd. Tekhnika, Kiev, 1970.
8. G. I. Zhungietu and G. N. Dorofeenko, *Usp. Khim.* **36**, 48 (1967) [*CA* **67**, 11638 (1967).
9. V. V. Mezheritskii, A. L. Vasserman, and G. N. Dorofeenko, *Heterocycles* **12**, 51 (1979).
10. H. C. Van der Plas, "Ring Transformations of Heterocycles," Vol. 2, p. 1. Academic Press, New York, 1973.
11. A. I. Meyers, "Heterocycles in Organic Synthesis," Wiley (Interscience), New York, 1974.
12. A. T. Balaban, *in* "New Trends in Heterocyclic Chemistry" (R. B. Mitra *et al.*, ed.), p. 79. Elsevier, Amsterdam, 1979.
13. A. R. Katritzky, *Tetrahedron Rep.* **36**, 679 (1980).
14. S. V. Krivun, O. F. Alferova, and S. V. Sayapina, *Usp. Khim,* **43**, 1739 (1974). [*CA* **82**, 43101 (1975)].
15. H. Perst, "Oxonium Ions in Organic Chemistry," Verlag Chemie, Weinheim, 1971;

369

J. Staunton, *in* "Comprehensive Organic Chemistry" (P. G. Sammes, ed.), Vol. 4, p. 607. Pergamon, Oxford, 1979.

16. H. Sugimoto, *Kagaku (Kyoto)* **25**, 829 (1970) [*CA* **74**, 87704 (1971)].
17. V. G. Kharachenko, S. N. Chalaya, and T. M. Konovalova, *Khim. Geterotsikl. Soedin.*, 147 (1975) [*CA* **82**, 170522 (1975)].
18. A. T. Balaban, *Math. Chem.* **8**, 159 (1980).
19. A. T. Balaban and F. Kerek, *Rev. Roum. Chim.* **19**, 631 (1974).
20. J. N. Collie and T. Tickle, *J. Chem. Soc.* **75**, 710 (1899).
21. M. M. Evstifeev, G. Kh. Aminova, E. P. Olekhnovich, G. N. Dorofeenko, V. P. Karmazin, and M. I. Knyazhanskii, *Zh. Obshch. Khim.* **46**, 2696 (1976) [*CA* **86**, 154828 (1977)].
22. M. M. Evstifeev, G. Kh. Aminova, G. N. Dorofeenko, and E. P. Olekhnovich, *Zh. Obshch. Khim.* **46**, 2693 (1976) [*CA* **86**, 62698 (1977)].
23. G. Wittig, F. Bangert, and H. E. Richter *Justus Liebigs Ann. Chem.* **446**, 155 (1925).
24. L. A. Flexser, L. P. Hammett, and A. Dingwall, *J. Am. Chem. Soc.* **57**, 2103 (1935).
25. M. E. Perel'son, Yu. N. Sheinker, and A. A. Savina, "Spectra and Structures of Coumarins, Chromones, and Xanthones" (in Russian). Izd. Meditsina, Moscow, 1975 [*CA* **85**, 93698 (1976)].
26. K. A. Hofmann, A. Metzler, and H. Lecher, *Ber. Dtsch. Chem. Ges.* **43**, 178 (1910).
27. F. Kehrmann and A. Duttenhöfer, *Ber. Dtsch. Chem. Ges.* **39**, 1299 (1906).
28. R. M. Anker and A. H. Cook, *J. Chem. Soc.*, 117 (1946).
29. A. Baeyer, *Ber. Dtsch. Chem. Ges.* **43**, 2337 (1910).
30. A. I. Kiprianov and A. I. Tolmachev, *Zh. Obshch. Khim.* **29**, 2868 (1959)[*CA* **54**, 12126 (1960)].
31. D. A. Griffin and J. Staunton, *J. C. S. Chem. Commun.*, 675 (1975).
32. W. H. Pirkle and M. Dines, *J. Heterocycl. Chem.* **6**, 313 (1969).
33. H. Meerwein, G. Hinz, P. Hofmann, E. Kroning, and E. Pfeil, *J. Prakt. Chem.* **147**, 257 (1936).
34. B. Föhlisch and D. Krockenberger, *Chem. Ber.* **101**, 3990 (1968).
35. V. A. Zagorevskii, Ph.D. Thesis, Moscow, 1972.
36. A. I. Buryak, Ph.D. Thesis, Donetsk, 1975.
37. G. Seitz and H. Moennighoff, *Arch. Pharm. (Weinheim, Ger.)* **306**, 389 (1973).
38. J. A. Van Allan, G. A. Reynolds, and D. P. Maier, *J. Org. Chem.* **33**, 4418 (1968).
39. S. V. Krivun, *Khim. Geterotsikl. Soedin.*, 764 (1976) [*CA* **85**, 123805 (1976)].
40. S. N. Baranov, A. I. Buryak, and S. V. Krivun, U.S.S.R. Patent 382,617 (1973) [*CA* **79**, 92008 (1973)].
41. S. V. Krivun, A. I. Buryak, and S. N. Baranov, *Khim. Geterotsikl. Soedin.*, 1199 (1973) [*CA* **80**, 47774 (1974)].
42. J. A. Van Allan, S. Chie Chang, and G. A. Reynolds, *J. Heterocycl. Chem.* **11**, 195 (1974).
43. D. M. McKinnon, *Can. J. Chem.* **48**, 3388 (1970).
44. A. Baeyer and J. Piccard, *Justus Liebigs Ann. Chem.* **384**, 208 (1911).
45. G. A. Reynolds and J. A. Van Allan, *J. Heterocycl. Chem.* **11**, 1075 (1974).
46. G. N. Dorofeenko, A. V. Koblik, B. A. Tertov, and T. I. Polyakova, *Khim. Geterotsikl. Soedin.*, 1580 (1972) [*CA* **78**, 58189 (1973)].
47. G. N. Dorofeenko, A. V. Koblik, B. A. Tertov, and T. I. Polyakova, *Khim. Geterotsikl. Soedin.*, 1016 (1973) [*CA* **79**, 137021 (1973)].
48. A. V. Koblik, T. I. Polyakova, B. A. Tertov, B. V. Mezhov, and G. N. Dorofeenko, *Zh. Org. Khim.* **11**, 2153 (1975) [*CA* **84**, 43782 (1976)].
49. T. I. Polyakova, Ph.D. Thesis, Rostov University, Rostov-on-Don, 1975.

50. G. N. Dorofeenko and V. V. Krasnikov, *Zh. Org. Khim.* **8**, 2620 (1972) [*CA* **78**, 97785 (1973)].
51. V. V. Krasnikov and G. N. Dorofeenko, *Khim. Geterotsikl. Soedin.*, 21 (1979) [*CA* **90**, 168702 (1979)].
52. S. N. Baranov, A. I. Buryak, S. V. Dulskaya, and S. V. Krivun, *Khim. Geterotsikl. Soedin.*, 280 (1972) [*CA* **76**, 153491 (1972)].
53. S. V. Krivun, A. I. Buryak, and S. N. Baranov, *Khim. Geterotsikl. Soedin.*, 1317 (1973) [*CA* **80**, 47782 (1974)].
54. R. Wizinger and H. Tobel, *Helv. Chim. Acta* **40**, 1305 (1957).
55. R. Wizinger and H. J. Angliker, *Helv. Chim. Acta* **49**, 2046 (1966).
56. F. Kröhnke and K. Dickoré, *Chem. Ber.* **92**, 46 (1959).
57. S. V. Krivun, *Dokl. Akad. Nauk SSSR* **180**, 615 (1968) [*CA* **69**, 106425 (1968)].
58. S. V. Krivun, G. N. Dorofeenko, and A. S. Kovalevskii, *Khim. Geterotsikl. Soedin.*, 733 (1970) [*CA* **73**, 98769 (1970)].
59. G. N. Dorofeenko, A. V. Koblik, T. I. Polyakova, and L. A. Murad'yan, *Khim. Geterotsikl. Soedin.*, 1045 (1980) [*CA* **94**, 47071 (1981)].
60. Yu. N. Porshnev, V. A. Churkina, and V. V. Titov, *Khim. Geterotsikl. Soedin.*, 459 (1978) [*CA* **89**, 59817 (1978)].
61. S. V. Krivun, A. I. Buryak, S. V. Sayapina, O. F. Voziyanova, and S. N. Baranov, *Khim. Geterotsikl. Soedin.*, 1004 (1973) [*CA* **80**, 49227 (1974)].
62. G. N. Dorofeenko and E. I. Sadekova, *Khim. Prir. Soedin.*, 714 (1977) [*CA* **88**, 121496 (1978)].
63. S. V. Krivun, O. F. Voziyanova, and S. N. Baranov, *Zh. Obshch. Khim.* **42**, 298 (1972) [*CA* **77**, 34224 (1972)].
64. S. V. Krivun, S. N. Baranov, and O. F. Voziyanova, *Dokl. Akad. Nauk SSSR* **196**, 600 (1971) [*CA* **75**, 35599 (1971)].
65. S. V. Krivun, O. F. Voziyanova, and S. N. Baranov, *Zh. Obshch. Khim.* **42**, 58 (1972) [*CA* **77**, 48587 (1972)].
66. S. N. Baranov, I. A. Dumbai, and S. V. Krivun, *Khim. Geterotsikl. Soedin.*, 1313 (1972) [*CA* **78**, 29544 (1973)].
67. G. W. Fischer and W. Schroth, *Tetrahedron* **32**, 2225 (1976).
68. W. Schroth and G. Fischer, *Z. Chem.* **4**, 27 (1964).
69. K. Undheim and E. T. Oestensen, *Acta Chem. Scand.* **27**, 1385 (1973).
70. K. Undheim and C. E. Carlberg, *Acta Chem. Scand., Ser. B* **B28**, 517 (1974).
71. G. N. Dorofeenko, A. V. Koblik, T. I. Polyakova, and B. A. Tertov, *Zh. Org. Khim.* **10**, 1998 (1974) [*CA* **81**, 169390 (1974)].
72. A. V. Koblik, *Noveishie Usp. Khim. Geterotsikl. Soedin. Kisloroda, Conf. Abstr.*, 29 (1977).
73. M. V. Nekhoroshev and O. Yu. Okhlobystin, *Zh. Org. Khim.* **13**, 1294 (1977) [*CA* **87**, 117749 (1977)].
74. O. V. Drygina, G. N. Dorofeenko, and O. Yu. Okhlobystin, *Khim. Geterotsikl. Soedin.*, 1172 (1977) [*CA* **88**, 23011 (1978)].
75. O. V. Drygina, G. N. Dorofeenko, and O. Yu. Okhlobystin, *Proc. Int. Conf. Organomet. Chem., 8th, 1977* Abstr. 4A, 34 (1977).
76. O. Yu. Okhlobystin, "Electron Transfer in Organic Reactions" (in Russian). Izd. Rostov. Univ., Rostov-on-Don, 1974 [*CA* **82**, 57452 (1975)].
77. N. T. Berberova, A. A. Bumber, M. V. Nekhoroshev, V. B. Panov, and O. Yu. Okhlobystin, *Dokl. Akad. Nauk SSSR* **246**, 108 (1979) [*CA* **91**, 80912 (1979)].
78. L. A. Polyakova, K. A. Bilevich, N. N. Bubnov, G. N. Dorofeenko, and O. Yu. Okhlobystin, *Dokl. Akad. Nauk SSSR* **212**, 370 (1973) [*CA* **79**, 145660 (1973)].

79. G. N. Dorofeenko, G. P. Safaryan, V. F. Voloshinova, and O. Yu. Okhlobystin, *Khim. Geterotsikl. Soedin.*, 999 (1976) [*CA* **85**, 159822 (1976)].
80. F. Klages and H. Träger, *Chem. Ber.* **86**, 1327 (1953).
81. F. Gordzeevich and S. Skrovachevska, *Noveishie Usp. Khim. Geterotsikl. Soedin. Kisloroda,* Conf. Abstr., 18 (1977).
82. A. Williams, *J. Am. Chem. Soc.* **93**, 2733 (1971).
83. J. W. Bunting, *Adv. Heterocycl. Chem.* **25**, 1 (1979).
84. H. Hartmann and D. Förster, G. D. R. Patent 91,668 (1972) [*CA* **78**, 58241 (1973)].
85. J. Liebscher and H. Hartmann, *Z. Chem.* **13**, 132, 342 (1973).
85a. J. Liebscher and H. Hartmann, *J. Prakt. Chem.* **318**, 705 (1976); G. D. R. Patent 106,831 (1974) [*CA* **82**, 97939 (1975)].
85b. W. Schroth and R. Spitzner, *Z. Chem.* (in press).
86. D. Farcasiu, *Tetrahedron* **25**, 1209 (1969).
87. J. Carretto and M. Simalty, *Tetrahedron Lett.*, 3445 (1973).
88. D. Farcasiu, A. Vasilescu, and A. T. Balaban, *Tetrahedron* **27**, 681 (1971).
89. V. G. Kharchenko, S. N. Chalaya, L. G. Chichenkova, and A. S. Tatarinov, *Zh. Org. Khim.* **11**, 444 (1975) [*CA* **82**, 170579 (1975)].
90. V. G. Kharchenko, S. K. Klimenko, M. N. Berezhnaya, and I. Ya. Evtushenko, *Zh. Org. Khim.* **10**, 1302 (1974) [*CA* **81**, 105201 (1974)].
91. S. K. Klimenko, N. M. Yartseva, M. N. Berezhnaya, M. E. Stankevich, and V. G. Kharchenko, *Zh. Org. Khim.* **10**, 2206 (1974) [*CA* **82**, 57522 (1975)].
92. G. N. Dorofeenko and E. V. Kuznetsov, *Zh. Org. Khim.* **5**, 191 (1969) [*CA* **70**, 87459 (1969)].
93. G. N. Dorofeenko, E. V. Kuznetsov, and V. E. Ryabinina, *Tetrahedron Lett.*, 711 (1969).
94. A. T. Balaban and C. D. Nenitzescu, *Justus Liebigs Ann. Chem.* **625**, 74 (1959).
95. G. N. Dorofeenko, S. M. Luk'yanov, E. P. Olekhnovich, and T. I. Davidenko, *Khim. Geterotsikl. Soedin.*, 735 (1973) [*CA* **79**, 91897 (1973)].
96. G. N. Dorofeenko, V. V. Mezheritskii, E. P. Olekhnovich, and A. L. Vasserman, *Zh. Org. Khim.* **9**, 395 (1973) [*CA* **78**, 124400 (1973)].
97. A. Rieche, H. Gross, and E. Höft, *Chem. Ber.* **93**, 88 (1960).
98. G. N. Dorofeenko and G. P. Safaryan, *Khim. Geterotsikl. Soedin.*, 278 (1970) [*CA* **72**, 111216 (1970)].
99. G. N. Dorofeenko and G. P. Safaryan, *Khim. Geterotṣikl. Soedin.*, 585 (1970) [*CA* **73**, 76986 (1970)].
100. R. Grewe, *Naturwissenchaften* **33**, 333 (1946).
101. R. Grewe and A. Mondon, *Chem. Ber.* **81**, 279 (1948).
102. G. N. Dorofeenko, G. P. Safaryan, and T. I. Polyakova, *Khim. Geterotsikl. Soedin.*, 1461 (1972) [*CA* **78**, 58188 (1973)].
103. G. P. Safaryan and G. N. Dorofeenko, *Khim. Geterotsikl. Soedin.*, 1323 (1976) [*CA* **86**, 72367 (1977)].
104. D. V. Pruchkin, E. V. Kuznetsov, and G. N. Dorofeenko, *Khim. Geterotsikl. Soedin.*, 275 (1978) [*CA* **88**, 169895 (1978)].
105. P. P. Hopf and R. J. W. Le Fèvre, *J. Chem. Soc.* 1989 (1938).
106. S. M. Luk'yanov and G. N. Dorofeenko, *Zh. Org. Khim.* **9**, 1360 (1973) [*CA* **79**, 105039 (1973)].
107. W. Dilthey and J. Fischer, *Ber. Dtsch. Chem. Ges.* **57**, 1653 (1924).
108. W. Schroth and G. Fischer, *Z. Chem.* **3**, 147, 277 (1963).
109. S. V. Krivun, Zh. V. Shiyan, and G. N. Dorofeenko, *Zh. Obshch. Khim.* **34**, 167 (1964) [*CA* **60**, 10641 (1964)].

110. G. N. Dorofeenko, L. N. Volovel'skii, and B. M. Savin, *Zh. Obshch. Khim.* **38,** 2686 (1968) [*CA* **70,** 88068 (1969)].

111. G. N. Dorofeenko, L. N. Volovel'skii, and B. M. Savin, *Zh. Obshch. Khim.* **39,** 656 (1969) [*CA* **71,** 50323 (1969)].

112. B. M. Savin, L. N. Volovel'skii, and G. N. Dorofeenko, *Khim. Geterotsikl. Soedin.,* 147 (1970) [*CA* **72,** 111679 (1970)].

113. B. M. Savin, G. N. Dorofeenko, and L. N. Volovel'skii, *Khim. Geterotsikl. Soedin., Sb.* **2,** 242 (1970) [*CA* **77,** 88781 (1972)].

114. V. V. Mezheritskii and G. N. Dorofeenko, *Khim. Geterotsikl. Soedin., Sb.* **2,** 232 (1970) [*CA* **76,** 140412 (1970)].

115. V. V. Mezheritskii, A. L. Vasserman, and G. N. Dorofeenko, *Khim. Geterotsikl. Soedin.,* 1163 (1972) [*CA* **77,** 164381 (1972)].

116. G. N. Dorofeenko and L. N. Etmetchenko, *Khim. Geterotsikl. Soedin., Sb.* **2,** 250 (1970).

117. G. N. Dorofeenko and E. P. Olekhnovich, *Zh. Org. Khim.* **6,** 192 (1970) [*CA* **72,** 90202 (1970)].

118. V. V. Mezheritskii, E. P. Olekhnovich, S. M. Luk'yanov, and G. N. Dorofeenko, "Ortho Esters in Organic Synthesis" (in Russian). Izd. Rostov. Univ., Rostov-on-Don, 1976 [*CA* **87,** 38241 (1977)].

119. G. N. Dorofeenko, E. P. Olekhnovich, and L. I. Laukhina, *Khim. Geterotsikl. Soedin.,* 435 (1971) [*CA* **76,** 34054 (1972)].

120. G. N. Dorofeenko, V. V. Mezheritskii, Yu. I. Ryabukhin, and E. P. Olekhnovich, *Khim. Geterosikl. Soedin.,* 1314 (1973) [*CA* **80,** 47777 (1974)].

121. G. N. Dorofeenko and V. V. Mezheritskii, *Khim. Geterotsikl. Soedin., Sb.* **2,** 217 (1970) [*CA* **76,** 140416 (1972)].

122. W. Schroth and G. W. Fischer, *Chem. Ber.* **102,** 1214 (1969).

123. G. I. Zhungietu and B. P. Sukhanyuk, *Khim. Geterotsikl. Soedin.,* 1030 (1972) [*CA* **77,** 152030 (1972)].

124. G. N. Dorofeenko, V. I. Volbushko, V. I. Dulenko, and E. N. Kornilova, *Khim. Geterotsikl. Soedin.,* 1181 (1976) [*CA* **86,** 43592 (1977)].

125. V. A. Chuiguk and N. N. Vlasova, *Khim. Geterotsikl. Soedin.,* 1484 (1977) [*CA* **88,** 89612 (1978)].

126. G. I. Zhungietu, B. P. Sukhanyuk, F. N. Chukhrii, and L. N. Volovel'skii, *Khim. Geterotsikl. Soedin.,* 219 (1973) [*CA* **78,** 136517 (1973)].

127. V. M. Vlasov, *Zh. Vses. Khim. O-va.* **15,** 708 (1970) [*CA* **74,** 53412 (1971)].

128. W. Schroth and G. Fischer, *Angew. Chem.* **75,** 574 (1963); *Angew. Chem. Int. Ed. Engl.* **2,** 394 (1963).

129. W. Schroth and G. W. Fischer, *Chem. Ber.* **102,** 575 (1969).

129a. G. W. Fischer and W. Schroth, *Chem. Ber.* **102,** 590 (1969).

129b. R. Gompper and W. Elser, *Justus Liebigs Ann. Chem.* **725,** 73 (1969).

129c. W. Schroth and U. Burkhardt, *Z. Chem.* (in press).

130. W. Schroth, G. W. Fischer, and J. Rottmann, *Chem. Ber.* **102,** 1202 (1969).

131. H. Teufel, *Chem.-Ztg.* **98,** 606 (1974).

132. M. Pulst and M. Weissenfels, *Z. Chem.* **16,** 337 (1976).

133. V. I. Minkin and G. N. Dorofeenko, *Usp. Khim.* **29,** 1301 (1960) [*CA* **55,** 12265 (1961)].

134. G. N. Dorofeenko and A. I. Pyshchev, *Zh. Org. Khim.* **9,** 1084 (1973) [*CA* **79,** 66123 (1973)].

135. G. N. Dorofeenko and A. I. Pyshchev, *Khim. Geterotsikl. Soedin.,* 1031 (1974) [*CA* **81,** 169392 (1974)].

136. J. Andrieux, J.-P. Battioni, M. Giraud, and D. Molho, *Bull. Soc. Chim. Fr.*, 2093 (1973).
137. A. I. Pyshchev, Ph.D. Thesis, Rostov University, Rostov-on-Don, 1977.
138. M. Weissenfels, P. Schneider, D. Schmiedl, and H. Altmann, *Z. Chem.* **12**, 263 (1972).
139. G. N. Dorofeenko, A. L. Shinkarenko, A. L. Kazakov, A. I. Pyshchev, and V. V. Mezheritskii, *Khim. Prir. Soedin.* **10**, 160 (1974) [*CA* **81**, 25496 (1974)].
140. G. N. Dorofeenko, V. V. Krasnikov, and A. I. Pyshchev, *Khim. Geterotsikl. Soedin.*, 599 (1977) [*CA* **87**, 85121 (1977)].
141. L. Yu. Ukhin, A. I. Pyshchev, V. V. Krasnikov, Zh. I. Orlova, and G. N. Dorofeenko, *Dokl. Akad. Nauk SSSR* **234**, 1351 (1977) [*CA* **87**, 168162 (1977)].
142. J. Liebscher and H. Hartmann, *Synthesis*, 241 (1979).
143. H. Hartmann and D. Förster, *J. Prakt. Chem.* **313**, 1110 (1971).
144. S. A. Barker and T. Riley, *J. C. S. Perkin I*, 809 (1972).
145. H. Strzelecka and M. Simalty, *Bull. Soc. Chim. Fr.*, 4122 (1968).
146. K. Dimroth, K. Vogel, and W. Krafft, *Chem. Ber.* **101**, 2215 (1968).
147. G. N. Dorofeenko and L. B. Olekhnovich, *Khim. Geterotsikl. Soedin.*, 883 (1972) [*CA* **77**, 164383 (1972)].
148. J. A. Van Allan and G. A. Reynolds, *J. Org. Chem.* **33**, 1102 (1968).
149. V. M. Vlasov, *Zh. Vses. Khim. O-va.* **15**, 476 (1970) [*CA* **75**, 35600 (1971)].
150. G. N. Dorofeenko, E. P. Olekhnovich, and L. I. Laukhina, *Zh. Org. Khim.* **7**, 1296 (1971) [*CA* **75**, 98392 (1971)].
151. Z. Csürös, G. Deák, and P. Sallay, *Acta Chim. Acad. Sci. Hung.* **70**, 123 (1971).
152. K. Dimroth and W. Mach, *Angew. Chem.* **80**, 489 (1968); *Angew. Chem. Int. Ed. Engl.* **7**, 460 (1968).
153. V. V. Mezheritskii and G. N. Dorofeenko, *Zh. Org. Khim.* **3**, 1533 (1967) [*CA* **68**, 68821 (1968)].
154. G. N. Dorofeenko and N. A. Lopatina, *Khim. Geterotsikl. Soedin.*, 160 (1971) [*CA* **75**, 48821 (1971)].
155. G. I. Zhungietu and G. V. Lazur'evskii, *Zh. Vses. Khim. O-va.* **13**, 597 (1968) [*CA* **70**, 47246 (1969)].
156. H. Strzelecka and M. Simalty, *Bull. Soc. Chim. Fr.*, 832 (1968).
157. K. Dimroth and P. Heinrich, *Angew. Chem.* **78**, 714 (1966); *Angew. Chem. Int. Ed. Engl.* **5**, 676 (1966).
158. G. N. Dorofeenko and S. M. Luk'yanov, *Zh. Org. Khim.* **7**, 419 (1971) [*CA* **74**, 125329 (1971)].
159. G. N. Dorofeenko, S. M. Luk'yanov, and E. S. Matskovskaya, *Zh. Org. Khim.* **8**, 1960 (1972) [*CA* **78**, 29545 (1973)].
160. G. N. Dorofeenko and S. M. Luk'yanov, *Khim. Geterotsikl. Soedin.*, 886 (1972) [*CA* **77**, 164379 (1972)].
161. G. N. Dorofeenko, S. M. Luk'yanov, and T. I. Davidenko, *Zh. Org. Khim.* **9**, 2433 (1973) [*CA* **80**, 70645 (1974)].
162. G. N. Dorofeenko, S. M. Luk'yanov, and T. I. Davidenko, *Zh. Org. Khim.* **11**, 163 (1975) [*CA* **82**, 170592 (1975)].
163. S. M. Luk'yanov, L. N. Etmetchenko, A. V. Koblik, O. A. Rakina, and G. N. Dorofeenko, *Zh. Org. Khim.* **11**, 908 (1975) [*CA* **83**, 9688 (1975)].
164. S. M. Luk'yanov, L. N. Etmetchenko, A. V. Koblik, O. A. Rakina, and G. N. Dorofeenko, *Zh. Org. Khim.* **11**, 1962 (1975) [*CA* **84**, 17071 (1976)].
165. G. N. Dorofeenko, E. I. Sadekova, and V. I. Beletskaya, *Zh. Org. Khim.* **6**, 1118 (1970) [*CA* **73**, 35158 (1970)].

166. G. N. Dorofeenko, S. M. Luk'yanov, L. N. Etmetchenko, A. V. Koblik, and O. A. Rakina, *Zh. Org. Khim.* **12**, 685 (1976) [*CA* **85**, 32758 (1976)].
167. S. M. Luk'yanov and G. N. Dorofeenko, *Zh. Org. Khim.* **11**, 1985 (1975) [*CA* **84**, 30807 (1976)].
168. G. N. Dorofeenko, S. V. Krivun, and V. V. Mezheritskii, *Zh. Obshch. Khim.* **35**, 632 (1965) [*CA* **63**, 2947 (1965)].
169. S. M. Luk'yanov, and G. N. Dorofeenko, *Zh. Org. Khim.* **12**, 684 (1976) [*CA* **85**, 32757 (1976)].
170. S. M. Luk'yanov, A. V. Koblik, and G. N. Dorofeenko, *Zh. Org. Khim.* **12**, 2267 (1976) [*CA* **86**, 72361 (1976)].
171. A. T. Balaban and C. D. Nenitzescu, *Justus Liebigs Ann. Chem.* **625**, 66 (1959).
172. P. F. G. Praill and A. L. Whitear, *Proc. Chem. Soc.*, 312 (1959).
173. P. F. G. Praill, *Chem. Ind. (London)*, 1123 (1959).
174. A. T. Balaban ánd C. D. Nenitzescu, *Org. Synth.* **44**, 98 (1964).
175. A. T. Balaban˙and C. D. Nenitzescu, *Org. Synth. Collect. Vol.* **5**, 1106 (1973).
176. A. T. Balaban and A. J. Boulton, *Org. Synth. Collect. Vol.* **5**, 1112 (1973).
177. A. Dinculescu and A. T. Balaban, *Org. Prep. Proc. Int.* (in press).
178. E. Gârd, F. Chiraleu, I. I. Stanoiu, and A. T. Balaban, *Rev. Roum. Chim.* **18**, 257 (1973).
179. A. Bota, A. T. Balaban, and F. Chiraleu, *Rev. Roum. Chim.* **21**, 101 (1976).
180. A. T. Balaban, A. Bota, and I. I. Stanoiu, *Rev. Roum. Chim.* **21**, 1183 (1976).
181. A. T. Balaban, A. Bota, F. Chiraleu, E. Sliam, A. Hanes, and C. Draghici, *Rev. Roum. Chim.* **22**, 1003 (1977).
182. M. Arnaud, C. Roussel, and J. Metzger, *Tetrahedron Lett.*, 1795 (1979).
182a. A. T. Balaban and A. Bota, *Org. Prep. Proced. Int.* (in press).
183. V. I. Dulenko, N. N. Alekseev, and V. M. Golyak, *Khim. Geterotsikl. Soedin.*, 1424 (1975) [*CA* **84**, 59106 (1976)].
184. A. T. Balaban and I. I. Badilescu, *Rev. Roum. Chim.* **21**, 1339 (1976).
185. A. T. Balaban, *Tetrahedron Lett.*, 4643 (1968).
186. A. T. Balaban, *Rev. Roum. Chim.* **18**, 1609 (1973).
187. C. Uncuta and A. T. Balaban, *Rev. Roum. Chim.* **21**, 251 (1976).
188. A. G. Anderson and P. J. Stang, *J. Org. Chem.* **41**, 3034 (1976).
189. A. T. Balaban, *Org. Prep. Proced. Int.* **9**, 125 (1977).
190. G. N. Dorofeenko, Yu. A. Zhdanov, and L. N. Etmetchenko, *Khim. Geterotsikl. Soedin.*, 781 (1969) [*CA* **72**, 111223 (1970)].
191. G. N. Dorofeenko, A. V. Koblik, and T. I. Polyakova, *Khim. Geterotsikl. Soedin.*, 878 (1973) [*CA* **79**, 126231 (1973)].
192. S. E. Earnest and D. B. Brown, *J. Heterocycl. Chem.* **12**, 815 (1975).
193. M. Arnaud, A. Pedra, C. Roussel, and J. Metzger, *J. Org. Chem.* **44**, 2972 (1979).
194. I. Tabushi, K. Fujita, and R. Oda, *Tetrahedron Lett.*, 4247, 5455 (1968).
195. C. D. Nenitzescu and A. T. Balaban, *in* "Friedel-Crafts and Related Reactions" (G. A. Olah, ed.) Vol. 3, Part 2, p. 1033. Wiley (Interscience), New York, 1964.
196. V. G. Kharchenko, N. M. Yartseva, and M. E. Stankevich, U.S.S.R. Patent 369,121 (1973) [*CA* **79**, 5263 (1973)].
197. A. G. Ismailov and G. I. Safarov, *Zh. Org. Khim.* **2**, 1624 (1966) [*CA* **66**, 65157 (1967)].
198. O. P. Shelyapin, I. V. Samartseva, and L. A. Pavolova, *Zh. Org. Khim.* **9**, 1987 (1973) [*CA* **79**, 137062 (1973)].
199. V. G. Kharchenko and V. I. Kleimenova, *Zh. Org. Khim.* **7**, 613 (1971) [*CA* **75**, 5634 (1971)].

200. J. Faust, *Z. Chem.* **8**, 171 (1968).
201. V. G. Kharchenko, V. I. Kleimenova, and A. R. Yakoreva, *Khim. Geterotsikl. Soedin.*, 900 (1970) [*CA* **74**, 76272 (1970)].
202. M. Simalty, J. Carretto, and R. Fugnitto, *Bull. Soc. Chim. Fr.*, 2959 (1966).
203. R. R. Schmidt, *Chem. Ber.* **98**, 334 (1965).
204. G. V. Boyd, *J. Chem. Soc.*, 1978 (1958).
205. G. V. Boyd, *J. Chem. Soc.*, 55 (1959).
206. G. N. Dorofeenko, G. P. Safaryan, and V. I. Dulenko, *Zh. Obshch. Khim.* **36**, 811 (1966) [*CA* **65**, 12159 (1966)].
207. A. T. Balaban, M. Mocanu, and Z. Simon, *Tetrahedron* **20**, 119 (1964).
208. T. C. Chadwick, *Anal. Chem.* **45**, 985 (1973).
209. M. C. Fabre, R. Fugnitto, and H. Strzelecka, *C. R. Acad. Sci., Ser. C* **282**, 175 (1976).
210. W. Dilthey and J. Fischer *Ber. Dtsch. Chem. Ges.* **56**, 1012 (1923).
211. G. N. Dorofeenko, Yu. A. Zhdanov, A. D. Semenov, V. A. Palchkov, and S. V. Krivun, *Zh. Obshch. Khim.* **36**, 1728 (1966) [*CA* **66**, 55327 (1967)].
212. J. Kelemen and R. Wizinger, *Helv. Chim. Acta* **45**, 1918 (1962).
213. G. N. Dorofeenko, O. E. Shelepin, Z. N. Nazarova, V. N. Novikov, and G. P. Tikhonova, *Zh. Obshch. Khim.* **35**, 570 (1965) [*CA* **63**, 1766 (1965)].
214. Yu. A. Zhdanov, G. N. Dorofeenko, V. A. Palchkov, and G. P. Safaryan, *Dokl. Akad. Nauk SSSR* **155**, 1115 (1964) [*CA* **61**, 3070 (1964)].
215. Yu. A. Zhdanov, G. N. Dorofeenko, and V. A. Palchkov, *Khim. Geterotsikl. Soedin.*, 812 (1965) [*CA* **65**, 2205 (1966)].
216. M. Simalty, H. Strzelecka, and H. Khedija, *Tetrahedron* **27**, 3503 (1971).
217. E. Gârd, A. Vasilescu, G. D. Mateescu, and A. T. Balaban, *J. Labelled Compd.* **3**, 196 (1967).
218. A. T. Balaban, E. Romas, and C. Rentia, *Tetrahedron* **22**, 1 (1966).
219. G. N. Dorofeenko, Yu. A. Zhdanov, A.D. Semenov, V. A. Palchkov, and E. P. Olekhnovich, *Zh. Org. Khim.* **2**, 1864 (1966) [*CA* **66**, 55326 (1967)].
220. H. Khedija, H. Strzelecka, and M. Simalty, *Bull. Soc. Chim. Fr.*, 3173 (1972).
221. S. V. Krivun and G. N. Dorofeenko, *Khim. Geterotsikl. Soedin.*, 656 (1966) [*CA* **67**, 64170 (1967)].
222. G. N. Dorofeenko, V. V. Mezheritskii, and B. I. Arbashev, *Zh. Org. Khim.* **3**, 1835 (1967) [*CA* **68**, 21776 (1968)].
223. H. Khedija, M. Simalty, H. Strzelecka, and B. Tchoubar, *C. R. Acad. Sci., Ser. C.* **272**, 1370 (1971).
224. H. Strzelecka, *Ann. Chim. (Paris)* [*14*] **1**, 201 (1966).
225. R. Wizinger and P. Ulrich, *Helv. Chim. Acta* **39**, 217 (1956).
226. A. I. Kiprianov and A. I. Tolmachev, *Zh. Obshch. Khim.* **30**, 638 (1960) [*CA* **54**, 24703 (1960)].
227. A. I. Tolmachev and M. A. Kudinova, *Khim. Geterotsikl. Soedin.*, 804 (1969) [*CA* **73**, 16250 (1970)].
228. J. A. Van Allan, G. A. Reynolds, and C. C. Petropoulos, *J. Heterocycl. Chem.* **9**, 783 (1972).
229. G. A. Reynolds and J. A. Van Allan, *J. Heterocycl. Chem.* **6**, 623 (1969).
230. G. A. Reynolds and J. A. Van Allan, *J. Org. Chem.* **34**, 2736 (1969).
231. J. A. Van Allan, G. A. Reynolds, D. P. Maier, and S. Chie Chang, *J. Heterocycl. Chem.* **9**, 1229 (1972).
232. H. Khedija, H. Strzelecka, and M. Simalty, *Bull. Soc. Chim. Fr.*, 218 (1973).
233. R. Michelot and H. Khedija, *Tetrahedron* **29**, 1031 (1973).

234. H. Brockmann, H. Junge, and R. Mühlmann, *Ber. Dtsch. Chem. Ges.* **77**, 529 (1944).
235. L. Roosens and R. Wizinger, *Bull. Soc. Chim. Belg.* **66**, 109 (1957).
236. R. Wizinger and W. Haldemann, *Chem. Ber.* **93**, 1533 (1960).
237. N. E. Shelepin, N. S. Loseva, L. E. Nivorozhkin, and V. I. Minkin, *Khim. Geterotsikl. Soedin.*, 733 (1971) [*CA* **76**, 25034 (1972)].
238. M. Simalty, H. Strzelecka, and H. Khedija, *Bull. Soc. Chim. Fr.*, 3603 (1971).
239. N. V. Khromov-Borisov and L. A. Gavrilova, *Zh. Obshch. Khim.* **32**, 3211 (1962) [*CA* **58**, 10163 (1963)].
240. H. Khedija, H. Strzelecka, and M. Simalty, *Tetrahedron* **28**, 3545 (1972).
241. R. A. Abramovich and E. P. Kyba, *Org. Prep. Proced. Int.* **3**, 127 (1971).
242. W. Stevens and R. Wizinger, *Helv. Chim. Acta* **44**, 1708 (1961).
243. H.-D. Kirner and R. Wizinger, *Helv. Chim. Acta* **44**, 1766 (1961).
244. E. A. Vimorozko, Diploma Work, Rostov University, Rostov-on-Don, 1971.
245. G. N. Dorofeenko, V. V. Mezheritskii, and A. L. Vasserman, *Khim. Geterotsikl. Soedin.*, 37 (1974) [*CA* **80**, 133177 (1974)].
246. A. L. Vasserman, V. V. Mezheritskii, and G. N. Dorofeenko, *Khim. Geterotsikl. Soedin.*, 892 (1974) [*CA* **81**, 120375 (1974)].
247. G. A. Reynolds and J. A. Van Allan, *J. Org. Chem.* **36**, 600 (1971).
248. G. A. Reynolds, J. A. Van Allan, and D. Daniel, *J. Heterocycl. Chem.* **7**, 1395 (1970).
249. H. Strzelecka, *C. R. Acad. Sci.* **255**, 731 (1962).
250. G. N. Dorofeenko, V. V. Mezheritskii, and A. L. Vasserman, *Khim. Geterotsikl. Soedin.*, 1338 (1974) [*CA* **82**, 139912 (1975)].
251. J. A. Van Allan and G. A. Reynolds, *J. Heterocycl. Chem.* **9**, 669 (1972).
252. J. A. Van Allan, S. Farid, G. A. Reynolds, and S. Chie Chang, *J. Org. Chem.* **38**, 2834 (1973).
253. A. T. Balaban, *Tetrahedron Lett.*, 599 (1978).
254. A. N. Narkevich, G. N. Dorofeenko, and Yu. A. Zhdanov, *Zh. Org. Khim.* **1**, 975 (1965) [*CA* **63**, 7007 (1965)].
255. W. Borsche and K. Wunder, *Justus Liebigs Ann. Chem.* **411**, 38 (1915).
256. M. Kamel and H. Shoeb, *Tetrahedron* **20**, 483 (1964).
257. F. Kehrmann, *Justus Liebigs Ann. Chem.* **372**, 287 (1910).
258. N. V. Kholodova, Diploma Work, Rostov University, Rostov-on-Don, 1973.
259. N. V. Khromov-Borisov and L. A. Gavrilova, *Zh. Obshch. Khim.* **32**, 86 (1962) [*CA* **57**, 12417 (1962)].
260. S. N. Baranov, V. F. Lipnitskii, and S. V. Krivun, U.S.S.R. Patent 546,614 (1977) [*CA* **87**, 53084 (1977)].
261. A. T. Balaban, to be published.
262. C. G. Le Fèvre and R. J. W. Le Fèvre, *J. Chem. Soc.*, 2894 (1932).
263. H. E. Johnston and R. J. W. Le Fèvre, *J. Chem. Soc.*, 2900 (1932).
264. A. T. Balaban, A. R. Katritzky, and B. M. Semple, *Tetrahedron* **23**, 4001 (1967).
265. S. N. Baranov, M. A. Lazovskaya, and S. V. Krivun, U.S.S.R. Patent 351,846 (1972) [*CA* **78**, 58242 (1972)].
266. M. Siemiatycki and R. Fugnitto, *Bull. Soc. Chim. Fr.*, 538 (1961).
267. V. V. Bessonov, O. Yu. Okhlobytsin, T. I. Panova, and L. Yu. Ukhin, *Teor. Eksp. Khim.* **12**, 829 (1976) [*CA* **86**, 154984 (1977)].
268. Yu. P. Andreichikov, N. V. Kholodova, and G. N. Dorofeenko, *Khim. Geterotsikl. Soedin.*, 1578 (1975) [*CA* **84**, 43980 (1976)].
269. V. V. Krasnikov, Yu. P. Andreichikov, N. V. Kholodova, and G. N. Dorofeenko, *Zh. Org. Khim.* **13**, 1566 (1977) [*CA* **87**, 152357 (1977)].

270. Yu. P. Andreichikov, N. V. Kholodova, and G. N. Dorofeenko, *Zh. Org. Khim.* **13**, 1565 (1977) [*CA* **87**, 184316 (1977)].
271. Yu. P. Andreichikov, N. V. Kholodova, and G. N. Dorofeenko, *Dokl. Akad. Nauk SSSR* **236**, 1364 (1977) [*CA* **88**, 89469 (1977)].
272. M. M. Evstifeev, G. Kh. Aminova, G. N. Dorofeenko, and E. P. Olekhnovich, *Zh. Obshch. Khim.* **44**, 657 (1974) [*CA* **81**, 9099 (1974)].
273. M. M. Evstifeev, G. N. Dorofeenko, E. P. Olekhnovich, and G. Kh. Aminova, *Zh. Obshch. Khim* **46**, 1334 (1976) [*CA* **85**, 93513 (1976)].
274. N. V. Kholodova, Yu. P. Andreichikov, G. N. Dorofeenko, Ya. R. Tymyanskii, and M. I. Knyazhanskii, U.S.S.R. Patent 570,608 (1977) [*CA* **88**, 22623 (1978)].
275. S.V. Krivun, *Khim. Geterotsikl. Soedin.,* 716 (1970) [*CA* **73**, 66381 (1970)].
276. S. Sib and M. Simalty, *Tetrahedron Lett.,* 3661 (1973).
277. S. Sib, J. Carretto, and M. Simalty, *Tetrahedron Lett.,* 217 (1972).
278. P. Beak, *Tetrahedron Lett.,* 863 (1963).
279. S. N. Baranov, A. I. Buryak, and S. V. Krivun, *Dopov. Akad. Nauk Ukr. RSR, Ser. B: Geol., Geofiz., Khim. Biol.* **33**, 629 (1971) [*CA* **75**, 88438 (1971)].
280. S. N. Baranov, A. I. Buryak, and S. V. Krivun, *Khim. Geterotsikl. Soedin.,* 279 (1971) [*CA* **75**, 48816 (1971)].
281. S. V. Krivun, A. I. Buryak, and S. N. Baranov, *Dopov. Akad. Nauk Ukr. RSR, Ser. B: Geol. Geofiz., Khim. Biol.* **34**, 931 (1972) [*CA* **78**, 159364 (1973); correction: *CA* **78**, 29543 (1973)].
282. S. V. Krivun, S. V. Sayapina, and S. N. Baranov, *Khim. Geterotsikl. Soedin.,* 873 (1973) [*CA* **79**, 126229 (1973)].
283. S. V. Krivun, S. N. Baranov, and A. I. Buryak, *Khim. Geterotsikl. Soedin.,* 1320 (1971) [*CA* **76**, 25030 (1972)].
284. S. V. Krivun, A. I. Buryak, O. F. Voziyanova, S. V. Sayapina, and S. N. Baranov, U.S.S.R. Patent 431,163 (1974) [*CA* **82**, 32470 (1975)].
285. S. N. Baranov, M. A. Lazovskaya, and S. V. Krivun, *Khim. Geterotsikl. Soedin.,* 565 (1971) [*CA* **76**, 25155 (1972)].
286. J. P. Griot, J. Royer, and J. Dreux, *Tetrahedron Lett.,* 2195 (1969).
287. H. Decker and T. Fellenberg, *Justus Liebigs Ann. Chem.* **364**, 1 (1909).
288. A. Löwenbein and B. Rosenbaum, *Justus Liebigs Ann. Chem.* **448**, 223 (1926).
289. D. W. Hill and R. R. Melhuish, *J. Chem. Soc.,* 1161 (1935).
290. H. Stetter and A. Reischl, *Chem. Ber.* **93**, 1253 (1960).
291. S. V. Krivun and S. V. Dul'skaya, *Khim. Geterotsikl. Soedin.,* 1454 (1970) [*CA* **74**, 53411 (1971)].
292. G. Salvadori and A. Williams, *J. C. S. Chem. Commun.,* 775 (1968).
293. G. Salvadori and A. Williams, *J. Am. Chem. Soc.* **93**, 2727 (1971).
294. G. Doddi, S. Fornarini, G. Illuminati, and F. Stegel, *J. Org. Chem.* **44**, 4496 (1979).
295. C. L. Pedersen, O. Buchardt, S. Larsen, and K. J. Watson, *Tetrahedron Lett.,* 2195 (1973).
296. C. L. Pedersen and O. Buchardt, *Acta Chem. Scand., Ser. B* **B29**, 285 (1975).
297. A. R. Katritzky, R. T. C. Brownlee, and G. Musumarra, *Heterocycles* **12**, 775 (1979).
298. A. T. Balaban and W. Silhan, *Tetrahedron* **26**, 743 (1970).
299. C. Jutz, R.-M. Wagner, A. Kraatz, and H.-G. Löbering, *Justus Liebigs Ann. Chem.,* 874 (1975).
300. S. Bersani, G. Doddi, S. Fornarini, and F. Stegel, *J. Org. Chem.* **43**, 4112 (1978).
301. G. W. Fischer, T. Zimmermann, and M. Weissenfels, *Z. Chem.* **21**, 260 (1981).
302. A. T. Balaban and V. Wray, *Org. Magn. Reson.* **9**, 16 (1977).
303. C. Schiele, A. Wilhelm, D. Hendriks, M. Stepec, and G. Paal, *Tetrahedron* **24**, 5029 (1968).

304. C. Schiele, A. Wilhelm, and G. Paal, *Justus Liebigs Ann. Chem.* **722**, 162 (1969).
305. G. Paal and A. Wilhelm, *Tetrahedron* **27**, 811 (1971).
306. N. E. Shelepin, L. E. Nivorozhkin, G. N. Dorofeenko, and V. I. Minkin, *Khim. Geterotsikl. Soedin.*, 1313 (1970).
307. G. W. Fischer, T. Zimmermann, and M. Weissenfels, *Z. Chem.* **21**, 282 (1981).
308. G. Märkl, F. Lieb, and A. Merz, *Angew. Chem.* **79**, 947 (1967); *Angew. Chem., Int. Ed. Engl.* **6**, 944 (1967).
309. S. V. Krivun, O. F. Voziyanova, and S. N. Baranov, *Dopov. Akad. Nauk Ukr. RSR, Ser. B: Geol., Geofiz., Khim. Biol.* **34**, 529 (1972) [*CA* **77**, 101765 (1972)].
310. S. V. Krivun, O. F. Voziyanova, and S. N. Baranov, *Zh. Obshch. Khim.* **42**, 298 (1972) [*CA* **77**, 34224 (1972)].
311. S. Hünig, B. J. Garner, G. Ruider, and W. Schenk, *Justus Liebigs Ann. Chem.*, 1036 (1973).
312. S. V. Krivun, *Dokl. Akad. Nauk SSSR* **182**, 347 (1968) [*CA* **70**, 29009 (1969)].
313. Yu. A. Zhdanov, S. V. Krivun, and V. A. Polenov, *Khim. Geterotsikl. Soedin.*, 368 (1969) [*CA* **71**, 22157 (1969)].
313a. V. I. Boev and A. V. Dombrovskii, *Zh. Obshch. Khim.* **50**, 467 (1980) [*CA* **92**, 215501 (1980)].
314. K. Dimroth and G. Neubauer, *Chem. Ber.* **92**, 2042 (1959).
315. K. Dimroth, K. Wolf, and H. Kroke, *Justus Liebigs Ann. Chem.* **678**, 183 (1964).
316. K. Dimroth, H. Kroke, and K. Wolf, *Justus Liebigs Ann. Chem.* **678**, 202 (1964).
317. A. Safieddine, J. Royer, and J. Dreux, *Bull. Soc. Chim. Fr.*, 703 (1972).
318. J. Royer and J. Dreux, *Bull. Soc. Chim. Fr.*, 707 (1972).
319. O. Chalvet, C. Decoret, J. Dreux, A. Safieddine, and J. Royer, *Bull. Soc. Chim. Fr.*, 716 (1972).
320. H. Wache, Ph.D. Thesis, University of Marburg, 1964.
321. W. Michel, Ph.D. Thesis, University of Marburg, 1967.
322. G. W. Fischer, T. Zimmermann, and M. Weissenfels, *Z. Chem.* **21**, 446 (1981).
323. K. Dimroth and G. Neubauer, *Angew. Chem.* **69**, 720 (1957).
324. K. Dimroth, W. Krafft, and K. H. Wolf, *in* "Nitro-Compounds" (T. Urbanski, ed.), p. 361. Pergamon, Oxford, 1964 [*CA* **63**, 17956 (1965)].
325. G. Büchi and N. C. Yang, *J. Am. Chem. Soc.* **79**, 2318 (1957).
326. S. V. Krivun, *Dokl. Akad. Nauk SSSR* **210**, 1098 (1973) [*CA* **79**, 92073 (1973)].
327. S. V. Krivun, *Dopov. Akad. Nauk Ukr. RSR, Ser. B: Geol., Geofiz., Khim. Biol.* **36**, 717 (1974) [*CA* **81**, 169463 (1974)].
328. E. T. Oestensen and K. Undheim, *Acta Chem. Scand.* **27**, 2184 (1973).
329. E. T. Oestensen, *Acta Chem. Scand., Ser. B* **B29**, 927 (1975).
329a. M. Ohta and H. Kato, *Bull. Chem. Soc. Jpn.* **32**, 707 (1959).
329b. J. A. Van Allan, C. C. Petropoulos, G. A. Reynolds, and D. P. Maier, *J. Heterocycl. Chem.* **7**, 1363 (1970).
329c. J. A. Van Allan and A. James, U. S. Patent 3,554,745 (1971).
329d. J. A. Van Allan, S. Chie Chang, G. A. Reynolds, and D. P. Maier, *J. Chem. Eng. Data* **20**, 210 (1975).
330. A. I. Tolmachev and V. P. Sribnaya, *Zh. Obshch. Khim.* **35**, 316 (1965) [*CA* **62**, 16416 (1965)].
331. S. V. Krivun, *Khim. Geterotsikl. Soedin.*, 757 (1976) [*CA* **85**, 123804 (1976)].
332. H. W. Whitlock and N. A. Carlson, *Tetrahedron* **20**, 2101 (1964).
333. Y. Suzuki, T. Toda, and T. Mukai, *Heterocycles* **4**, 739 (1976).
334. A. T. Balaban, G. Mihai, and C. D. Nenitzescu, *Tetrahedron* **18**, 257 (1962).
335. A. Safieddine, J. Royer, and J. Dreux, *Bull. Soc. Chim. Fr.*, 2510 (1972).

336. V. G. Kharchenko, N. M. Yartseva, and N. I. Kozhevnikova, *Zh. Org. Khim.* **7,** 1551 (1971) [*CA* **77,** 5283 (1972)].
337. V. G. Kharchenko, N. M. Yartseva, and N. I. Kozhevnikova, *Zh. Org. Chim.* **9,** 189 (1973) [*CA* **78,** 97435 (1973)].
338. I. Degani, R. Fochi, and C. Vincenzi, *Boll. Sci. Fac. Chim. Ind. Bologna* **23,** 21 (1965) [*CA* **63,** 8137 (1965)].
339. K. Undheim, and E. T. Oestensen, *Acta Chem. Scand.* **27,** 1385 (1973).
340. C. Bratu and A. T. Balaban, *Rev. Roum. Chim.* **10,** 1001 (1965).
341. A. T. Balaban and M. D. Georghiu, *Rev. Roum. Chim.* **23,** 1065 (1978).
342. G. V. Boyd, *Chem. Ind. (London),* 1244 (1957).
343. G. V. Boyd and F. W. Clark, *J. Chem. Soc. C,* 859 (1966).
344. G. V. Boyd and A. W. Ellis, *J. Chem. Soc. B,* 349 (1966).
345. G. Fischer, Ph.D. Thesis, University of Leipzig, 1965.
346. G. Fischer and W. Schroth, *Z. Chem.* **3,** 191 (1963).
347. W. Schroth and G. W. Fischer, *Tetrahedron* **32,** 2219 (1976).
348. H.-J. Timpe and A. V. El'kov, *Z. Chem.* **15,** 172 (1975).
349. R. Borsdorf, G. W. Fischer, and W. Schroth, *J. Prakt. Chem.* **320,** 463 (1978).
350. H. D. Kirner and R. Wizinger, *Helv. Chim. Acta* **44,** 1778 (1961).
351. J. A. Van Allan and G. A. Reynolds, *J. Heterocycl. Chem.* **13,** 577 (1976).
352. J. A. Van Allan, G. A. Reynolds, and C. C. Petropoulos, *J. Heterocycl. Chem.* **15,** 365 (1978).
353. G. A. Reynolds and J. A. Van Allan, *J. Heterocycl. Chem.* **14,** 539 (1977).
354. G. A. Reynolds and K. H. Drexhage, *J. Org. Chem.* **42,** 885 (1977).
355. A. A. Ishchenko, N. A. Derevyanko, M. A. Kudinova, G. G. Dyadyusha, and A. I. Tolmachev, *Teor. Eksp. Khim.* **13,** 828 (1977) [*CA* **88,** 75290 (1978)].
356. A. I. Tolmachev, G. G. Dyadyusha, E. F. Karaban, A. A. Ishchenko, and N. A. Derevyanko, *Khim. Geterotsikl. Soedin.,* 739 (1978) [*CA* **89,** 146726 (1978)].
357. A. I. Tolmachev, M. Yu. Kornilov, and E. F. Karaban, *Teor. Eksp. Khim.* **12,** 817 (1976) [*CA* **86,** 107984 (1977)].
358. A. I. Tolmachev and M. A. Kudinova, *Dopov. Akad. Nauk Ukr. RSR, Ser. B: Geol., Khim. Biol. Nauki,* 48 (1977) [*CA* **86,** 157007 (1977)].
359. R. Neidlein and I. Körber, *Arch. Pharm. (Weinheim, Ger.)* **311,** 170 (1978).
360. W. Dilthey, *Ber. Dtsch. Chem. Ges.* **52,** 1195 (1919).
361. W. Dilthey and R. Taucher, *Ber. Dtsch. Chem. Ges.* **53,** 252 (1920).
362. A. T. Balaban, M. Gavat, P. T. Frangopol, M. Mocanu, and C. D. Nenitzescu, *Rev. Roum. Chim.* **9,** 79 (1964).
363. A. T. Balaban, G. D. Mateescu, and M. Elian, *Tetrahedron* **18,** 1083 (1962).
364. A. T. Balaban, E. Gârd, and C. N. Rentea, *Abh. Dtsch. Akad. Wiss. Berlin, Kl. Chem., Geol., Bio.,* 659 (1964).
365. I. I. Stanoiu, E. Gârd, C. Uncuta, F. Chiraleu, and A. T. Balaban, *Rev. Roum. Chim.* **24,** 209 (1979).
366. I. I. Stanoiu, F. Chiraleu, E. Gârd, and A. T. Balaban, *Rev. Roum. Chim.* **22,** 117 (1977).
367. I. I. Stanoiu, A. T. Balaban, and F. Chiraleu, *Rev. Roum. Chim.* **22,** 1499 (1977).
368. I. I. Stanoiu, E. Gârd, C. Uncuta, F. Chiraleu, and A. T. Balaban, *Rev. Roum. Chim.* **22,** 1359 (1977).
369. A. T. Balaban, I. I. Stanoiu, F. Chiraleu, and I. Motoc, *Rev. Roum. Chim.* **23,** 143 (1978).
369a. A. T. Balaban, C. Uncuta, and F. Chiraleu, submitted for publication.
369b. A. T. Balaban, M. Elian and P. Filip, submitted for publication.

370. E. Gârd, I. Bally, A. Vasilescu, A. Arsene, and A. T. Balaban, *J. Labelled Compd.* **1**, 182 (1965).

371. E. Gârd, I. I. Stanoiu, A. T. Balaban, and F. Chiraleu, *Rev. Roum. Chim.* **14**, 247 (1969).

372. A. T. Balaban, E. Gârd, A. Vasilescu, and A. Barabas, *J. Labelled Compd.* **1**, 266 (1965).

373. A. Barabas, E. Gârd, A. Vasilescu, and A. T. Balaban, *J. Labelled Compd.* **2**, 359 (1966).

374. A. T. Balaban, I. I. Stanoiu, and F. Chiraleu, *J. C. S. Chem. Commun.*, 984 (1976); *Rev. Roum. Chim.* **23**, 187 (1978).

375. A. T. Balaban, A. Bota, D. Oniciu, J. Metzger, C. Roussel, and G. Klatte, *J. Chem. Res.* (in press).

375a. A. T. Balaban, *J. Labelled Compd. Radiopharm.* **18**, 1627 (1981).

376. D. L. Pavia, *Diss. Abstr. Int. B* **30**, 570 (1969) [*CA* **73**, 35149 (1970)].

377. H. H. Wasserman and D. L. Pavia, *J. C. S. Chem. Commun.*, 1459 (1970).

378. Z. Yoshida, T. Sugimoto, and S. Yoneda, *Tetrahedron Lett.*, 4259 (1971).

379. A. T. Balaban, C. Bratu, and C. N. Rentea, *Tetrahedron* **20**, 265 (1964).

380. V. A. Palchkov, Yu. A. Zhdanov, and G. N. Dorofeenko, *Zh. Org. Khim.* **1**, 1171 (1965) [*CA* **65**, 11276 (1965)].

381. J. B. Conant and A. W. Sloan, *J. Am. Chem. Soc.* **45**, 2466 (1923).

382. W. T. Bowie and M. Feldman, *J. Phys. Chem.* **71**, 3696 (1967).

383. K. Conrow and P.C. Radlick, *J. Org. Chem.* **26**, 2260 (1961).

384. R. B. Woodward and R. Hoffmann, "The Conservation of Orbital Symmetry," Verlag Chemie, Weinheim, 1970.

385. N. T. Anh, "Die Woodward-Hoffmann-Regeln und ihre Anwendung." Verlag Chemie, Weinheim, 1972.

386. T. L. Gilchrist and R. C. Storr, "Organic Reactions and Orbital Symmetry." Cambridge Univ. Press, London and New York, 1971.

387. G. Maier, "Valenzisomerisierungen." Verlag Chemie, Weinheim, 1972.

388. J. A. Berson, *J. Am. Chem. Soc.* **74**, 358 (1952).

389. A. T. Balaban, unpublished.

390. A. Baeyer and J. Piccard, *Justus Liebigs Ann. Chem.* **407**, 332 (1914).

391. E. Gârd, A. Runge, A. Barabas, and A. T. Balaban, *J. Labelled Compd.* **3**, 151 (1967).

392. J.-J. Basselier, *C. R. Acad. Sci.* **248**, 700 (1959); *Ann. Chim. (Paris)* **6**, 1131 (1961).

393. G. Rio and Y. Fellion, *Tetrahedron Lett.*, 1213 (1962).

394. W. Dilthey and T. Böttler, *Ber. Dtsch. Chem. Ges.* **52**, 2040 (1919).

395. N. G. Bokii and Yu. T. Struchkov, *Cryst. Struct. Commun.* **6**, 317 (1977).

396. A. I. Pyshchev, N. G. Bokii, and Yu. T. Struchkov, *Tetrahedron* **34**, 2131 (1978) [*CA* **90**, 151923 (1979)].

397. L. Yu. Ukhin, V. V. Bessonov, A. I. Yanovskii, T. V. Timofeeva, N. G. Furmanova, and Yu. T. Struchkov, *Khim. Geterotsikl. Soedin.*, 461 (1980) [*CA* **93**, 168061 (1980)].

398. W. Dilthey, *J. Prakt. Chem.* **95**, 107 (1917).

399. W. Dilthey, *J. Prakt. Chem.* **101**, 177 (1921).

400. A. T. Balaban, *Tetrahedron* **26**, 743 (1970).

401. A. T. Balaban and C. Toma, *Tetrahedron, Suppl.* **7**, 1 (1966).

402. M. Dupré, M.-L. Filleux-Blanchard, M. Simalty, and H. Strzelecka, *C. R. Acad. Sci., Ser. C* **268**, 1611 (1969).

403. M. Simalty, H. Strzelecka, and M. Dupré, *C. R. Acad. Sci., Ser. C* **266**, 1306 (1968).

404. R. Lombard and J.-P. Stéphan, *Bull. Soc. Chim. Fr.*, 1458 (1958).

405. R. Lombard and J.-P. Stéphan, *C. R. Acad. Sci.* **237**, 333 (1953).
406. R. Lombard and A. Kress, *Bull. Soc. Chim. Fr.,* 1528 (1960).
407. C. Toma and A. T. Balaban, *Tetrahedron, Suppl.* **7,** 9 (1966).
408. A. B. Susan and A. T. Balaban, *Rev. Roum. Chim.* **14,** 111 (1969).
409. A. R. Katritzky, R. T. C. Brownlee, and G. Musamarra, *Tetrahedron* **36,** 1643 (1980).
410. A. R. Katritzky, R. H. Manzo, J. M. Lloyd, and R. C. Patel, *Angew Chem.* **92,** 315 (1980); *Angew Chem., Int. Ed. Engl.* **19,** 306 (1980).
411. O. Diels and K. Alder, *Ber. Dtsch. Chem. Ges.* **60,** 716 (1927).
412. G. Köbrich, *Justus Liebigs Ann. Chem.* **648,** 114 (1961).
413. K. Hafner and K. D. Asmus, *Justus Liebigs Ann. Chem.* **671,** 31 (1964).
414. A. T. Balaban, *Tetrahedron* **24,** 5059 (1968).
415. W. Schneider, *Justus Liebigs Ann. Chem.* **438,** 115 (1924).
416. W. Schneider and W. Müller, *Justus Liebigs Ann. Chem.* **438,** 147 (1924).
417. W. Schneider and K. Weiss, *Ber. Dtsch. Chem. Ges.* **61,** 2445 (1928).
418. W. Schneider and W. Riedel, *Ber. Dtsch. Chem. Ges.* **74,** 1252 (1941).
419. A. T. Balaban, P. T. Frangopol, G. D. Mateescu, and C. D. Nenitzescu, *Bull. Soc. Chim. Fr.,* 298 (1962).
420. A. T. Balaban, *Tetrahedron* **26,** 739 (1970).
421. A. T. Balaban and C. D. Nenitzescu, *J. Chem. Soc.,* 3566 (1961).
422. A. T. Balaban, T. H. Crawford, and R. H. Wiley, *J. Org. Chem.* **30,** 879 (1965).
423. G. Köbrich, *Angew. Chem.* **72,** 348 (1960).
424. G. Köbrich and D. Wunder, *Justus Liebigs Ann. Chem.* **654,** 131 (1962).
425. J. Royer, Ph.D. Thesis, University of Lyon, 1969.
426. J. Royer and J. Dreux, *C. R. Acad. Sci.* **258,** 5895 (1964).
427. G. V. Boyd and S. R. Dando, *J. C. S. Perkin I,* 1142 (1972).
428. G. Märkl, *Angew. Chem.* **74,** 696 (1962); *Angew. Chem., Int. Ed. Engl.* **1,** 511 (1962).
429. C. Crisan and H. Normant, *Bull. Soc. Chim. Fr.,* 1451 (1957).
430. E. N. Marvell, G. Caple, T. Gosink, and G. Zimmer, *J. Am. Chem. Soc.* **88,** 619 (1966).
431. T. Gosink, Ph.D. Thesis, University of Oregon, 1966.
432. E. N. Marvell, T. Chadwick, G. Caple, T. Gosink, and G. Zimmer, *J. Org. Chem.* **37,** 2992 (1972).
433. E. N. Marvell, T. Gosink, P. Churchley, and T. H. Li, *J. Org. Chem.* **37,** 2989 (1972).
434. H. C. Van der Plas, *Acc. Chem. Res.* **11,** 462 (1978).
435. A. T. Balaban and C. D. Nenitzescu, *Chem. Ber.* **93,** 599 (1960).
436. A. T. Balaban, *Org. Prep. Proced. Int.* **1,** 63 (1969).
437. A. T. Balaban, M. Frangopol, P. T. Frangopol, and E. Gârd, *Prep. Bio-Med. Appl. Labelled Mol., Proc. Symp.,* 45 (1964) [*CA* **64,** 15823 (1966)].
438. E. Gârd, A. Vasilescu, A. T. Balaban, and A. Barabas, *Proc. Int. Conf. Methods Prep. Stor. Labelled Comp.,* 2nd, 1966, 649 (1968) [*CA* **71,** 38674 (1969)].
439. A. T. Balaban, M. D. Gheorghiu, and C. Draghici, *Isr. J. Chem.* **20,** 168 (1980).
440. N. I. Shuikin, I. F. Bel'skii, A. T. Balaban, and C. D. Nenitzescu, *Izv. Akad. Nauk SSSR, Otd. Khim. Nauk,* 491 (1962) [*CA* **57,** 15058 (1962)].
441. C. L. Pedersen, *Acta Chem. Scand., Ser. B* **B29,** 791 (1975).
442. F. Quint, R. Pütter, and W. Dilthey, *Ber. Dtsch. Chem. Ges.* **71,** 356 (1938).
443. V. I. Dulenko, N. N. Alekseev, V. M. Golyak, and L. V. Dulenko, *Khim. Geterotsikl. Soedin.,* 1135 (1977) [*CA* **87,** 201220 (1977)].
444. N. N. Alekseev, V. M. Golyak, L. V. Dulenko, V. M. Marchenko, and V. I. Dulenko, U.S.S.R. Patent 539,882 (1976) [*CA* **87,** 5789 (1977)].

445. F. Feist, *Ber. Dtsch. Chem. Ges.* **34,** 1992 (1901).

446. I. E.-S. El-Kholy, F. K. Rafla, and M. M. Mishrikey, *J. Chem. Soc. C,* 1950 (1969).

447. F. Feist, *Ber. Dtsch. Chem. Ges.* **26,** 747 (1893).

448. I. Degani, R. Fochi, and C. Vincenzi, *Gazz. Chim. Ital.* **97,** 397 (1967) [*CA* **67,** 53994 (1967)].

449. N. N. Alekseev, V. M. Golyak, and V. I. Dulenko, U.S.S.R. Patent 640,991 (1979) [*CA* **90,** 137667 (1979)].

450. G. Traverso, *Ann. Chim. (Rome)* **46,** 821 (1956) [*CA* **51,** 6622 (1957)].

451. G. Traverso and M. Sanesi, *Ann. Chim. (Rome)* **43,** 795 (1953) [*CA* **49,** 13981 (1955)].

452. G. Traverso, *Ann. Chim. (Rome)* **44,** 1018 (1954) [*CA* **50,** 366 (1956)].

453. G. Traverso, *Ann. Chim. (Rome)* **45,** 687 (1955) [*CA* **51,** 5761 (1957)].

454. F. Arndt, P. Nachtwey, and J. Pusch, *Ber. Dtsch. Chem. Ges.* **58,** 1633 (1925).

455. G. Guillouzo, *Bull. Soc. Chim. Fr.,* 1316 (1958).

456. S. Bezzi, C. Garbuglio, M. Mammi, and G. Traverso, *Gazz. Chim. Ital.* **88,** 1226 (1958) [*CA* **53,** 22007 (1959)].

457. S. Bezzi, M. Mammi, and C. Garbuglio, *Nature (London)* **182,** 247 (1958).

458. G. Traverso, *Ann. Chim. (Rome)* **47,** 3 (1957) [*CA* **51,** 10543 (1957)].

459. N. Lozac'h and J. Vialle, *Chem. Org. Sulfur Compd.* **2,** 257 (1966).

460. N. Lozac'h, *in* "Organosulfur Chemistry" (M. J. Janssen, ed.), pp. 179ff. Wiley (Interscience), New York, 1967.

461. J. P. Marino, *Top. Sulfur Chem.* **1,** 86 (1970).

462. P. L. Kumler, C. L. Pedersen, and O. Buchardt, *Acta Chem. Scand.* **22,** 2719 (1968).

463. C. L. Pedersen, N. Harrit, and O. Buchardt, *Acta Chem. Scand.* **24,** 3435 (1970).

464. W. Schneider and F. Seebach, *Ber. Dtsch. Chem. Ges.* **54,** 2285 (1921).

465. G. N. Dorofeenko, A. N. Narkevich, and Yu. A. Zhdanov, *Khim. Geterotsikl. Soedin.,* 1130 (1967) [*CA* **69,** 67175 (1968)].

466. G. N. Dorofeenko, A. N. Narkevich, Yu. A. Zhdanov, O. E. Shelepin, and T. G. Soroka, *Khim. Geterotsikl. Soedin., Sb.* **2,** 223 (1970) [*CA* **76,** 140454 (1972)].

467. V. Snieckus and G. Kan, *J. C. S. Chem. Commun.,* 1208 (1970).

468. C. L. Pedersen and O. Buchardt, *Acta Chem. Scand.* **24,** 834 (1970).

469. M. Lempert-Sréter and K. Lempert, *Acta Chim. Acad. Sci. Hung.* **65,** 443 (1970).

470. M. Lempert-Sréter and K. Lempert, *Acta Chim. Acad. Sci. Hung.* **88,** 189 (1976).

471. A. R. Katritzky, J. Lewis, and P.-L. Nie, *J. C. S. Perkin I,* 446 (1979).

472. R. G. Jones and M. J. Mann, *J. Am. Chem. Soc.* **75,** 4048 (1953).

473. C. Ainsworth and R. G. Jones, *J. Am. Chem. Soc.* **76,** 3172 (1954).

474. T. Toda, H. Morino, Y. Suzuki, and T. Mukai, *Chem. Lett.,* 155 (1977) [*CA* **87,** 5854 (1977)].

475. A. R. Katritzky, S. Q. A. Rizivi, and J. W. Suwinski, *Heterocycles* **3,** 379 (1975).

476. A. R. Katritzky, S. Q. A. Rizivi, and J. W. Suwinski, *J. Chem. Soc. C,* 2489 (1975).

477. A. T. Balaban, M. Marculescu-Frangopol, and P. T. Frangopol, *Isotopentechnik* **2,** 235 (1962).

478. V. I. Dulenko, S. N. Baranov, G. N. Dorofeenko, I. G. Katts, and L. V. Dulenko, *Dokl. Akad. Nauk SSSR* **195,** 607 (1970) [*CA* **74,** 76351 (1971)].

479. S. V. Krivun, V. I. Dulenko, L. V. Dulenko, and G. N. Dorofeenko, *Dokl. Akad. Nauk SSSR* **166,** 359 (1966) [*CA* **64,** 11153 (1966)].

480. V. I. Dulenko, N. N. Alekseev, and L. I. Kapkan, *Khim. Geterotsikl. Soedin.,* 1342 (1973) [*CA* **80,** 27132 (1974)].

481. V. I. Dulenko and N. N. Alekseev, *Khim. Geterotsikl. Soedin.,* 1212 (1973) [*CA* **79,** 146441 (1973)].

482. G. N. Dorofeenko, V. I. Dulenko, and V. I. Volbushko, *Khim. Geterotsikl. Soedin.,* 450 (1973) [*CA* **79,** 18609 (1973)].

483. E. V. Kuznetsov, I. V. Shcherbakova, and G. N. Dorofeenko, *Khim. Geterotsikl. Soedin.*, 1176 (1977) [*CA* **88**, 37569 (1978)].
484. C. Buehler and D. Pearson, "Survey of Organic Syntheses," p. 280. Wiley (Interscience), New York, 1970.
485. H. G. Rajoharison, H. Soltani, M. Arnaud, C. Roussel, and J. Metzger, *Synth. Commun.* **10**, 195 (1980).
486. A. T. Balaban and C. D. Nenitzescu, *J. Chem. Soc.*, 3553 (1961).
486a.   A. T. Balaban, M. D. Gheorghiu, and A. Bota, submitted for publication.
487. R. Wizinger and P. Ulrich, *Helv. Chim. Acta* **39**, 207 (1956).
488. G. Suld and C. C. Price, *J. Am. Chem. Soc.* **83**, 1770 (1961).
489. G. Suld and C. C. Price, *J. Am. Chem. Soc.* **84**, 2094 (1962).
490. K. Kanai, M. Umehara, H. Kitano, and K. Fukui, *Nippon Kagaku Zasshi* **84**, 432 (1963) [*CA* **59**, 13934 (1963)].
491. H. Yasuba, T. Imai, K. Okamoto, S. Kusabayashi, and H. Mikawa, *Bull. Chem. Soc. Jpn.* **43**, 3101 (1970).
492. C. C. Price, J. Follweiler, H. Pirelahi, and M. Sistin, *J. Org. Chem* **36**, 791 (1971).
493. M. Siemiatycki, *Ann. Chim. (Paris)* **2**, 189 (1957) [*CA* **53**, 366 (1959)].
494. G. A. Reynolds, *Synthesis*, 638 (1975).
495. G. A. Reynolds and J. A. Van Allan, *J. Heterocycl. Chem.* **9**, 1105 (1972).
495a.   G. M. Senkler, J. Stackhouse, B. E. Maryanoff, and K. Mislow, *J. Am. Chem. Soc.* **96**, 5648, 5650, 5651 (1974).
496. H. Kato, T. Ogawa, and M. Ohta, *Bull. Chem. Soc. Jpn.* **33**, 1467 (1960).
497. F. Arndt, R. Schwarz, C. Martius, and E. Aron, *Rev. Fac. Sci. Univ. Istanbul, Ser.* A13, 57 (1948) [*CA* **42**, 4176 (1948)].
498. M. Rolla, M. Sanesi, and G. Traverso, *Ann. Chim. (Rome)* **42**, 673 (1952) [*CA* **47**, 11934 (1953)].
499. W. Dilthey, *J. Prakt. Chem.* **94**, 53 (1916).
500. G. N. Dorofeenko, E. A. Zvezdina, M. P. Zhdanova, V. V. Derbenev, and E. S. Matskovskaya, *Khim. Geterotsikl. Soedin.*, 1036 (1974) [*CA* **81**, 169408 (1974)].
501. I. I. Stanoiu, M. Paraschiv, E. Gârd, and A. T. Balaban, *Rev. Roum. Chim.* **22**, 865 (1977).
502. G. N. Dorofeenko, G. A. Korol'chenko, and S. V. Krivun, *Khim. Geterotsikl. Soedin.*, 817 (1965) [*CA* **64**, 19548 (1966)].
503. G. N. Dorofeenko and S. V. Krivun, *Ukr. Khim. Zh.* **29**, 1058 (1963) [*CA* **60**, 7977 (1964)].
504. S. V. Krivun and G. N. Dorofeenko, *Zh. Obshch. Khim.* **34**, 2091 (1964) [*CA* **61**, 6982 (1964)].
505. S. V. Krivun and G. N. Dorofeenko, *Chem. Heterocycl. Compd. (Engl. Transl.)* **2**, 499 (1966).
506. S. Hünig and G. Ruider, *Justus Liebigs Ann. Chem.*, 1415 (1974).
507. L. C. King. and F. J. Ozog, *J. Org. Chem.* **20**, 448 (1955).
508. G. I. Zhungietu, G. N. Dorofeenko, B. P. Sukhanyuk, and D. D. Buburuz, *Khim. Geterotsikl. Soedin.*, 1437 (1970) [*CA* **74**, 76251 (1971)].
509. A. T. Balaban and C. D. Nenitzescu, *J. Chem. Soc.*, 3561 (1961).
510. A. T. Balaban, C. D. Nenitzescu, M. Gavat, and G. Mateescu, *J. Chem. Soc.*, 3564 (1961).
511. A. T. Balaban, M. Gavat, and C. D. Nenitzescu, *Tetrahedron* **18**, 1079 (1962).
512. P. F. G. Praill and A. L. Whitear, *J. Chem. Soc.*, 3573 (1961).
513. B. K. Blount and R. Robinson, *J. Chem Soc.*, 555 (1933).
514. M. Vajda and F. Ruff, *Acta Chim. Acad. Sci. Hung.* **40**, 217 (1964).

515. M. Vajda, *Acta Chim. Acad. Sci. Hung.* **40**, 295 (1964).
516. S. V. Krivun, V. I. Dulenko, L. V. Dulenko, and G. N. Dorofeenko, *Dokl. Akad. Nauk SSSR* **166**, 359 (1966) [*CA* **64**, 11153 (1966)].
517. G. N. Dorofeenko, A. D. Semenov, V. I. Dulenko, and S. V. Krivun, *Zh. Org. Khim.* **2**, 1492 (1966) [*CA* **66**, 46286 (1967)].
518. G. N. Dorofeenko, E. V. Kuznetsov, and S. V. Krivun, *Zh. Org. Khim.* **2**, 1499 (1966) [*CA* **66**, 46287 (1967)].
519. S. V. Krivun, G. N. Dorofeenko, and E. I. Sadekova, *Zh. Obshch. Khim.* **40**, 1429 (1970) [*CA* **74**, 53435 (1971)].
520. G. N. Dorofeenko, G. P. Safaryan, and E. V. Kuznetsov, *Khim. Geterotsikl. Soedin.*, 1013 (1970) [*CA* **74**, 76295 (1971)].
521. G. N. Dorofeenko, E. I. Sadekova, and L. K. Pyshkina, *Khim. Geterotsikl. Soedin.*, Sb. **2**, 196 (1970) [*CA* **77**, 114177 (1972)].
522. G. N. Dorofeenko, V. G. Korobkova, and S. V. Krivun, *Khim. Geterotsikl. Soedin.*, Sb. **2**, 200 (1970) [*CA* **76**, 140460 (1972)].
523. K. Dimroth and H. Odenwälder, *Chem. Ber.* **104**, 2984 (1971).
524. G. N. Dorofeenko and V. G. Korobkova, *Khim. Geterotsikl. Soedin.*, 1601 (1971) [*CA* **77**, 19846 (1972)].
525. E. V. Kuznetsov, D. V. Pruchkin, and G. N. Dorofeenko, *Khim. Geterotsikl. Soedin.*, 1479 (1977) [*CA* **88**, 62275 (1978)].
526. E. V. Kuznetsov, I. V. Shcherbakova, and G. N. Dorofeenko, *Khim. Geterotsikl. Soedin.*, 1481 (1977) [*CA* **88**, 89499 (1978)].
527. G. N. Dorofeenko, L. V. Dulenko, V. I. Dulenko, and S. V. Krivun, *Zh. Org. Khim.* **1**, 1171 (1965) [*CA* **63**, 11483 (1965)].
528. G. N. Dorofeenko and L. V. Dulenko, *Khim. Geterotsikl. Soedin.*, 417 (1969) [*CA* **72**, 3412 (1970)].
529. Yu. A. Zhdanov, V. I. Kornilov, and G. N. Dorofeenko, *Dokl. Akad. Nauk SSSR* **178**, 849 (1968) [*CA* **69**, 19442 (1968)].
530. G. N. Dorofeenko and V. G. Korobkova, *Chem. Ind. (London)*, 1848 (1968).
531. G. N. Dorofeenko, V. G. Korobkova, and E. A. Guzhina, *Khim. Geterotsikl. Soedin.*, 345 (1971) [*CA* **76**, 14384 (1972)].
532. L. V. Dulenko, G. N. Dorofeenko, S. N. Baranov, I. G. Katts, and V. I. Dulenko, *Khim. Geterotsikl. Soedin.*, 320 (1971) [*CA* **76**, 14378 (1972)].
533. R. L. Shriner and W. R. Knox, *J. Org. Chem.* **16**, 1064 (1951).
534. G. N. Dorofeenko, G. V. Lazur'evskii, and G. I. Zhungietu, *Dokl. Akad. Nauk SSSR* **161**, 355 (1965) [*CA* **63**, 569 (1965)].
535. G. N. Dorofeenko and G. I. Zhungietu, *Zh. Obshch. Khim.* **35**, 589 (1965) [*CA* **63**, 569 (1965)].
536. R. L. Shriner, H. W. Johnston, and C. E. Kaslow, *J. Org. Chem.* **14**, 204 (1949).
537. G. N. Dorofeenko and V. I. Dulenko, *Zh. Obshch. Khim.* **32**, 3445 (1962) [*CA* **58**, 9013 (1963)].
538. G. N. Dorofeenko and V. I. Dulenko, *Dokl. Akad. Nauk SSSR* **157**, 361 (1964) [*CA* **61**, 9458 (1964)].
539. L. V. Dulenko, V. I. Dulenko, and G. N. Dorofeenko, *Zh. Obshch. Khim.* **34**, 3588 (1964) [*CA* **62**, 5246 (1965)].
540. G. N. Dorofeenko, V. I. Dulenko, and L. V. Dulenko, *Zh. Obshch. Khim.* **34**, 3116 (1964) [*CA* **62**, 521 (1965)].
541. G. N. Dorofeenko, Yu. A. Zhdanov, V. I. Dulenko, V. A. Palchkov, and N. V. Kovalenko, *Khim. Geterotsikl. Soedin.*, 172 (1966) [*CA* **65**, 2206 (1966)].
542. A. T. Balaban and N. S. Barbulescu, *Rev. Roum. Chim.* **11**, 109 (1966).

543. G. N. Dorofeenko, Yu. A. Zhdanov, G. I. Zhungietu, and S. V. Krivun, *Tetrahedron* **22**, 1821 (1966).

544. G. N. Dorofeenko, Yu. A. Zhdanov, G. I. Zhungietu, and S. V. Krivun, *Tetrahedron* **23**, 1565 (1967).

545. N. Barbulescu and G. Nicolae, *Rev. Chim. (Bucharest)* **22**, 368 (1971) [*CA* **75**, 129638 (1971)].

546. G. I. Zhungietu, G. N. Dorofeenko, and G. V. Lazur'evskii, *Dokl. Akad. Nauk SSSR* **163**, 372 (1965) [*CA* **63**, 11661 (1965)].

547. A. T. Balaban, A. Dinculescu, and A. Zlota, (in preparation).

548. H. Meislich, *in* "Pyridine and its Derivatives" (E. Klingsberg, ed.), Ch. 12, p. 509. Wiley (Interscience), New York, 1962.

549. W. Borsche and W. Peter, *Justus Liebigs Ann. Chem.* **453**, 148 (1925).

550. R. H. Wiley, N. R. Smith, and L. H. Knabeschuh, *J. Am. Chem. Soc.* **75**, 4482 (1953).

551. R. H. Wiley, P. Beasley, and L. H. Knabeschuh, *J. Am. Chem. Soc.* **76**, 311 (1954).

552. H. Stetter and C. W. Schellhammer, *Chem. Ber.* **90**, 775 (1957).

553. I. E.-S. El-Kholy, F. K. Rafla, and G. Soliman, *J. Chem. Soc.*, 4490 (1961).

554. C.-S. Wang, *J. Heterocycl. Chem.* **7**, 389 (1970).

555. M. Conrad and M. Guthzeit, *Ber. Dtsch. Chem. Ges.* **20**, 154 (1887).

556. W. Borsche and I. Bonacker, *Ber. Dtsch. Chem. Ges.* **54**, 2678 (1921).

557. M. A. F. Elkaschef and M. H. Nosseir, *J. Am. Chem. Soc.* **82**, 4344 (1960).

558. M. A. F. Elkaschef, M. H. Nosseir, and A. Abdel-Kader, *J. Chem. Soc.*, 4647 (1963).

559. H. Pechmann and W. Welsh, *Ber. Dtsch. Chem. Ges.* **17**, 2384 (1884).

560. H. Pechmann, *Justus Liebigs Ann. Chem.* **264**, 261 (1891).

561. W. T. Caldwell, F. T. Tyson, and L. Lauer, *J. Am. Chem. Soc.* **66**, 1479 (1944).

562. H. Gault, J. Gilbert, and D. Briaucourt, *C. R. Acad. Sci., Ser. C* **266**, 131 (1968).

563. H. Ost, *J. Prakt. Chem.* **29**, 57 (1884).

564. H. Ost, *J. Prakt. Chem.* **27**, 257 (1883).

565. L. Haitinger and A. Lieben, *Monatsh. Chem.* **6**, 279 (1885).

566. J. W. Armit and T. J. Nolan, *J. Chem. Soc.*, 3023 (1931).

567. K. Heyns and G. Vogelsang, *Chem. Ber.* **87**, 1377 (1954).

568. G. N. Dorofeenko, V. G. Korobkova, and V. I. Volbushko, *Khim. Geterotsikl. Soedin.*, 553 (1977) [*CA* **87**, 84858 (1977)].

569. A. T. Balaban, P. T. Frangopol, A. R. Katritzky, and C. D. Nenitzescu, *J. Chem. Soc.*, 3889 (1962).

570. H. Kato, T. Ogawa, and M. Ohta, *Chem. Ind. (London)*, 1300 (1960).

571. A. R. Katritzky, J. Lloyd, and R. C. Patel, unpublished.

572. M. P. Sammes and K. L. Yip, *J. C. S. Perkin I*, 1373 (1978).

573. C. Toma and A. T. Balaban, *Tetrahedron, Suppl.* **7**, 27 (1966).

574. A. Dinculescu and A. T. Balaban, *Rev. Roum. Chim.* **25**, 505 (1980).

575. A. N. Narkevich, G. N. Dorofeenko, and Yu. A. Zhdanov, *Zh. Obshch. Khim.* **36**, 819 (1966) [*CA* **65**, 10665 (1966)].

576. Yu. A. Zhdanov, G. N. Dorofeenko, and A. N. Narkevich, *Zh. Obshch. Khim.* **33**, 2418 (1963) [*CA* **59**, 14105 (1963)].

577. A. N. Narkevich, G. N. Dorofeenko, and Yu. A. Zhdanov, *Dokl. Akad. Nauk SSSR* **176**, 103 (1967) [*CA* **68**, 78605 (1968)].

578. E. A. Zvezdina, M. P. Zhdanova, V. A. Bren, and G. N. Dorofeenko, *Khim. Geterotsikl. Soedin.*, 944 (1978) [*CA* **89**, 179817 (1978)].

579. Yu. A. Zhdanov, A. N. Narkevich, and G. N. Dorofeenko, *Sint. Prir. Soedin., Ikh Analogov Fragm.*, 160 (1965) [*CA* **65**, 7264 (1966)].
580. M. H. O'Leary and G. A. Samberg, *J. Am. Chem. Soc.* **93**, 3530 (1971).
581. G. N. Dorofeenko, E. A. Zvezdina, and V. V. Derbenev, *Zh. Org. Khim.* **9**, 1079 (1973) [*CA* **79**, 53145 (1973)].
582. N. S. Zefirov, G. N. Dorofeenko, and T. M. Pozdnyakova, *Zh. Org. Khim.* **9**, 387 (1973) [*CA* **78**, 124413 (1973)].
583. A. R. Katritzky, K. Horvath, and B. Plau, *J. C. S. Chem. Commun.*, 300 (1979).
584. G. N. Dorofeenko, A. N. Narkevich, Yu. A. Zhdanov, and T. G. Soroka, *Khim. Geterotsikl. Soedin.*, 315 (1970) [*CA* **73**, 66502 (1970)].
585. A. R. Katritzky and J. W. Suwinski, *Tetrahedron Lett.*, 4123 (1974).
586. A. R. Katritzky and J. W. Suwinski, *Tetrahedron* **31**, 1549 (1975).
587. R. Neidlein and P. Witerzens, *Monatsh. Chem.* **106**, 643 (1975).
588. A. Camerman, L. H. Jensen, and A. T. Balaban, *Acta Crystallogr., Sect. B* **B25**, 2623 (1969).
589. A. T. Balaban, A. Dinculescu, H. N. Koutrakis, and F. Chiraleu, *Tetrahedron Lett.*, 437 (1979).
590. H. A. P. de Jongh and H. Wynberg, *Tetrahedron* **21**, 515 (1965).
591. A. T. Balaban, C. Uncuta, A. Dinculescu, M. Elian, and F. Chiraleu, *Tetrahedron Lett.*, 1553 (1980).
592. W. Dilthey, *Ber. Dtsch. Chem. Ges.* **55**, 57 (1922).
593. W. Dilthey, C. Ammon, and E. Ebert, *J. Prakt. Chem.* **107**, 7 (1924).
594. W. Dilthey and A. Schaefer, *J. Prakt. Chem.* **108**, 332 (1924).
595. W. Dilthey and H. Dierichs, *J. Prakt. Chem.* **144**, 1 (1936).
596. R. Wizinger and H. Wenning, *Helv. Chim. Acta* **23**, 247 (1940).
597. K. Dimroth, C. Reichardt, T. Siepmann, and F. Bohlmann, *Justus Liebigs Ann. Chem.* **661**, 1 (1963).
598. K. Dimroth, C. Reichardt, and A. Schweig, *Justus Liebigs Ann. Chem.* **669**, 95 (1969).
599. K. Dimroth and C. Reichardt, *Z. Anal. Chem.* **215**, 344 (1966).
600. C. Reichardt and K. Dimroth, *Fortschr. Chem. Forsch.* **11**, 1 (1968).
601. K. Dimroth and C. Reichardt, *Justus Liebigs Ann. Chem.* **727**, 93 (1969).
602. C. Reichardt, "Lösungsmittel-Effekte in der organischen Chemie," Verlag Chemie, Weinheim, 1969; "Solvent Effects in Organic Chemistry," Verlag Chemie, Weinheim, 1979.
603. K. Dimroth and H. Odenwälder, *Tetrahedron Lett.*, 553 (1971).
604. Ya. R. Tymanskii, M. I. Knyazhanskii, Yu. P. Andreichikov, G. E. Trukhan, and G. N. Dorofeenko, *Zh. Org. Khim.* **12**, 1126 (1976) [*CA* **85**, 102252 (1976)].
605. A. R. Katritzky, Z. Zakaria, E. Lunt, P. G. Jones, and O. Kennard, *J. C. S. Chem. Commun.*, 268 (1979).
605a. A. R. Katritzky, Z. Zakaria, and E. Lunt, *J. C. S. Perkin I*, 1879 (1980).
606. K. Ziegler and F. A. Fries, *Ber. Dtsch. Chem. Ges.* **59**, 242 (1926).
607. N. F. Eweiss, A. R. Katritzky, P.-L. Nie, and C. A. Ramsden, *Synthesis*, 634 (1977).
608. A. R. Katritzky, N. F. Eweiss, and P.-L. Nie, *J. C. S. Perkin I*, 433 (1979).
609. A. R. Katritzky, U. Gruntz, A. A. Ikizler, D. H. Kenny, and B. P. Leddy, *J. C. S. Perkin I*, 436 (1979).
610. A. R. Katritzky, F. Al Omran, R. C. Patel, and S. S. Thind, *J. C. S. Perkin I*, 1890 (1980).

611. A. R. Katritzky, K. Horvath, and B. Plau, *Synthesis,* 437 (1979).
612. A. R. Katritzky, A. Chermprapai, and R. C. Patel, *J. C. S. Chem. Commun.,* 238 (1979).
613. A. R. Katritzky, A. Chermprapai, and R. C. Patel, *J. C. S. Perkin I,* 2901 (1980).
614. U. Gruntz, A. R. Katritzky, D. H. Kenney, M. C. Rezende, and H. Sheikh, *J. C. S. Chem. Commun.,* 701 (1977).
615. A. R. Katritzky, U. Gruntz, D. H. Kenny, M. C. Rezende, and H. Sheikh, *J. C. S. Perkin I,* 430 (1979).
616. A. R. Katritzky and L. Mazorati, *J. Org. Chem.* **45,** 2515 (1981).
617. A. R. Katritzky, J. B. Bapat, R. J. Blade, B. P. Leddy, P.-L. Nie, C. A. Ramsden, and S. S. Thind, *J. S. C. Perkin I,* 418 (1979).
618. A. R. Katritzky, M. F. Abdel-Megeed, G. Lhommet, and C. A. Ramsden, *J. C. S. Perkin I,* 426 (1979).
619. A. R. Katritzky, A. Saba, and R. C. Patel, *J. C. S. Perkin I,* 1492 (1981).
620. A. R. Katritzky, M. J. Cook, A. Ikizler, and G. H. Millet, *J. C. S. Perkin I,* 2500 (1979).
621. A. R. Katritzky, U. Gruntz, N. Mongelli, and M. C. Rezende, *J. C. S. Chem. Commun.,* 133 (1978).
622. A. R. Katritzky, U. Gruntz, N. Mongelli, and M. C. Rezende, *J. C. S. Perkin I,* 1953 (1979).
623. A. R. Katritzky and S. S. Thind, *J. C. S. Chem. Commun.,* 138 (1979).
624. A. R. Katritzky and S. S. Thind, *J. C. S. Perkin I,* 865 (1980).
625. A. R. Katritzky and S. S. Thind, unpublished.
626. A. R. Katritzky, A. M. El-Mowafy, and R. C. Patel, *Recl. Trav. Chim. Pays-Bas* **98,** 302 (1979).
627. A. R. Katritzky, G. Liso, E. Lunt, R. C. Patel, S. S. Thind, and A. Zia, *J. C. S. Perkin I,* 849 (1980).
627a. A. R. Katritzky and S. S. Thind, *J. C. S. Perkin I,* 1870 (1980).
628. A. R. Katritzky, G. de Ville, and R. C. Patel, *J. C. S. Chem. Commun.,* 602 (1979).
629. A. J. Boulton, J. Epsztajn, A. R. Katritzky, and P.-L. Nie, *Tetrahedron Lett.,* 2689 (1976).
630. A. R. Katritzky, J. Lewis, and P.-L. Nie, *J. C. S. Perkin I,* 442 (1979).
631. A. R. Katritzky, S. Bravo, and R. C. Patel, unpublished.
632. A. R. Katritzky, J. B. Bapat, R. M. Claramunt-Elguero, F. S. Yates, A. Dinculescu, A. T. Balaban, and F. Chiraleu, *J. Chem. Res., Synop.,* 395 (1978); *J. Chem. Res., Miniprint,* 4783 (1978).
633. G. V. Boyd and S. R. Dando, *J. Chem. Soc. C,* 3873 (1971).
634. H. Ost, *J. Prakt. Chem.* **29,** 378 (1884).
635. E. Mennel, *J. Prakt. Chem.* **32,** 176 (1885).
636. J. P. Wibaut and R. J. C. Kleipool, *Recl. Trav. Chim. Pays-Bas* **66,** 24 (1947).
637. D. Klostermans, *Recl. Trav. Chim. Pays-Bas* **66,** 93 (1947).
638. R. J. C. Kleipool and J. P. Wibaut, *Recl. Trav. Chim. Pays-Bas* **69,** 1041 (1950).
639. S. Hünig and G. Köbrich, *Justus Liebigs Ann. Chem.* **617,** 181 (1958).
640. K. N. Campbell, I. F. Ackerman, and B. K. Campbell, *J. Org. Chem.* **15,** 221 (1950).
641. R. Adams and J. L. Johnson, *J. Am. Chem. Soc.* **71,** 705 (1949).
642. J. U. Lerch, *Monatsh. Chem.* **5,** 367 (1884).
643. A.-P. Smirnoff, *Helv. Chim. Acta* **4,** 599 (1921).
644. V. Ettel and J. Hebky, *Collect. Czech. Chem. Commun.* **15,** 639 (1950).
645. R. H. Wiley, L. H. Knabeschuh, A. L. Duckwall, and N. R. Smith, *J. Am. Chem. Soc.* **76,** 625 (1954).

646. O. Fischer and K. Demeler, *Ber. Dtsch. Chem. Ges.* **32**, 1307 (1899).
647. D. Vorländer, *Ber. Dtsch. Chem. Ges.* **58**, 1893 (1925).
648. K. Dimroth and K. H. Wolf, *Angew. Chem.* **72**, 777 (1960).
649. F. Eiden, *Naturwissenschaften* **47**, 60 (1960).
650. H. Kato, T. Ogawa, and M. Ohta, *Bull. Chem. Soc. Jpn.* **33**, 1468 (1960).
651. F. Eiden, *Arch. Pharm. (Weinheim, Ger.)* **293**, 404 (1960).
652. F. Eiden, *Arch. Pharm. (Weinheim, Ger.)* **295**, 667 (1962).
653. I. Belsky, H. Dodiuk, and Y. Shvo, *J. Org. Chem.* **39**, 989 (1974).
654. N. N. Alekseev, V. M. Golyak, and V. I. Dulenko, U.S.S.R. Patent 659,562 (1979) [*CA* **91**, 91332 (1979)].
655. E. Schmitz, *Chem. Ber.* **91**, 1488 (1958).
656. A. Peratoner, *Gazz. Chim. Ital.* **41**, 619 (1912) [*CA* **6**, 993 (1912)].
657. G. Soliman and I. E.-S. El-Kholy, *J. Chem. Soc.*, 1755 (1954).
658. I. E.-S. El-Kholy, F. K. Rafla, and M. M. Mishrikey, *J. Chem. Soc. C*, 1578 (1970).
659. M. J. Cook, A. R. Katritzky, and G. H. Millet, *Heterocycles* **7**, 227 (1977).
660. K. Dimroth, G. Arnoldy, S. Eicken, and G. Schiffler, *Justus Liebigs Ann. Chem.* **604**, 221 (1957).
661. R. A. Abramovitch and T. Takaya, *J. Org. Chem.* **38**, 3311 (1973).
662. P. L. Kumler and O. Buchardt, *J. C. S. Chem. Commun.*, 1321 (1968).
663. J. Streith and C. Sigwalt, *Bull. Soc. Chim. Fr.*, 1157 (1970).
664. V. Snieckus, *J. C. S. Chem. Commun.*, 831 (1969).
665. A. R. Katritzky and J. M. Lagowski, "Chemistry of the Heterocyclic N-Oxides," p. 321. Academic Press, New York, 1971.
666. A. Roedig, H. A. Renk, M. Schlosser, and T. Neukam, *Justus Liebigs Ann. Chem.*, 1206 (1974).
667. J. B. Bapat, R. J. Blade, A. J. Boulton, J. Epsztajn, A. R. Katritzky, J. Lewis, P. Molina-Buendia, P.-L. Nie, and C. A. Ramsden, *Tetrahedron Lett.*, 2691 (1976).
668. A. R. Katritzky and P. Molina-Buendia, *J. C. S. Perkin I*, 1957 (1979).
669. A. R. Katritzky, P.-L. Nie, A. Dondoni, and D. Tassi, *Synth. Commun.* **7**, 387 (1977).
670. A. R. Katritzky, P.-L. Nie, A. Dondoni, and D. Tassi, *J. C. S. Perkin I*, 1961 (1979).
671. I. Ichimoto, K. Fujii, and C. Tatsumi, *Agric. Biol. Chem.* **31**, 979 (1967).
672. A. Marxer and A. F. Thomas, *Angew. Chem.* **72**, 270 (1960).
673. M. P. Zhdanova, E. A. Zvezdina, and G. N. Dorofeenko, *Khim. Geterotsikl. Soedin.*, 277 (1975) [*CA* **82**, 156212 (1975)].
674. M. P. Zhdanova, E. A. Zvezdina, and G. N. Dorofeenko, *Khim. Geterotsikl. Soedin.*, 456 (1978) [*CA* **89**, 43321 (1978)].
675. E. A. Zvezdina, M. P. Zhdanova, and G. N. Dorofeenko, *Khim. Geterotsikl. Soedin.*, 324 (1979) [*CA* **91**, 5080 (1979)].
676. E. A. Zvezdina, G. N. Dorofeenko, M. P. Zhdanova, and A. M. Simonov, U.S.S.R. Patent 490,801 (1975) [*CA* **84**, 74304 (1976)].
677. E. A. Zvezdina, M. P. Zhdanova, A. M. Simonov, and G. N. Dorofeenko, *Khim. Geterotsikl. Soedin.*, 1180 (1975) [*CA* **84**, 30994 (1976)].
678. T. Jaworski and S. Kwiatkowski, *Rocz. Chem.* **44**, 555 (1970) [*CA* **73**, 130845 (1970)].
679. R. W. J. Carney, J. Wojtkunski, B. Fechting, R. T. Puckett, B. Biffar, and G. DeStevens, *J. Org. Chem.* **36**, 2602 (1971).
680. G. Märkl, *Angew. Chem.* **78**, 907 (1966); *Angew. Chem., Int. Ed. Engl.* **5**, 846 (1966).
681. K. Dimroth, N. Greif, W. Städe, and F. W. Steuber, *Angew. Chem.* **79**, 727 (1967); *Angew. Chem., Int. Ed. Engl.* **6**, 711 (1967).

682. A. I. Tolmachev and E. S. Koslov, *Zh. Obshch. Khim.* **37**, 1922 (1967) [*CA* **68**, 105298 (1968)].
683. G. Märkl, F. Lieb, and A. Merz, *Angew. Chem.* **79**, 475 (1967); *Angew. Chem., Int. Ed. Engl.* **6**, 458 (1967).
684. K. Dimroth, *Top. Curr. Chem.* **38**, 1 (1973).
685. G. Märkl, D. E. Fischer, and H. Olbrich, *Tetrahedron Lett.*, 645 (1970).
686. C. C. Price, T. Parasaran, and T. V. Lakshminarayan, *J. Am. Chem. Soc.* **88**, 1034 (1966).
687. G. Märkl, A. Merz, and H. Rausch, *Tetrahedron Lett.*, 2989 (1971).
688. G. Märkl, F. Lieb, and A. Merz, *Angew. Chem.* **79**, 59 (1967); *Angew. Chem., Int. Ed. Engl.* **6**, 87 (1967).
689. G. Märkl, J. Heterocycl. Chem. **9**, S-69 (1972).
690. N. I. Shvetsov-Shilovskii, R. G. Bobkova, N. P. Ignatova, and N. N. Mel'nikov, *Usp. Khim.* **46**, 967 (1977) [*CA* **87**, 135478 (1977)].
691. R. Gompper and O. Christmann, *Chem. Ber.* **94**, 1795 (1961).
692. R. Gompper and O. Christmann, *Angew. Chem.* **71**, 32 (1959); *Chem. Ber.* **94**, 1784 (1961).
693. K. Dimroth and K. H. Wolf, *Angew. Chem.* **72**, 777 (1960).
694. K. Dimroth, H. Kroke, and K. Wolf, *Justus Liebigs Ann. Chem.* **678**, 202 (1964).
695. A. T. Balaban and A. Barabas, *Chem. Ind. (London)*, 404 (1967).
696. A. Barabas and A. T. Balaban, *Tetrahedron* **27**, 5495 (1971).
697. P. Canonne and L. C. Leitch, *Can. J. Chem.* **45**, 1761 (1967).
698. P. Canonne and H. Bilodau, *Can. J. Chem.* **44**, 2849 (1966); P. Canonne, P. Holm, and L. C. Leitch, *ibid.* **45**, 2151 (1967).
699. N. K. Cuong, F. Fournier, and J.-J. Basselier, *Bull. Soc. Chim. Fr.*, 2117 (1974).
700. K. Dimroth and G. Bräuniger, *Angew. Chem.* **68**, 519 (1956).
701. K. Dimroth, G. Bräuniger, and G. Neubauer, *Chem. Ber.* **90**, 1634 (1957).
702. K. Dimroth, G. Neubauer, H. Möllenkamp, and G. Oosterloo, *Chem. Ber.* **90**, 1668 (1957).
703. K. Dimroth, F. Kalk, and G. Neubauer, *Chem. Ber.* **90**, 2058 (1957).
704. ' K. Dimroth, F. Kalk, R. Sell, and K. Schlömer, *Justus Liebigs Ann. Chem.* **624**, 51 (1959).
705. K. Dimroth, G. Laubert, and K. H. Blöcher, *Justus Liebigs Ann. Chem.* **765**, 133 (1972).
706. G. Schill and H. Zollenkopf, *Justus Liebigs Ann. Chem.* **721**, 53 (1969).
707. K. Dimroth, W. Umbach, and K. H. Blöcher, *Angew. Chem.* **75**, 860 (1963).
708. K. Dimroth, A. Berndt, and R. Volland, *Chem. Ber.* **99**, 3040 (1966).
709. K. Dimroth and H. Wache, *Chem. Ber.* **99**, 399 (1966).
710. K. Dimroth and G. Laubert, *Angew. Chem.* **81**, 392 (1969); *Angew. Chem., Int. Ed. Engl.* **8**, 370 (1969).
711. K. Dimroth and G. Neubauer, *Angew. Chem.* **69**, 95 (1957).
712. G. A. Reynolds and J. A. Van Allan, *J. Heterocycl. Chem.* **8**, 301 (1971).
713. K. Dimroth, K. H. Wolf, and H. Wache, *Angew. Chem.* **75**, 860 (1963).
714. Y. Tamura, K. Sumoto, and H. Ikeda, *Chem. Ind. (London)*, 498 (1972).
715. G. Märkl and H. Baier, *Tetrahedron Lett.*, 4379 (1968).
716. G. Märkl and H. Baier, *Tetrahedron Lett.*, 4439 (1972).
717. K. T. Potts, A. J. Elliot, and M. Šorm, *J. Org. Chem.* **37**, 3838 (1972).
718. A. T. Balaban, *Rev. Roum. Chim.* **21**, 241 (1976).
719. O. Buchardt, C. L. Pedersen, U. Svanholm, A. M. Duffield, and A. T. Balaban, *Acta Chem. Scand.* **23**, 3125 (1969).

720. E. Klingsberg, *Abstr. Meet., Am. Chem. Soc., 1965*, 66S (1965); see also F. D. Popp and A. C. Noble, *Adv. Heterocycl. Chem.* **8**, 21 (1967).

721. E. A. Zvezdina, V. V. Derbenev, V. A. Bren, A. N. Popova, and G. N. Dorofeenko, *Khim. Geterotsikl. Soedin.*, 1025 (1976) [*CA* **86**, 4742 (1977)].

722. G. I. Zhungietu, I. V. Shantsevoi, and D. D. Buburuz, *Khim. Geterotsikl. Soedin.*, 281 (1971) [*CA* **75**, 35972 (1971)].

723. G. I. Zhungietu, I. V. Shantsevoi, and S. V. Krivun, *Khim. Geterotsikl. Soedin.*, 45 (1973) [*CA* **78**, 111271 (1973)].

724. G. I. Zhungietu, I. V. Shantsevoi, V. M. Kurilenko, and Zh. N. Khlienko, U.S.S.R. Patent 372,220 (1973) [*CA* **79**, 42571 (1973)].

725. G. I. Zhungietu, E. A. Revenko, and F. N. Chukrii, *Khim. Geterotsikl. Soedin.*, 347 (1973) [*CA* **78**, 147927 (1973)].

726. W. Tochtermann, *Top. Curr. Chem.* **15**, 378 (1970).

727. J.-P. Le Roux, J.-C. Cherton, and P.-L. Desbene, *C. R. Acad. Sci., Ser. C.* **280**, 37 (1975).

728. K. Hafner, *Angew. Chem.* **69**, 393 (1957).

729. K. Hafner and H. Kaiser, *Justus Liebigs Ann. Chem.* **618**, 140 (1958).

730. K. Hafner and H. Kaiser, *Org. Synth. Collect. Vol.* **5**, 1088 (1973).

731. G. N. Dorofeenko, A. V. Koblik, and T. I. Polyakova, U.S.S.R. Patent 534,446 (1976) [*CA* **87**, 5729 (1977)].

732. K. Hafner, *Justus Liebigs Ann. Chem.* **606**, 79 (1957).

733. J.-E. Mansson, M. Nilsson, and O. Wennerström, *Acta Chem. Scand., Ser. B* **31**, 47 (1977).

734. E. J. Corey and J. Streith, *J. Am. Chem. Soc.* **86**, 950 (1964).

735. W. H. Pirkle and L. H. McKendry, *J. Am. Chem. Soc.* **91**, 1179 (1969).

736. E. J. Corey and W. H. Pirkle, *Tetrahedron Lett.*, 5255 (1967).

737. E. Paterno, G. Chieffi, and G. Perret, *Gazz. Chim. Ital.* **44**, 151 (1914) [*CA* **8**, 2687 (1914)].

738. M. Guia and M. Civera, *Gazz. Chim. Ital.* **81**, 875 (1951) [*CA* **47**, 4335 (1953)].

739. P. Yates and M. J. Jorgenson, *J. Am. Chem. Soc.* **80**, 6150 (1958).

740. P. Yates and M. J. Jorgenson, *J. Am. Chem. Soc.* **85**, 2956 (1963).

741. P. Yates and I. W. J. Still, *J. Am. Chem. Soc.* **85**, 1208 (1963).

742. A. Padwa and R. Hartman, *J. Am. Chem. Soc.* **88**, 1518 (1966).

743. P. Yates, E. S. Hand, P. Singh, S. K. Roy, and I. W. J. Still, *J. Org. Chem.* **34**, 4046 (1969).

744. J. W. Pavlik and J. Kwong, *J. Am. Chem. Soc.* **95**, 7914 (1973).

745. N. Ishibe, M. Odani, and M. Sunami, *J. C. S. Chem. Commun.*, 1034 (1971).

746. N. Ishibe, M. Sunami, and M. Odani, *J. Am. Chem. Soc.* **95**, 463 (1973).

747. J. W. Pavlik and L. T. Pauliukonis, *Tetrahedron Lett.*, 1939 (1976).

748. E. B. Keil and J. W. Pavlik, *J. Heterocycl. Chem.* **13**, 1149 (1976).

749. D. H. R. Barton and L. A. Hulshof, *J. C. S. Perkin I*, 1103 (1977).

750. J. A. Barltrop, A. C. Day, and C. J. Samuel, *J. Am. Chem. Soc.* **101**, 7521 (1979).

751. J. W. Pavlik, D. R. Bolin, K. C. Bradford, and W. G. Anderson, *J. Am. Chem. Soc.* **99**, 2816 (1977).

752. J. W. Pavlik and E. L. Clennan, *J. Am. Chem. Soc.* **95**, 1697 (1973).

753. J. A. Barltrop, J. C. Barrett, R. W. Carder, A. C. Day, J. R. Harding, W. E. Long, and C. J. Samuel, *J. Am. Chem. Soc.* **101**, 7510 (1979).

754. J. A. Barltrop, A. C. Day, and C. J. Samuel, *J. C. S. Chem. Commun.*, 823 (1976).

755. J. A. Barltrop, K. Dawes, A. C. Day, and A. J. H. Summers, *J. C. S. Chem. Commun.*, 1240 (1972).

756. J. A. Barltrop, K. Dawes, A. C. Day, and A. J. H. Summers, *J. Am. Chem. Soc.* **95**, 2406 (1973).
757. M. Shiozaki and T. Hiraska, *Tetrahedron Lett.*, 4655 (1972).
758. A. T. Balaban, *Rev. Roum. Chim.* **11**, 1097 (1966).
759. A. T. Balaban, *Rev. Roum. Chim.* **15**, 463 (1970).
760. J. A. Barltrop, R. Carder, A. C. Day, J. R. Harding, and C. Samuel, *J. C. S. Chem. Commun.*, 729 (1975).
761. J. A. Barltrop and A. C. Day, *J. C. S. Chem. Commun.*, 177 (1975).
762. J. A. Barltrop, A. C. Day, P. D. Moxon, and R. W. Ward, *J. C. S. Chem. Commun.*, 786 (1975).
763. E. F. Ullman, *J. Am. Chem. Soc.* **85**, 3529 (1963).
764. E. F. Ullman and W. A. Henderson, *J. Am. Chem. Soc.* **86**, 5050 (1964).
765. J. M. Dunston and P. Yates, *Tetrahedron Lett.*, 505 (1964).
766. E. F. Ullman and J. E. Milks, *J. Am. Chem. Soc.* **84**, 1315 (1962); **86**, 3814 (1964).
767. E. N. Marvell and T. Gosink, *J. Org. Chem.* **37**, 3036 (1972).
768. G. Maier and M. Wiessler, *Tetrahedron Lett.*, 4987 (1969).
769. W. Schroth and R. Spitzner, *Z. Chem.*, in preparation.
770. P. Schiess, H. L. Chia, and C. Suter, *Tetrahedron Lett.*, 5747 (1968).
771. P. Schiess and H. L. Chia, *Helv. Chim. Acta* **53**, 485 (1970).
772. P. Schiess, R. Seeger, and C. Suter, *Helv. Chim. Acta* **53**, 1713 (1970).
773. P. Schiess and C. Suter, *Helv. Chim. Acta* **54**, 2636 (1971).
774. P. Schiess, *Helv. Chim. Acta* **55**, 2365 (1972).
775. A. Roedig, K. Fleischmann, F. Frank, and R. Rettenberger, *Justus Liebigs Ann. Chem.*, 2091 (1977).
776. E. T. Oestensen and M. M. Mishrikey, *Acta Chem. Scand., Ser. B* **30**, 635 (1976).
777. E. T. Oestensen, A. Abdel-Azeen Abdallah, S. H. Skarre, and M. M. Mishrikey, *Acta Chem. Scand., Ser. B* **31**, 496 (1977).
778. A. Mustafa, *Chem. Rev.* **43**, 509 (1948).
779. J. H. Day, *Chem. Rev.* **63**, 65 (1963).
780. G. N. Dorofeenko, Ph.D. Habil. Thesis, Rostov University, Rostov-on-Don, 1966.
781. A. T. Balaban, C. N. Rentea, and M. Bacescu-Roman, *Rev. Roum. Chim.* **10**, 863 (1965).
782. A. T. Balaban, *in* "La nature et les propriétés des liaisons de coordination," p. 233, CNRS, Paris, 1970.
783. A. T. Balaban, *Tetrahedron Lett.*, 5055 (1978).
784. L. Yu. Ukhin, V. J. Il'in, Zh. I. Orlova, N. G. Bokii, and Yu. T. Struchkov, *J. Organomet. Chem.* **113**, 167 (1976).
785. L. Yu. Ukhin, V. V. Krasnikov, and G. N. Dorofeenko, *Koord. Khim.* **4**, 455 (1978) [*CA* **89**, 24512 (1978)].
786. L. Yu. Ukhin, Zh. I. Orlova, V. I. Il'in, A. I. Pyshchev, and G. N. Dorofeenko, *Koord. Khim.* **4**, 772 (1978) [*CA* **89**, 109911 (1978)].
787. R. P. Hughes, *J. Organomet. Chem.* **141**, C29 (1977).
788. L. Yu. Ukhin, V. I. Il'in, Zh. I. Orlova, N. G. Bokii, and Yu. T. Struchkov, U.S.S.R. Patent 530,884 (1976) [*CA* **86**, 106788 (1977)].
789. L. Yu. Ukhin and E. P. Onokolova, *Dokl. Akad. Nauk SSSR* **241**, 858 (1978) [*CA* **89**, 197699 (1978)].
790. L. Yu. Ukhin, A. I. Pyshchev, V. V. Krasnikov, Zh. I. Orlova, and G. N. Dorofeenko, *Dokl. Akad. Nauk SSSR* **234**, 1351 (1977) [*CA* **87**, 168162 (1977)].
791. T. L. Gilchrist, R. Livingston, C. W. Rees, and E. Angerer, *J. C. S. Perkin I*, 2535 (1973).

792. A. Hantzsch, *Ber. Dtsch. Chem. Ges.* **52**, 1535, 1544 (1919).
793. R. C. Gibbs, J. R. Johnson, and E. C. Hughes, *J. Am. Chem. Soc.* **52**, 4895 (1930).
794. R. Wizinger and K. Wagner, *Helv. Chim. Acta* **34**, 2290 (1951).
795. R. Wizinger, A. Grüne, and E. Jacobi, *Helv. Chim. Acta* **39**, 1 (1956).
796. R. Wizinger, S. Losinger, and P. Ulrich, *Helv. Chim. Acta* **39**, 5 (1956).
797. A. T. Balaban, V. E. Sahini, and E. Keplinger, *Tetrahedron* **9**, 163 (1960).
798. A. T. Balaban, *Stud. Cercet. Chim.* **7**, 257 (1959) [*CA* **54**, 7521 (1960)].
799. A. T. Balaban and Z. Simon, *Tetrahedron* **18**, 315 (1962).
800. A. T. Balaban and Z. Simon, *Rev. Roum. Chim.* **10**, 1059 (1965).
801. A. T. Balaban and C. D. Nenitzescu, *Izv. Akad. Nauk SSSR, Otd. Khim. Nauk,* 2064 (1960) [*CA* **55**, 16139 (1961)].
802. I. Degani and C. Vincenzi, *Boll. Sci. Fac. Chim. Ind. Bologna* **23**, 249 (1965) [*CA* **63**, 13025 (1965)].
803. I. Degani, R. Fochi, and C. Vincenzi, *Gazz. Chim. Ital.* **94**, 203 (1964) [*CA* **61**, 5611 (1964)].
804. H. Suzuki, "Electronic Absorption Spectra and Geometry of Organic Molecules," p. 196. Academic Press, New York, 1967.
805. A. E. Gillam and E. S. Stern, "An Introduction to Electronic Absorption Spectroscopy in Organic Chemistry," 2nd ed., p. 153. Arnold, London, 1957.
806. R. A. Friedel and M. Orchin, "Ultraviolet Spectra of Aromatic Compounds." Wiley, New York, 1951.
807. J. N. Murrel, "The Theory of the Electronic Spectra of Organic Molecules." Wiley, New York, 1963.
808. H. H. Joffé and M. Orchin, "Theory and Application of Ultraviolet Spectroscopy," p. 299. Wiley, New York, 1962.
809. G. N. Lewis and M. Calvin, *Chem. Rev.* **25**, 273 (1939).
810. G. N. Lewis and J. Bigeleisen, *J. Am. Chem. Soc.* **65**, 2102, 2107 (1943).
811. Cf. also W. D. Kumler, *J. Am. Chem. Soc.* **68**, 1184 (1946).
812. R. N. Jones, *Chem. Rev.* **41**, 353 (1947).
813. J. R. Platt, *J. Chem. Phys.* **17**, 484 (1949).
814. J. R. Platt, *J. Chem. Phys.* **19**, 101 (1951).
815. H. B. Klevens and J. R. Platt, *J. Chem. Phys.* **17**, 470 (1949).
816. M. G. Mayer and A. L. Sklar, *J. Chem. Phys.* **6**, 64 (1938).
817. W. Moffit, *J. Chem. Phys.* **22**, 320 (1954).
818. L. Doub and J. M. Vandenbelt, *J. Am. Chem. Soc.* **69**, 2714 (1947).
819. E. A. Braude, *Annu. Rep. Prog. Chem.* **42**, 105 (1945) [*CA* **42**, 8773 (1948)].
820. E. Clar, "Aromatische Kohlenwasserstoffe," 2nd ed., p. 36. Springer-Verlag, Berlin and New York, 1952.
821. E. Clar, *Chem. Ber.* **82**, 495 (1949).
822. M. D. Gheorghiu and A. T. Balaban, *Rev. Roum. Chim.* **21**, 1513 (1976).
823. Y. Maroni-Barnaud, P. Maroni, M. Simalty, and Y. Madaule, *Bull. Soc. Chim. Fr.,* 1398 (1970).
824. M. Simalty, J. Carretto, and S. Sib, *Bull. Soc. Chim. Fr.,* 3920 (1970).
825. M. Simalty, J. Carretto, and S. Sib, *Bull. Soc. Chim. Fr.,* 3926 (1970).
826. A. T. Balaban, M. Gavat, P. T. Frangopol, M. Mocanu, and C. D. Nenitzescu, *Stud. Cercet. Chim.* **12**, 71 (1964).
827. J. R. Wilt, G. A. Reynolds, and J. A. Van Allan, *Tetrahedron* **29**, 795 (1973).
828. S. Kostanecki and G. Rossbach, *Ber. Dtsch. Chem. Ges.* **29**, 1488 (1896).
829. W. Dilthey, G. Bauriedel, B. Burger, G. Geisselbrecht, F. Ibach, K. Kiefer, A. Seeger, O. Simon, R. Taucher, and J. Winkler, *J. Prakt. Chem.* **102**, 209 (1921).

830. H. Kautsky and G. Müller, *Naturwissenschaften* **29**, 150 (1941).
831. E. G. Protsenko, V. G. Tishchenko, and B. G. Distanov, *Stsintill. Org. Lyuminofory,* 117 (1972) [*CA* **79**, 145563 (1973)].
832. S. K. Chakrabarti, P. T. Frangopol, and M. Frangopol, *Rev. Roum. Phys.* **17**, 1053 (1972).
833. V. P. Karmazin, M. I. Knyazhanskii, E. P. Olekhnovich, and G. N. Dorofeenko, *Zh. Prikl. Spektrosk.* **22**, 234 (1975) [*CA* **83**, 17984 (1975)].
834. A. A. Gyurov, *God. Mash.-Elektrotekh. Inst.* **2**, 15 (1955) [*CA* **54**, 22606 (1960)].
835. K. Kiciak, *Rocz. Chem.* **37**, 225 (1963) [*CA* **59**, 2971 (1963)].
836. F. P. Schäfer, W. Schmidt, and K. Marth, *Phys. Lett. A* **24A**, 280 (1967).
837. J. L. R. Williams and G. A. Reynolds, *J. Appl. Phys.* **39**, 5327 (1968).
838. G. A. Reynolds, S. A. Tuccio, O. G. Peterson, and D. P. Specht, Ger. Offen. 2,109,040 (1971) [*CA* **76**, 40148 (1972)].
839. G. Briegleb, "Elektronen-Donator-Acceptor-Komplexe." Springer-Verlag, Berlin and New York, 1961.
840. J. Rose, "Molecular Complexes." Pergamon, Oxford, 1967.
841. R. S. Mulliken and W. B. Person, "Molecular Complexes." Wiley (Interscience), New York, 1969.
842. R. Foster, "Organic Charge-Transfer Complexes." Academic Press, New York, 1969.
843. E. M. Kosower, "Molecular Biochemistry." McGraw-Hill, New York, 1962; "An Introduction to Physical Organic Chemistry." Wiley, New York, 1968.
844. A. Lablache-Combier, *Bull. Soc. Chim. Fr.,* 4791 (1972).
845. M. Feldman and S. Winstein, *Tetrahedron Lett.,* 853 (1962).
846. A. T. Balaban, *C. R. Acad. Sci.* **256**, 4041 (1963).
847. R. A. Mackay, J. R. Landolph, and E. J. Poziomek, *J. Am. Chem. Soc.* **93**, 5026 (1971).
848. R. A. Mackay and E. J. Poziomek, *J. Am. Chem. Soc.* **94**, 4167 (1972).
849. A. T. Balaban and M. Paraschiv, *Rev. Roum. Chim.* **19**, 1731 (1974).
850. S. Badilescu and A. T. Balaban, *Spectrochim. Acta, Part A* **32A**, 1311 (1976).
851. T. Tamamura, M. Yokoyama, S. Kusabayashi, and H. Mikawa, *Bull. Chem. Soc. Jpn.* **47**, 442 (1974).
852. T. Tamamura, H. Yasuba, K. Okamoto, T. Imai, S. Kusabayashi, and H. Mikawa, *Bull. Chem. Soc. Jpn.* **47**, 448 (1974).
853. T. Tamamura, T. Yamare, N. Yasuoka, and N. Kasai, *Bull. Chem. Soc. Jpn.* **47**, 832 (1974).
854. S. Badilescu, L. Manu, and A. T. Balaban, *Rev. Roum. Chim.* **24**, 947 (1979).
855. E. Le Goff and R. B. La Count, *J. Am. Chem. Soc.* **85**, 1354 (1963).
856. W. J. Siemons, P. E. Bierstedt, and R. G. Kepler, *J. Chem. Phys.* **39**, 3528 (1963).
857. A. Chyla and Z. Romaszewski, *Lecture Notes in Physics* **65**, 521 (1977).
858. C. Parkanyi and G. J. Leu, *Z. Naturforsch., B: Anorg. Chem., Org. Chem.* **30**, 984 (1975).
859. G. Briegleb, G. Betz, and W. Herre, *Z. Phys. Chem. (Wiesbaden)* **64**, 85 (1969).
860. M. Avram and G. D. Mateescu, "Infrared Spectroscopy." Wiley (Interscience), New York, 1972.
861. A. R. Katritzky, *Q. Rev. (London)* **13**, 353 (1959).
862. A. R. Katritzky (ed.), "Physical Methods in Heterocyclic Chemistry," Vol. 2. Academic Press, New York, 1963; Vol. IV, 1971.
863. I. I. Stanoiu, M. Paraschiv, E. Romas, and A. T. Balaban, *Spectrochim. Acta, Part A* **28A**, 1001 (1972).
864. Z. Yoshida, H. Sugimoto, and S. Yoneda, *Tetrahedron* **30**, 2099 (1974).

865. A Roedig and H. A. Renk, *Chem. Ber.* **106**, 3877 (1973).
866. A. D. Semenov, G. N. Dorofeenko, and V. I. Dulenko, *Khim. Geterotsikl. Soedin.*, 14 (1966) [*CA* **65**, 5333 (1966)].
867. Y. Maroni-Barnaud, P. Maroni, M. Simalty, and Y. Madaule, *Bull. Soc. Chim. Fr.*, 546 (1971).
868. A. T. Balaban, G. R. Bedford, and A. R. Katritzky, *J. Chem. Soc.*, 1646 (1964).
869. I. Degani, F. Taddei, and C. Vincenzi, *Boll. Sci. Fac. Chim. Ind. Bologna* **25**, 61 (1967) [*CA* **68**, 48853 (1968)].
870. I. Degani, L. Lunazzi, and F. Taddei, *Boll. Sci. Fac. Chim. Ind. Bologna* **23**, 131 (1965) [*CA* **63**, 15750 (1965)].
871. L. Radics and J. Kardos, *Org. Magn. Reson.* **5**, 251 (1973).
872. A. A. Bothner-By and S. M. Castellano, *Comput. Programs Chem.* **1**, 10 (1968).
873. V. M. S. Gil and A. J. L. Pinto, *Mol. Phys.* **19**, 573 (1970).
874. C. C. Rentia, A. T. Balaban, and Z. Simon, *Rev. Roum. Chim.* **11**, 1193 (1966).
875. A. Camerman and L. H. Jensen, *Acta Crystallogr., Sect. B* **25**, 12 (1969).
876. C. Uncuta and A. T. Balaban, *Rev. Roum. Chim.* **21**, 251 (1976).
877. A. I. Tolmachev, L. M. Shulezhko, and M. Yu. Kornilov, *Ukr. Khim. Zh.* **40**, 287 (1974) [*CA* **81**, 3730 (1974)].
878. M. Yu. Kornilov, L. M. Shulezhko, and A. I. Tolmachev, *Ukr. Khim. Zh.* **40**, 212 (1974) [*CA* **80**, 132264 (1974)].
879. Y. Yamamota, K. Kuno, and H. Nozaki, *Bull. Chem. Soc. Jpn.* **44**, 2265 (1971).
880. E. T. Oestensen, *Acta. Chem. Scand., Ser. B* **B28**, 1107 (1974).
881. A. T. Balaban and V. Wray, *Z. Naturforsch., B: Anorg. Chem., Org. Chem.* **30**, 654 (1975).
882. A. T. Balaban and V. Wray, unpublished.
883. M. T. Chenon, S. Sib, and M. Simalty, *Org. Magn Reson.* **12**, 71 (1979).
884. J. B. Conant, L. F. Small, and B. S. Taylor, *J. Am. Chem. Soc.* **47**, 1959 (1925); J. B. Conant and H. B. Cutter, *ibid.* **48**, 1016 (1926).
885. L. A. Polyakova, K. A. Bilevich, N. N. Bubnov, G. N. Dorofeenko, and O. Yu. Okhlobystin, *Dokl. Akad. Nauk SSSR* **212**, 370 (1973) [*CA* **79**, 145660 (1973)].
886. I. Degani and C. Vincenzi, *Boll. Sci. Fac. Chim. Ind. Bologna* **25**, 77 (1967) [*CA* **68**, 114369 (1968)].
887. I. Degani, L. Lunazzi, and G. F. Pedulli, *Mol. Phys.* **14**, 217 (1968).
888. F. W. Steuber and K. Dimroth, unpublished results, quoted after Ref. 684, namely on p. 45.
889. V. B. Panov, M. V. Nekhoroshev, and O. Yu. Okhlobystin, *Dokl. Akad. Nauk SSSR* **243**, 372 (1978) [*CA* **90**, 86384 (1979)].
890. M. Farcasiu and D. Farcasiu, *Chem. Ber.* **102**, 2294 (1969).
891. C. Hacquard and A. Rassat, *Mol. Phys.* **30**, 1935 (1975).
892. E. Krumbholds and F. W. Steuber, *Angew. Chem.* **87**, 588 (1975); *Angew. Chem., Int. Ed. Engl.* **14**, 553 (1975).
893. V. A. Samaraskii, V. V. Panov, M. V. Nekhoroshev, V. A. Khizhny, O. Yu. Okholobystin, and V. D. Pokhodenko, *Zh. Org. Khim.* **14**, 1643 (1978) [*CA* **89**, 214538 (1978)].
894. J. Alizon, J. Gallice, H. Robert, G. Delplanque, C. Weyl, C. Fabre, and H. Strzelecka, *Mol. Cryst. Liq. Cryst.* **33**, 91 (1976).
895. N. T. Berberova, G. N. Dorofeenko, and O. Yu. Okhlobystin, *Khim. Geterotsikl. Soedin.*, 318 (1977) [*CA* **87**, 67593 (1977)].
896. N. G. Bokii, R. V. Vedrinskii, V. V. Kitaev, N. A. Lopatina, and Yu. T. Struchkov, *Koord. Khim.* **2**, 103 (1976) [*CA* **84**, 114608 (1976)].
897. F. Bohlmann and C. Zdero, *Chem. Ber.* **101**, 3941 (1678).

898. J. Collin, *Bull. Soc. Chim. Belg.* **69**, 575 (1960).
899. W. K. Rohwedder, A. F. Marbrouk, and E. Selke, *J. Phys. Chem.* **69**, 1711 (1965).
900. A. M. Duffield, C. Djerassi, and A. T. Balaban. *Org. Mass Spectrom.* **5**, 87 (1971).
901. G. Hvistendahl, P. Gyorösi, and K. Undheim, *Org. Mass Spectrom.* **9**, 80 (1974).
902. P. Ellingsen, G. Hvistendahl, and K. Undheim, *Org. Mass Spectrom.* **13**, 455 (1978).
903. M. G. B. Drew, G. W. A. Fowles, D. A. Rice, and K. J. Shanton, *J. C. S. Chem. Commun.*, 614 (1974).
904. N. Serpone and P. H. Bird, *J. C. S. Chem. Commun.*, 284 (1975).
905. E. P. Kirpichev and Yu. I. Rubtsov, *Zh. Fiz. Khim.* **43**, 2025 (1969) [*CA* **72**, 36560 (1970)].
906. P. Beak, D. S. Mueller, and J. Lee, *J. Am. Chem. Soc.* **96**, 3867 (1974).
907. S. W. Benson, "Thermochemical Kinetics." Wiley, New York, 1968.
908. C. L. Norris, R. C. Benson, P. Beak, and W. H. Flygare, *J. Am. Chem. Soc.* **95**, 2766 (1973).
909. W. Haberditzl, *Wiss. Z. Tech. Hochsch. Chem. "Carl Schorlemmer" Leuna-Merseburg* **3**, 401 (1960/61) [*CA* **57**, 1720 (1962)].
910. R. Havemann, W. Haberditzl, and H. Köppel, *Z. Phys. Chem. (Leipzig)* **218**, 277 (1961).
911. E. H. Poindexter, J. A. Potenza, D. D. Thompson, Nguyen Van Nghia, and R. H. Webb, *Mol. Phys.* **14**, 385 (1968).
912. R. M. Bowyer, A. Ledwith, and D. C. Sherrington, *J. Chem. Soc. B,* 1511 (1971).
913. G. Schwarzenbach and K. Lutz, *Helv. Chim. Acta* **23**, 1147 (1940).
914. E. Gârd and A. T. Balaban, *J. Electroanal. Chem.* **4**, 48 (1962).
915. F. Pragst, *Electrochim. Acta* **21**, 407 (1976).
916. F. Pragst and U. Seydewitz, *J. Prakt. Chem.* **319**, 952 (1977).
917. F. Pragst, M. Janda, and I. Stibor, *Electrochim. Acta* **25**, 779 (1980).
918. N. T. Berberova, G. N. Dorofeenko, and O. Yu. Okhlobystin, *Khim. Geterotsikl. Soedin.*, 1574 (1976) [*CA* **86**, 88996 (1977)].
919. M. M. Evstifeev, L. L. Pyshcheva, A. I. Pyshchev, and G. N. Dorofeenko, *Zh. Obshch. Khim.* **46**, 1340 (1976) [*CA* **85**, 93514 (1976)].
920. M. V. Nekhoroshev and O. Yu. Okhlobystin, *Zh. Org. Khim.* **13**, 1294 (1977) [*CA* **87**, 117749 (1977)].
921. L. L. Pyshcheva and G. Kh. Aminova, *Molodye Uch.-Naucho-Tekh. Prog.*, 95 (1973) [*CA* **82**, 179968 (1975)].
922. E. Hopîrtean, *Rev. Roum. Chim.* **22**, 1385 (1977).
923. E. Hopîrtean, M. Preda, and C. Liteanu, *Fresenius' Z. Anal. Chem.* **286**, 65 (1977).
924. F. D. Saeva and G. R. Olin, *J. Am. Chem. Soc.* **102**, 299 (1980).
925. Yu. A. Zhdanov, G. N. Dorofeenko, and S. V. Zelenskaya, *Zh. Obshch. Khim.* **36**, 210 (1966) [*CA* **64**, 16601 (1966)].
926. J. Koutecky, *Collect. Czech. Chem. Commun.* **24**, 1608 (1959).
927. Z. Simon and C. Volanschi, *Stud. Cercet. Chim.* **8**, 641 (1960) [*CA* **55**, 19458 (1961)].
928. Z. Simon, *Opt. Spektrosk.* **12**, 22 (1962) [*CA* **57**, 1757 (1962)].
929. Z. Simon and A. T. Balaban, *Rev. Roum. Chim.* **9**, 339 (1964).
930. Z. Simon and A. T. Balaban, *Stud. Cercet. Chim.* **12**, 345 (1964).
931. L. Goodman and H. Shull, *J. Chem. Phys.* **22**, 1138 (1954).
932. G. V. Boyd, *Rev. Roum. Chim.* **12**, 1133 (1967).
933. E. A. Zvezdina, M. P. Zhdanova, V. A. Bren, and G. N. Dorofeenko, *Khim. Geterotsikl. Soedin.*, 1484 (1976) [*CA* **86**, 106304 (1977)].
934. G. V. Boyd and A. T. Balaban, *Rev. Roum. Chim.* **14**, 1575 (1969).
935. G. V. Boyd and N. Singer, *Tetrahedron* **21**, 1263 (1965).
936. N. Singer, P. R. Whittington, and G. V. Boyd, *Tetrahedron* **26**, 3731 (1970).

937. M. J. S. Dewar and G. J. Gleicher, *J. Chem. Phys.* **44**, 768 (1966).
938. Z. Yoshida, H. Sugimoto, and S. Yoneda, *Tetrahedron* **28**, 5873 (1972).
939. C. Decoret, J. Royer, and B. Tinland, *Bull. Soc. Chim. Fr.,* 2235 (1972).
940. G. Karlsson and O. Mårtensson, *Theor. Chim. Acta* **13**, 195 (1969).
941. O. Mårtensson and C. H. Warren, *Acta Chem. Scand.* **24**, 2745 (1970).
942. O. Mårtensson, *Acta Chem. Scand.* **24**, 3417 (1970).
943. R. W. Bigelow, *J. Chem. Phys.* **67**, 4498 (1977).
944. J. Fabian, A. Mehlhorn, and R. Zahradnik, *Theor. Chim. Acta* **12**, 247 (1968).
945. A. Mistr, M. Vávra, J. Skroupý, and R. Zahradnik, *Collect. Czech. Chem. Commun.* **37**, 1520 (1972).
946. A. Mistr and R. Zahradnik, *Collect. Czech. Chem. Commun.* **38**, 1668 (1973).
947. M. H. Palmer, R. H. Findlay, W. Moyes, and A. J. Gaskell, *J. C. S. Perkin II,* 841 (1975).
948. S. K. Chakrabarti, P. T. Frangopol, and M. Frangopol, *Rev. Roum. Phys.* **17**, 1053 (1972).
949. N. Vlahovici and P. T. Frangopol, *Rev. Roum. Chim.* **22**, 1379 (1977).
950. A. F. Pronin and V. G. Kharchenko, *Khim. Geterotsikl. Soedin.,* 1206 (1977) [*CA* **88**, 21875 (1978)].
951. K. Hafner and H. Kaiser, *Org. Synth. Collect. Vol.* **5**, 1108 (1973).
952. A. T. Balaban and A. J. Boulton, *Org. Synth. Collect. Vol.* **5**, 1114 (1973).
953. L. G. S. Brooker, D. W. Heseltine, and D. S. Daniel, French Demande 2,004,639 (1969) [*CA* **73**, 30656 (1970)].
954. B. D. Illingsworth and J. E. Jones, French Patent 1,522,354 (1968) [*CA* **72**, 17279 (1970)].
955. J. E. Jones and W. E. Yoerger, U.S. Patent 3,958,991 (1976) [*CA* **85**, 169691 (1976)].
956. J. F. Van Besauw and A. L. Poot, Ger. Offen. 2,506,445 (1975) [*CA* **84**, 52160 (1976)].
957. T. Tani, U.S. Patent 3,970,459 (1976) [*CA* **86**, 99007 (1977)].
958. J. K. Lindsay, U.S. Patent 3,577,238 (1969) [*CA* **75**, 50436 (1971)].
959. F. J. Rauner and C. G. Houle, Australian Patent 288,153 (1969) [*Ref. Zh., Khim.,* 17 H 7061 (1969)].
960. L. G. S. Brooker, D. W. Heseltine, and D. S. Daniel, U.S. Patent 3,579,346 (1971) [*Ref. Zh., Khim.,* 8 H 690 (1972)].
961. J. A. Van Allan, C. C. Natale, and F. J. Rauner, Belgian Patent 623,972 (1963) [*CA* **63**, 10102 (1965)].
962. G. A. Reynolds and J. A. Van Allan, French Patent 2,055,690 (1971) [*CA* **76**, 87190 (1972)].
963. J. Rochlitz and R. Lehner, Ger. Offen. 2,160,812 (1973) [*CA* **79**, 120881 (1973)].
964. M. Ikeda, H. Sato, E. Torii, K. Morimoto, and Y. Hasegawa, Ger. Offen. 2,149,293 (1973) [*CA* **79**, 47839 (1973)].
965. S. H. Merrill and L. E. Contois, Ger. Offen. 2,013,506 (1970) [*CA* **75**, 103700 (1970)].
966. C. J. Fox, U.S. Patent 3,784,376 (1970) [*CA* **80**, 114823 (1974)].
967. R. W. Stahr and T. H. Morse, U.S. Patent 3,810,759 (1974) [*CA* **81**, 56640 (1974)].
968. G. A. Reynolds and J. A. Van Allan, French Patent 2,105,830 (1972) [*CA* **79**, 11993 (1973)].
969. L. E. Contois and S. H. Merrill, French Demande 2,016,435 (1969) [*CA* **74**, 93486 (1971)].
970. C. J. Fox, French Patent 1,523,960 (1968) [*CA* **72**, 49676 (1970)].
971. D. R. Davis, C. C. Natale, and C. J. Fox, Belgian Patent 626,528 (1963) [*CA* **60**, 8822 (1964)].
972. K. Ueno and T. Tsunoda, Japanese Kokai 74/135,616 (1974) [*CA* **83**, 106227 (1975)].

973. G. A. Reynolds, J. A. Van Allan, and L. E. Contois, U.S. Patent 3,938,994 (1976) [CA **84,** 152236 (1976)].
974. M. Kuroda and K. Morimoto, Japanese Patent 77143957 (1971) [*Ref. Zh., Khim.,* 13 H 951 (1972)].
975. M. A. Berwick, C. J. Fox, and W. A. Light, Ger. Offen. 2,557,430 (1976) [CA **86,** 131100 (1977)].
976. R. F. Bartlett and L. K. Case, U.S. Patent 3,982,935 (1976) [CA **86,** 148806 (1977)].
977. T. Yamaoka, K. Ueno, T. Tsunoda, and K. Torige, *Polymer* **18,** 81 (1977).
978. D. M. Sturmer, U.S. Patent 4,028,113 (1977) [CA **87,** 54536 (1977)]; U.S. Patent 3,984,248 (1976) [CA **86,** 18376 (1977)].
979. M. T. Regan, G. A. Reynolds, D. P. Specht, and J. A. Van Allan, Ger. Offen. 2,733,911 (1978) [CA **88,** 144335 (1978)].
980. M. Okazaki, A. Yamaguchi, A. Kozima, and M. Sasaki, Ger. Offen. 2,717,007 (1977) [CA **88,** 43755 (1978)].
981. C. F. Fox, Ger. Offen. 2,631,629 (1977) [CA **86,** 163620 (1977)].
982. K. Emoto and K. Futaki, Japanese Kokai 77/52,637 (1977) [CA **88,** 97395 (1978)]; Japanese Patent 76/81,622 (1976) [CA **86,** 131104 (1977)].
983. Kodak Soc., Belgian Patent 649,986 (1965) [CA **64,** 15230 (1966)].
984. F. J. Rauner and C. G. Houle, French Patent 1,387,433 (1965) [CA **64,** 9136 (1966)].
985. J. G. McNally, Ger. Offen. 2,035,392 (1971) [CA **74,** 149244 (1971)].
986. S. S. Fico and J. W. Manthey, U.S. Patent 3,772,028 (1973) [CA **80,** 151233 (1974)].
987. F. J. Rauner and C. G. Houle, U.S. Patent 3,300,314 (1967) [*Ref. Zh., Khim.,* 12 H 436 (1975)].
988. G. A. Reynolds, U.S. Patent 3,148,067 (1964) [CA **61,** 15571 (1964)].
989. M. J. Alsup and A. R. Guevara, U.S. Patent 3,703,373 (1972) [CA **78,** 36321 (1973)].
990. J. G. McNally, U.S. Patent 3,679,415 (1972) [CA **77,** 146217 (1972)].
991. J. J. De Palma and A. W. Johnson, French Patent 1,391,547 (1965) [CA **64,** 1510 (1966)].
992. D. Gallois and P. Carlu, French Patent 1,461,640 (1966) [CA **66,** 120791 (1967)].
993. C. G. Houle and T. J. Masseth, U.S. Patent 3,671,251 (1972) [CA **77,** 133216 (1972)].
994. D. G. Borden, Ger. Offen. 1,815,868 (1969) [CA **73,** 40481 (1970)].
995. O. Michihiro, T. Mayuzumi, T. Fujino, and A. Noshiro, Ger. Offen. 2,233,514 (1973) [CA **78,** 137311 (1973)].
996. R. C. De Selms and C. V. Wilson, U.S. Patent 3,503,740 (1970) [CA **72,** 138325 (1970)].
997. E. P. Gramza and D. D. Schreiber, U.S. Patent 3,684,502 (1972) [CA **77,** 171226 (1972)].
998. E. J. Seus, U.S. Patent 3,591,374 (1971) [*Ref. Zh., Khim.,* 6 H 731 (1972)].
999. W. A. Light, S. African Patent 70101,473 (1970) [CA **75,** 13556 (1971)].
1000. K. Maruyama, K. Kojima, T. Kubota, M. Charada, and M. Oda, Japanese Patent 7,615,786 (1976) [*Ref. Zh., Khim.,* 2 H 329 (1977)].
1001. S. Inoue and Y. Sumimoto, Japanese Patent 7,428,457 (1974) [CA **82,** 178197 (1975)].
1002. P. J. Grisdale, Fr. Patent 2,083,980 (1972) [CA **77,** 103347 (1972)].
1003. K. Maruyama, T. Kubota, K. Kojima, and H. Tamura, Japanese Patent 74,105,345 (1974) [CA **87,** 171363 (1974)].
1004. F. J. Kryman and W. J. Staudenmayer, U.S. Patent 3,679,408 (1972) [CA **77,** 158751 (1972)].
1005. C. J. Fox and W. A. Light, U.S. Patent 3,706,554 (1972) [CA **78,** 50580 (1973)].
1006. H. Hartmann, *J. Prakt. Chem.* **313,** 1113 (1971).
1007. L. G. S. Brooker, D. W. Heseltine, and D. S. Daniel, French Demande 2,004,640 (1969) [CA **73,** 30660 (1970)].

1008. Y. Murakami, Y. Hasegana, and K. Morimoto, U.S. Reissue Patent 28,698 (1976) [CA 85, 169686 (1976)].
1009. V. P. Grigor'ev and V. V. Ekilik, Zh. Prikl. Khim. 41, 2770 (1968) [CA 70, 73486 (1969)].
1010. V. P. Grigor'ev and V. V. Ekilik, Zh. Prikl. Khim. 42, 1295 (1969) [CA 72, 8612 (1970)].
1011. V. P. Grigor'ev and I. M. Gershanova, Zh. Prikl. Khim. 42, 2135 (1969) [CA 72, 38140 (1969)].
1012. A. N. Nikolaev, Zh. Prikl. Khim. 44, 449 (1971, [CA 75, 57892 (1971)].
1013. V. P. Grigor'ev, I. M. Gershanova, and V. V. Ekilik, Zh. Prikl. Khim. 44, 1037 (1971) [CA 75, 79503 (1971)].
1014. V. P. Grigor'ev, V. V. Kuznetsov, V. V. Ekilik, and O. E. Shelepin, Zh. Prikl. Khim. 42, 804 (1969) [CA 71, 44010 (1969)].
1015. J. D. Bode, U.S. Patent 3,434,973 (1969) [CA 70, 108552 (1969)].
1016. V. V. Ekilik and V. P. Grigor'ev, Zashch. Met. 13, 690 (1977) [CA 88, 112361 (1977)].
1017. V. P. Grigor'ev, G. N. Ekilik, and V. V. Ekilik, Izv. Vyssh. Uchebn. Zaved., Khim. Khim. Tekhnol. 20, 1171 (1977) [CA 88, 43071 (1978)].
1018. E. M. Golubchik, Zashch. Met. 12, 605 (1976) [CA 86, 80719 (1977)].
1019. I. P. Krainov, B. I. Shapiro, I. V. Mangubi, and B. G. Distanov, U.S.S.R. Patent 609,083 (1978) [CA 89, 76470 (1978)].
1020. C. E. H. Bawn, R. Carruthers, and A. Ledwith, J. C. S. Chem. Commun., 522 (1965).
1021. E. F. Melega, A. Barabas, P. Blenke, and V. Toniu, Roumanian Patent 50,964 (1968) [CA 70, 20843 (1969)].
1022. A. Dinculescu, A. T. Balaban, A. Blaga, and E. Dinculescu, Roumanian Patent 76, 892 (1981).
1023. C. Georgoulis, J. Landais, C. Prévost, and M. Siemiatycki, C. R. Acad. Sci. 250, 3168 (1960).
1024. A. T. Balaban, Rev. Roum. Chim. 14, 1331 (1969).
1025. T. C. Chadwick, Anal. Chem. 47, 933 (1975).
1026. E. Hopîrtean, Rev. Roum. Chim. (in press).
1027. G. Arnold, G. Paal, W. Alfred, K. Halfar, and H. P. Vollmer, Ger. Offen. 2,346,278 (1975) [CA 83, 88759 (1975)].
1028. A. Bloom and J. W. Burke, Appl. Opt. 16, 2614 (1977).
1029. L. Strzelecki, Bull. Soc. Chim. Fr., 2666 (1967).
1030. G. A. Reynolds and J. A. Van Allan, U.S. Patent 3,417,083 (1969) [Ref. Zh. Khim., 11 H 395 (1970)].
1031. E. P. Olekhnovich, G. N. Tregub, G. N. Dorofeenko, V. P. Karmazin, and M. I. Knyazhanskii, U.S.S.R. Patent 514,833 (1976) [CA 85, 102217 (1976)].
1032. V. G. Tishchenko, U.S.S.R. Patent 167,866 (1965) [CA 63, 2519 (1965)].
1033. A. T. Balaban, I. I. Stanoiu, and E. Gârd, Rev. Roum. Chim. 22, 1191 (1977).
1034. T. E. Young and P. H. Scott, U.S. Patent 3,388,133 (1968) [CA 69, 59213 (1968)]; U.S. Patent 3,388,134 (1968) [CA 69, 67357 (1968)].
1035. N. S. Semenov, Yu. A. Nikolyukin, S. N. Baranov, and V. I. Dulenko, U.S.S.R. Patent 382,619 (1973) [CA 79, 66339 (1973)].
1036. N. Barbulescu, G. Nicolae, and G. Brotea, Roumanian Patent 59,006 (1975) [CA 88, 169973 (1978)].
1037. A Dinculescu, S. Cilianu, and C. Draghici, Roumanian Patent 69,761 (1979); A. Dinculescu and A. Ardeleanu, Roumanian Patent 76,897 (1981).

1038. Yu. D. Beletskii, A. N. Narkevich, G. N. Dorofeenko, and Yu. A. Zhdanov, *Zh. Vses. Khim. O-va.* **11**, 359 (1966) [*CA* **65**, 9355 (1966)].
1039. Yu. D. Beletskii, A. N. Narkevich, G. N. Dorofeenko, and Yu. A. Zhdanov, *Genetika*, 118 (1966) [*CA* **66**, 549 (1967)].
1040. A. N. Narkevich, Yu. D. Beletskii, G. N. Dorofeenko, Yu. A. Zhdanov, and E. K. Razoriteleva, *Genetika*, 33 (1968) [*CA* **69**, 74545 (1968)].
1041. E. P. Gus'kov, Ph.D. Thesis, Rostov University, Rostov-on-Don, 1970.
1042. A. N. Narkevich, Yu. D. Beletskii, Yu. G. Suchkov, G. N. Dorofeenko, and Yu. A. Zhdanov, *Genetika*, 165 (1970) [*CA* **73**, 127857 (1970)].
1043. E. P. Gus'kov, L. A. Plugina, L. L. Gumanov, and Yu. D. Beletskii, *Dokl. Akad. Nauk SSSR* **194**, 1214 (1970) [*CA* **74**, 39675 (1970)].
1044. L. B. Olekhnovich, Ph.D. Thesis, Rostov University, Rostov-on-Don, 1973.
1045. A. N. Narkevich, Ph.D. Thesis, Rostov University, Rostov-on-Don, 1970.
1046. A. P. Ordyntseva, *Farmakol. Tsentr. Kholinolitikov Drugikh Neirotropnykh Sredstv*, 325 (1969) [*CA* **73**, 118965 (1970)].
1047. V. V. Babin, L. I. Isakova, V. V. Morozovskii, S. G. Khlistovskaya, V. I. Shepelev, G. N. Shibanov, V. A. Palchkov, and V. F. Ermolova, U.S.S.R. Patent 704,575 (1979) [*CA* **92**, 89322 (1980)].
1048. J. D. Roberts and M. C. Caserio, "Modern Organic Chemistry," p. 308. Benjamin, New York, 1967.
1049. W. Schneider and A. Ross, *Ber. Dsch. Chem. Ges.* **55**, 2775 (1922).
1050. G. N. Dorofeenko, V. V. Mezheritskii, and A. L. Vasserman, *Khim. Geterotsikl. Soedin.*, 570 (1974) [*CA* **81**, 49516 (1974)].
1051. V. V. Mezheritskii, A. L. Vasserman, and G. N. Dorofeenko, *Khim. Geterotsikl. Soedin.*, 897 (1974) [*CA* **81**, 120376 (1974)].
1052. J. A. Van Allan and G. A. Reynolds, *J. Heterocycl. Chem.* **6**, 29 (1969).
1053. F. Eiden and H. Fenner, *Chem. Ber.* **101**, 3403 (1968).
1054. W. Dilthey and E. Floret, *Justus Liebigs Ann. Chem.* **440**, 89 (1924).
1055. W. Dilthey and B. Burger, *Ber. Dsch. Chem. Ges.* **54**, 825 (1921).
1056. J. A. Van Allan and G. A. Reynolds, *J. Heterocycl. Chem.* **11**, 395 (1974).
1057. S. V. Krivun, U.S.S.R. Patent 410,016 (1974) [*CA* **80**, 120907 (1974)].
1058. D. Lloyd and F. I. Wasson, *Chem. Ind. (London)*, 1559 (1963).
1059. L. Douglas and F. I. Wasson, *J. Chem. Soc. C*, 1086 (1966).
1060. F. Bardone-Gaudemar, *Ann. Chim. (Paris)* **3**, 52 (1958) [*CA* **52**, 20038 (1958)].
1061. W. C. Dovey and R. Robinson, *J. Chem. Soc.*, 1389 (1935).
1062. Z. S. Ariyan and H. Suschitzky, *J. Chem. Soc.*, 2242 (1961).
1063. C. Gastaldi, *Gazz. Chim. Ital.* **51**, 289 (1921) [*CA* **16**, 1410 (1922)].
1064. W. Dilthey, G. Fröde, and H. Koenen, *J. Prakt. Chem.* **114**, 153 (1926).
1065. A. Treibs and H. Bader, *Chem. Ber.* **90**, 789 (1957).
1066. W. Dilthey, *Ber. Dtsch. Chem. Ges.* **55**, 1275 (1922).
1067. A. I. Tolmachev, *Zh. Obshch Khim.* **33**, 1864 (1963) [*CA* **60**, 689 (1964)].
1068. J. Bolle and G. Tomaszewski, French Patent 1,340,970 (1963) [*CA* **60**, 5463 (1964)].
1069. D. Farcasiu and E. Gârd, *Tetrahedron* **24**, 4741 (1968).
1070. L. C. King, F. J. Ozog, and J. Moffat, *J. Am. Chem. Soc.* **73**, 300 (1951).
1071. A. T. Balaban and C. D. Nenitzescu, *Tetrahedron Lett.*, No. 2, 7 (1960).
1072. G. N. Dorofeenko, V. I. Dulenko, and N. V. Kovalenko, *Zh. Obshch. Khim.* **34**, 332 (1964) [*CA* **60**, 10641 (1964)].
1073. A. T. Balaban, D. Farcasiu, and C. D. Nenitzescu, *Tetrahedron* **18**, 1075 (1962).
1074. A. T. Balaban and C. D. Nenitzescu, *Tetrahedron* **10**, 55 (1960).
1075. H. Strzelecka, M. Simalty-Siemiatycki, and C. Prévost, *C. R. Acad. Sci.* **258**, 6167 (1964).

1076. G. N. Dorofeenko and G. I. Zhungietu, *Zh. Obshch. Khim.* **35**, 963 (1965) [*CA* **63**, 9909 (1965)].
1077. E. V. Kuznetsov, I. V. Shcherbakova, and G. N. Dorofeenko, *Khim. Geterotsikl. Soedin.*, 745 (1976) [*CA* **85**, 177200 (1976)].
1078. G. I. Zhungietu and E. M. Perepelitsa, *Zh. Obshch. Khim.* **36**, 1858 (1966) [*CA* **66**, 55362 (1967)].
1079. R. L. Letsinger and J. D. Jamison, *J. Am. Chem. Soc.* **83**, 193 (1961).
1080. M. K. Georgi and J. Rétey, *J. C. S. Chem. Commun.*, 32 (1971).
1081. M. Simalty-Siemiatycki, *Bull. Soc. Chim. Fr.*, 1944 (1965).
1082. G. N. Dorofeenko, E. I. Demidenko, and S. V. Krivun, *Izv. Vyssh. Uchebn. Zaved., Khim. Khim. Tekhnol.* **10**, 304 (1967) [*CA* **68**, 29529 (1968)].
1083. J. A. Durden and D. G. Crosby, *J. Org. Chem.* **30**, 1684 (1965).
1084. C. Gastaldi, *Gazz. Chim. Ital.* **52**, 169 (1922) [*CA* **16**, 2515 (1922)].
1085. C. Gastaldi and G. L. Peyretti, *Gazz. Chim. Ital.* **53**, 11 (1923) [*CA* **17**, 2284 (1923)].
1086. G. N. Dorofeenko and V. V. Tkachenko, *Khim. Geterotsikl. Soedin.*, 176 (1974) [*CA* **81**, 13347 (1974)].
1087. G. I. Zhungietu, Ph.D. Thesis, Rostov University, Rostov-on-Don, 1965.
1088. A. G. Ismailov and M. R. Atakishieva, *Khim. Geterotsikl. Soedin.*, 777 (1967) [*CA* **68**, 114352 (1968)].
1089. G. N. Dorofeenko, Z. N. Nazarova, and V. N. Novikov, *Zh. Obshch. Khim.* **34**, 3918 (1964) [*CA* **62**, 9099 (1965)].
1090. H. E. Johnston and R. J. W. Le Fèvre, *J. Chem. Soc.*, 2900 (1932).
1091. W. Dilthey, *J. Prakt. Chem.* **104**, 28 (1922).
1092. L. Amoros-Marin and R. B. Carlin, *J. Am. Chem. Soc.* **81**, 733 (1959).
1093. K. Hafner, H. Pelster, and H. Patzelt, *Justus Liebigs Ann. Chem.* **650**, 80 (1961).
1094. W. Dilthey and W. Radmacher, *J. Prakt. Chem.* **111**, 153 (1925).
1095. G. N. Dorofeenko and S. V. Krivun, *Zh. Obshch. Khim.* **34**, 105 (1964) [*CA* **60**, 10641 (1964)].
1096. G. N. Dorofeenko and S. V. Krivun, *Zh. Obshch. Khim.* **32**, 2386 (1962) [*CA* **58**, 7904 (1963)].
1097. T. Eicher and S. Böhm, *Chem. Ber.* **107**, 2238 (1974).
1098. W. Dilthey and C. Berres, *J. Prakt. Chem.* **111**, 340 (1925).
1099. W. Dilthey, *J. Prakt. Chem.* **102**, 209 (1921).
1100. H. Meerwein, K. Bodenbrenner, P. Borner, F. Kunert, and K. Wunderlich, *Justus Liebigs Ann. Chem.* **632**, 38 (1960).
1101. M. Simalty-Siemiatycki, *Bull. Soc. Chim. Fr.*, 1944 (1965).
1102. V. A. Palchkov, Ph.D. Thesis, Rostov University, Rostov-on-Don, 1966.
1103. E. V. Kuznetsov, Ph.D. Thesis, Rostov University, Rostov-on-Don, 1970.
1104. R. Neidlein and P. Witerzens, *Arch. Pharm. (Weinheim, Ger.)* **309**, 649 (1976).
1105. A. Dinculescu, Ph.D. Thesis, Polytechnique, Bucharest, 1979.
1106. A. Dinculescu and A. T. Balaban, Roumanian Patent 72,176 (1979).
1107. A. Dinculescu and A. T. Balaban, Roumanian Patent 73,505 (1980).
1108. A. Dinculescu and A. T. Balaban, Roumanian Patent 71,871 (1979).
1109. A. Dinculescu and A. T. Balaban, Roumanian Patent 72,174 (1979).
1110. A. Dinculescu, H. N. Koutrakis, and A. T. Balaban, *Rev. Roum. Chim.* **24**, 439 (1979).
1111. A. Dinculescu and A. T. Balaban, Roumanian Patent 71,869 (1979).
1112. A. Dinculescu and A. T. Balaban, Roumanian Patent 74,733 (1980).
1113. J. A. Van Allan, G. A. Reynolds, J. T. Alessi, S. Chie Chang, and R. C. Joines, *J. Heterocycl. Chem.* **8**, 919 (1971).
1114. A. Dinculescu and A. T. Balaban, Roumanian Patent 74,547 (1980).

1115.   G. Seitz, *Angew. Chem.* **79**, 96 (1967); *Angew. Chem., Int. Ed. Engl.* **6**, 82 (1967).
1116.   A. Dinculescu and A. T. Balaban, Roumanian Patent 72,175 (1979).
1117.   A. Dinculescu, V. Voroneanu, A. Ardeleanu, and A. T. Balaban, Roumanian Patent **101**, 468 (1980).
1118.   G. N. Dorofeenko, Yu. P. Andreichikov, and G. E. Trukhan, *Khim. Geterotsikl. Soedin.*, 1344 (1974) [*CA* **82**, 139908 (1975)].
1119.   E. A. Zvezdina, M. P. Zhdanova, V. A. Bren, and G. N. Dorofeenko, *Khim. Geterotsikl. Soedin.*, 1461 (1974) [*CA* **82**, 97303 (1975)].
1120.   W. Schneider and A. Sack, *Ber. Dsch. Chem. Ges.* **56**, 1786 (1923).
1121.   A. Dinculescu and A. T. Balaban, Roumanian Patent 71,870 (1979).
1122.   N. Barbulescu and G. Nicolae, *Rev. Chim. (Bucharest)* **23**, 69 (1972).
1123.   C. Uncuta and A. T. Balaban, unpublished.
1124.   A. T. Balaban, A. Zlota, and A. Dinculescu, unpublished.
1125.   G. N. Dorofeenko, E. A. Zvezdina, M. P. Zhdanova, and I. A. Barchan, *Khim. Geterotsikl. Soedin.*, 1682 (1973) [*CA* **80**, 82805 (1974)].
1126.   A. Dinculescu and A. T. Balaban, Roumanian Patent 74,546 (1980).
1127.   J. Kelemen and R. Wizinger, *Helv. Chim. Acta* **45**, 1908 (1962).
1127a. A. Dinculescu and A. T. Balaban, Roumanian Patent 73,503 (1980).
1128.   N. V. Khromov-Borisov and L. A. Gavrilova, *Zh. Obshch. Khim.* **31**, 2192 (1961) [*CA* **56**, 2415 (1962)].
1129.   J. A. Van Allan and G. A. Reynolds, *J. Heterocycl. Chem.* **8**, 803 (1971).
1130.   F. Eiden and M. Peglow, *Arch. Pharm. (Weinheim, Ger.)* **303**, 71 (1970).
1131.   N. Barbulescu, G. Nicolae, and V. Niculaita, *Ann. Univ. Bucuresti, Chim.* **20**, 37 (1971) [*CA* **79**, 66146 (1973)].
1132.   V. A. Kaminski and M. N. Tilichenko, *Zh. Org. Khim.* **5**, 186 (1969) [*CA* **70**, 87541 (1969)].
1133.   A. Dinculescu and A. T. Balaban, *Chem. Scripta* (in press).
1134.   G. Seitz, H.-G. Lehmann, and H. Mönnighoff, *Justus Liebigs Ann. Chem.* **757**, 93 (1972).
1135.   A. R. Katritzky and M. Shanta, *J. C. S. Chem. Commun.*, 552 (1979).
1136.   A. R. Katritzky, G. Musumarra, K. Sakizadeh, S. M. M. El-Shafie, and B. Jovanovic, *Tetrahedron Lett.*, 2697 (1980).
1137.   A. R. Katritzky, M. J. Cook, A. Ikizler, and G. H. Millet, *J. C. S. Perkin I*, 2500 (1979).
1138.   R. Wizinger and H. Sontag, *Helv. Chim. Acta* **38**, 363 (1955).
1138a. L. G. S. Brooker, D. S. Daniel, and R. C. Taber, U.S. Patent 3,639,127 (1972) [*CA* **77**, 116064 (1972)].
1139.   A. R. Katritzky, A. Krutošíková, C. A. Ramsden, and J. Lewis, *Collect. Czech. Chem. Commun.* **43**, 2046 (1978).
1140.   A. R. Katritzky, C. A. Ramsden, Z. Zakaria, R. L. Harlow, and H. Simsen, *J. C. S. Chem. Commun.*, 363 (1979).
1141.   G. Mutz, Ph.D. Thesis, University of Marburg, 1960.
1142.   A. T. Balaban and F. A. Urseanu, Roumanian Patent 57,177 (1974) [*CA* **83**, 195216 (1975)].
1143.   A. I. Tolmachev, N. A. Derevyanko, E. F. Karaban, and M. A. Kudinova, *Khim. Geterotsikl. Soedin.*, 612 (1975) [*CA* **83**, 99157 (1975)].
1144.   K. Tamura, Y. Ogo, and T. Imoto, *Bull. Chem. Soc. Jpn.* **46**, 2988 (1973).
1145.   G. N. Dorofeenko, Yu. P. Andreichikov, E. A. Zvezdina, V. A. Bren, G. E. Trukhan, V. V. Derbenev, and A. N. Popova, *Khim. Geterotsikl. Soedin.*, 1349 (1974) [*CA* **82**, 111386 (1975)].

1145a. E. Stepan and A. T. Balaban, Roumanian Patent (appl.).
1146. G. Märkl, "Chimie Organique du Phosphore." 1969.
1147. J. C. J. Bart and J. J. Daly, *Angew. Chem.* **80**, 843 (1968); *Angew. Chem., Int. Ed. Engl.* **7**, 811 (1968).
1148. G. Märkl, F. Lieb, and A. Merz, *Angew. Chem.* **79**, 947 (1967); *Angew. Chem., Int. Ed. Engl.* **6**, 944 (1967).
1149. H. H. Pohl, Diploma Work, University of Marburg, 1971.
1150. W. Mach, Ph.D. Thesis, University of Marburg, 1968.
1151. M. Schoelm, Diploma Work, University of Marburg, 1968.
1152. G. Märkl, *Angew. Chem.* **78**, 907 (1966); *Angew. Chem., Int. Ed. Engl.* **5**, 846 (1966).
1153. A. Chatzidakis, Ph.D. Thesis, University of Marburg, 1969.
1154. G. I. Zhungietu, F. N. Chukhrii, and A. I. Tolmachev, *Zh. Vses. Khim. O-va.* **15**, 590 (1970) [*CA* **74**, 22956 (1971)].
1155. N. Greif, Ph.D. Thesis, University of Marburg, 1967.
1156. K. Worschech, Ph.D. Thesis, University of Marburg, 1960.
1157. W. Krafft, Ph.D. Thesis, University of Marburg, 1962.
1158. G. Oosterloo, Ph.D. Thesis, University of Marburg, 1958.
1159. W. Umbach, Ph.D. Thesis, University of Marburg, 1963.
1160. K. H. Blöcher, Ph.D. Thesis, University of Marburg, 1960.
1161. K. Schlömer, Ph.D. Thesis, University of Marburg, 1961.
1162. K. Dimroth, A. Berndt, and C. Reichardt, *Org. Synth. Collect.* **5**, 1128 (1973).
1163. K. Dimroth and W. Umbach, unpublished.
1164. D. Hammel, Diploma Work, University of Marburg, 1963.
1165. K. H. Wolf, Ph.D. Thesis, University of Marburg, 1961.
1166. K. Hafner, C. Bernhard, and R. Müller, *Justus Liebigs Ann. Chem.* **650**, 35 (1961).
1167. Yu. N. Porshnev, E. M. Tereshchenko, and M. J. Cherkashin, *Zh. Org. Khim.* **14**, 263 (1978) [*CA* **88**, 190452 (1978)].
1168. N. V. Kholodova, Yu. P. Andreichikov, and G. N. Dorofeenko, *Khim. Geterotsikl. Soed.*, 162 (1981).
1169. O. V. Dyrgina, G. N. Dorofeenko, and O. Yu. Okhlobystin, *Khim. Geterotsikl. Soed.*, 189 (1980); 454 (1981).
1170. A. Pelter, *Tetrahedron Lett.*, 22 (1981).
1170a. Yu. I. Ryabukhin, V. V. Mezheritskii, V. I. Dulenko, N. I. Basina, and G. N. Dorofeenko, *Khim. Geterotsikl. Soed.*, 1027 (1979).
1171. V. I. Boev and A. V. Dombrovskii, *Khim. Geterotsikl. Soed.*, 881, 887 (1979).
1171a. A. T. Balaban, M. Fahmy, and M. D. Gheorghiu (in preparation).
1172. R. Aveta, G. Doddi, G. Illuminati, and F. Stegel, *J. Am. Chem. Soc.* **103**, 6148 (1981).
1173. R. Aveta, G. Doddi, N. Insam, and F. Stegel, *J. Org. Chem.* **45**, 5160 (1980).
1174. A. R. Katritzky, G. Musumarra, K. Sakizadeh, and M. Misic-Vukovic, *J. Org. Chem.* **46**, 3280 (1981).
1175. A. R. Katritzky, A. M. El-Mowafy, G. Musumarra, K. Sakizadeh, C. Sana-Ullah, M. M. S. El-Shafie, and S. S. Thind, *J. Org. Chem.* **46**, 3823 (1981).
1176. A. R. Katritzky, G. Musumarra, and K. Sakizadeh, *J. Org. Chem.* **46**, 3831 (1981).
1177. A. R. Katritzky, G. Z. De Ville, and R. C. Patel, *Tetrahedron Lett.*, 1723 (1980).
1178. A. R. Katritzky and M. C. Rezende, *J. Chem. Res., Synop.*, 312 (1980).
1179. A. R. Katritzky, S. S. Thind, and S. Sukhpal, *J. Chem. Soc. Pakistan* **2**, 51 (1980).
1180. A. R. Katritzky, A. Banerji, B. S. El-Osta, J. R. Parker, and C. A. Ramsden, *J. C. S. Perkin I*, 690 (1979).

1181.   A. R. Katritzky and S. S. Thind, *J. C. S. Perkin I*, 661 (1981).
1182.   A. R. Katritzky and A. M. El-Mowafi, *J. Chem. Soc. Chem. Commun.*, 96 (1981).
1183.   A. R. Katritzky, M. C. Rezende, and S. S. Thind, *J. Chem. Res., Synop.*, 309 (1980).
1184.   A. R. Katritzky, A. Saba, and R. C. Patel, *J. C. S. Perkin I*, 1492 (1981).
1185.   A. R. Katritzky, R. C. Patel, and M. S. Shanta. *J. C. S. Perkin I*, 1888 (1980).
1186.   A. R. Katritzky, A. S. Afridi, and C. A. Ramsden, *Pakistan J. Sci. Ind. Res.* **21**, 1 (1978).
1187.   A. Arques, A. Lorenzo, P. Molina, and A. Soler, *An. Quim.* **75**, 118 (1979).
1188.   E. A. Zvezdina, M. P. Zhdanova, and G. N. Dorofeenko, *Khim. Geterotsikl. Soed.*, 321 (1979).
1189.   E. A. Zvezdina, M. P. Zhdanova, and G. N. Dorofeenko, *Khim. Geterotsikl. Soed.*, 748 (1980).
1190.   D. J. Harris, G. Y. P. Kan, T. Tschamber, and V. Snieckus, *Canad. J. Chem.* **58**, 494 (1980).
1191.   A. R. Katritzky and P. Ballesteros, *J. Chem. Res., Synop.*, 172 (1981).
1192.   A. R. Katritzky, P. Ballesteros, and A. T. Thomas, *J. C. S. Perkin I*, 1495 (1981).
1193.   A. T. Balaban, M. D. Gheorghiu, and T. S. Balaban (in press).
1194.   A. T. Balaban, A. Dinculescu, and M. Fahmy (in preparation).
1195.   S. Kurokawa and A. G. Anderson, *Bull. Chem. Soc. Jpn.* **52**, 257 (1979).
1196.   J. A. Barltrop, A. W. Baxter, A. C. Day, and E. Irving, *J. Chem. Soc. Chem. Commun.*, 606 (1980).
1197.   A. T. Balaban, S. Badilescu, and V. Ciorba (in press).
1198.   F. Pragst, R. Ziebig, U. Seydewitz, and G. Driesel, *Electrochim. Acta* **25**, 341 (1980).
1199.   P. Sandor and L. Radics, *Org. Magn. Reson.* **16**, 148 (1981).
1200.   V. B. Panov, M. V. Nekhoroshev, and O. Yu. Okhlobystin, *Zhurn. Obshch. Khim.* **49**, 234 (1979).
1201.   A. S. Mokovnik and O. Yu. Okhlobystin, *Khim. Geterotsikl. Soed.*, 1041 (1980).
1202.   R. W. Bigelow, *J. Chem. Phys.* **73**, 3864 (1980).
1203.   R. W. Bigelow, R. J. Weagley, and H. J. Freund, *J. Electr. Spectr. Rel. Phenom.* (in press).
1204.   B. Kopainsky, W. Kaiser, and K. H. Drexhage, *Opt. Commun.* **32**, 451 (1980).

Author Index

Author names are followed by an italicized number (in parentheses) representing a reference number and then by the page number(s) where this reference or name is cited or by an S which denotes the foregoing Note Added in Proof. No distinction has been made between a, ä, å, ă or â; between o or ö; and between u or ü. The page numbers where the references are listed have not been cited.

A

Abdel-Azeen Abdallah, A., (*177*) 170
Abdel-Kader, A., (*558*) 113
Abdel-Megeed, M. F., (*618*) 122, 123, 298, 333
Abramovich, R. A., (*241*) 35, 117, 235, 236; (*661*) 132, 293
Ackerman, I. F., (*640*) 126
Adams, R., (*641*) 126
Afridi, A. S., (*1186*) S
Ainsworth, C., (*473*) 100
Alder, K., (*411*) 76, 127, 258, 328, 329, 331, 332
Alekseev, N. N., (*183*) 28, 334; (*443*) 93; (*444*) 93; (*449*) 95; (*480*) 103; (*481*) 103; (*654*) 129
Alessi, J. T., (*1113*) 297, 299
Alferova, O. F., (*14*) 3, 26, 207, 216
Alfred, W., (*1027*) 217
Alizon, J., (*894*) 198
Alomran, F., (*610*) 122
Alsup, M. J., (*989*) 215
Altmann, H., (*138*) 22
Aminova, G., Kh. (*21*) 9, 208, 209; (*22*) 9, 208; 209; (*272*) 39, 209; (*273*) 39, 208; (*921*) 210
Ammon, C.. (*593*) 119
Amoros-Marin, L., (*1092*) 261, 262
Anderson, A. G., (*188*) 29, 108; (*1195*) S
Anderson, W. G., (*751*) 160, 161
Andreichikov, Yu. P., (*268*) 39, 120, 321, 326; (*270*) 270, 30, 40; (*271*) 39; (*74*) 39, 208; (604) 120; (1118) 299, 300, 323, 333–336, 346, 347; (*1145*) 355, 357, 358; (*1168*) S
Andrieux, J., (*136*) 22

Angerer, E., (*791*) 173
Angliker, H. J., (*55*) 13
Ahn, N. T., (*385*) 67
Anker, R. M., (*28*) 9, 33, 54, 110, 113, 115, 116, 249, 251, 252, 292, 293, 313
Arbashev, B. I., (*222*) 32, 128, 212
Ardeleanu, A., (*1117*) 298, 321, 355, 357, 358
Ariyan, Z. S., (*1062*) 240, 263, 265
Armit, J. W., (*566*) 113, 126
Arnaud, M., (*182*) 28; (*193*) 29, 244–249; (*485*) 103
Arndt, F., (*454*) 95; (*497*) 106
Arnold, G., (*1027*) 217
Arnoldy, G., (*660*) 132, 255, 322, 323, 329, 332, 339
Aron, E., (*497*) 106
Arques, A., (*1187*) S
Arsene, A., (*370*) 64
Asmus, K. D., (*413*) 76, 157, 366
Atakishieva, M. R., (*1088*) 257
Aveta, R., (*1172*) S; (*1173*) S
Avram, M., (*860*) 184

B

Babin, V. V., (*1047*) 219
Bacescu-Roman, M., (*781*) 171
Bader, H., (*1065*) 242, 269
Badilescu, I. I., (*184*) 28, 110, 188, 252
Badilescu, S., (*850*) 183; (*854*) 183; (*1197*) S
Baeyer, A., (*29*) 9, 73, 107, 109, 245; (*244*) 11, 69, 73, 107, 114, 240, 281, 303; (*390*) 69, 73, 102, 107, 240
Baier, H., (*715*) 152; (*716*) 153

(648) 126, 142; *(660)* 132, 255, 322, 323, 329, 332, 339; *(681)* 137, 359–362; *(684)* 138, 139; *(693)* 142; *(694)* 142; *(700)* 143, 363, 364; *(701)* 143, 363, 364; *(702)* 143, 363, 364; *(703)* 144; *(704)* 144; *(705)* 144, 145; *(707)* 144; *(708)* 144, 218; *(709)* 145; *(710)* 145; *(711)* 146; *(713)* 150, 157; *(888)* 197; *(1162)* 364; *(1163)* 365

Dinculescu, A., *(177)* 27, 215; *(547)* 113; *(574)* 116, 121, 126, 293, 294, 296–299, 301–307, 310–312, 317, 319, 320, 324, 326–328, 330, 333, 339–341, 343, 344, 347, 348, 355, 356; *(589)* 117, 187, 294, 296, 297, 303–306, 309, 311, 315; *(591)* 118, 295, 298, 307, 312, 315, 321; *(632)* 125, 293–296, 310, 315–319, 325–332, 346; *(1022)* 216, 297; *(1037)* 218; 301; *(1105)* 293–298, 301–307, 309–312, 315–320, 322, 324–333, 338, 340, 341, 343–348, 353, 355, 356; *(1106)* 293, 315, 325; *(1107)* 293; *(1108)* 293, 315, 325; *(1109)* 294, 295, 310, 316, 318, 319, 330, 332, 346; *(1110)* 294, 296, 309, 315, 318, 322, 327, 330, 331, 338, 346, 348; *(1111)* 295, 317, 327; *(1112)* 296, 297, 302, 317, 319, 324; *(1114)* 297, 320, 331; *(1116)* 298, 306, 320, 328, 333, 343; *(1117)* 298, 321, 355, 357, 358; *(1121)* 304, 327, 339–341, 343, 344, 348; *(1124)* 305, 309, 312; *(1126)* 307, 345; *(1127 a)* 312, 333, 347; *(1133)* 319; *(1194)* S

Dinculescu, E., *(1022)* 216, 297

Dines, M., *(32)* 9, 12, 339, 342, 346

Dingwall, W., *(24)* 9

Distanov, B. G., *(831)* 180; *(1019)* 216

Djerassi, C., *(900)* 199

Doddi, G., *(294)* 44, 170; *(300)* 45, 170; *(1172)* S; *(1173)* S

Dodiuk, H., *(653)* 127

Dombrovskii, A. V., *(313 a)* 49; *(1171)* S

Dondoni, A., *(669)* 134, 322, 350, 351; *(670)* 134, 349–351, 353

Dorofeenko, G. N., *(5)*, 3, 216; *(6)* 3, 216; *(7)* 3, 25, 216; *(8)* 3, 216; *(9)* 3, 31, 216; *(21)* 9, 208, 209; *(22)* 9, 208, 29; *(46)* 12; *(47)* 12, 97, 99, 103, 108, 218, 219, 251, 299, 312, 315; *(48)* 12, 311;

(50) 12, 25, 108; *(51)* 12, 25, 108; *(58)* 13, 108; *(59)* 13; *(62)* 13; *(71)* 14; *(74)* 14, 171; *(75)* 14; *(78)* 15, 66; *(79)* 15; *(92)* 17; *(93)* 17; *(95)* 17, 18; *(96)* 17, 20, 25; *(98)* 17, 110, 254; *(99)* 17, 110, 246, 247; *(102)* 18, 19, 288, 289; *(103)* 18, 111; *(104)* 18; *(106)* 18, 260, 275, 276; *(109)* 19, 244, 245, 247, 248, 250, 254, 256; *(110)* 19, 111, 290, 291; *(111)* 19, 112; *(112)* 19, 112; *(113)* 19; *(114)* 19, 25, 250, 254, 257, 259, 273; *(115)* 20; *(116)* 20, 29, 111, 264, 269; *(117)* 20, 111, 280, 281; *(118)* 20, 26; *(119)* 29; *(120)* 20, 253; *(121)* 20, 245, 247, 249–251; *(124)* 21, 22, 25, 39, 253, 254, 258, 269, 271; *(113)* 22; *(134)* 22; *(135)* 22; *(139)* 23; *(140)* 23, 108, 172; *(141)* 23, 108, 172; *(147)* 24; *(150)* 24; *(153)* 25, 250; *(154)* 25; *(158)* 26; *(159)* 26, 211; *(160)* 26, 30; *(161)* 26; *(163)* 26; *(163)* 26; *(164)* 26; *(165)* 26; *(166)* 26, 323; *(167)* 26, 27; *(168)* 27, 261, 262, 266, 268; *(169)* 27; *(170)* 27; *(190)* 29, 30, 110, 111, 282; *(191)* 29, 108; *(206)* 30, 111, 283–286; *(211)* 31, 32, 286; *(213)* 31, 32, 286; *(213)* 31, 32, 111, 280, 284–286; *(214)* 31, 32, 111, 285, 286; *(215)* 31, 32, 110, 283, 284; *(219)* 32, 280, 281, 287; *(221)* 32, 109, 278; *(222)* 32, 128, 212; *(245)* 36, 225; *(246)* 36; *(250)* 36, 225, 226, 259, 315; *(254)* 37; *(269)* 39; *(270)* 39, 40; *(271)* 39; *(272)* 39, 209; *(273)* 39, 208; *(274)* 39, 208; *(306)* 46; *(380)* 66, 196; *(465)* 98, 292, 303; *(466)* 98, 303; *(478)* 103; *(479)* 103, 111; *(482)* 103; *(483)* 103, 249; *(500)* 107, 116, 132, 320, 332, 333, 335, 336, 339, 340, 342, 345–349, 353; *(502)* 108, 256, 258, 259, 267, 269, 271, 276, 277; *(503)* 109, 251, 258, 260, 278; *(504)* 109, 278; *(505)* 109; *(508)* 110, 128; *(516)* 110; *(517)* 110; *(518)* 110; *(519)* 110; *(520)* 110; *(521)* 110; *(22)* 110; *(524)* 110; *(525)* 110; *(526)* 110; *(527)* 110; *(528)* 110; *(529)* 110; *(530)* 110; *(531)* 110; *(532)* 110; *(534)* 111; 280, 287; *(535)* 111, 280, 287; *(537)* 111, 282–284; *(538)* 111, 282–285; *(539)* 111, 282–284; *(540)* 111,

Emoto, K., (*982*) 215
Epsztajn, J., (*629*) 125, 327, 336, 337, 340; (*667*) 133, 302, 326, 331, 333, 337–340, 342, 343, 345
Ermolova, V. F., (*1047*) 219
Etmetchenko, L. N., (*116*) 20, 29, 111, 264, 269; (*163*) 26; (*164*) 26; (*166*) 26, 323; (*190*) 29, 30, 110, 111, 282
Ettel, V. (*644*) 126
Evstifeev, M. M., (*21*) 9, 208, 209; (*22*) 9, 208, 209; (*272*) 39, 209; (*273*) 39, 208; (*919*) 209
Evtushenko, I. Ya., (*90*) 16, 17, 30
Eweiss, N., (*607*) 122, 331, 332, 337, 339, 340, 343, 357; (*608*) 122, 326–328, 331, 332, 334, 335, 337–340, 343, 345, 357, 358

F

Fabian, J., (*944*) 213
Fabre, C., (*894*) 198
Fabre, M. C., (*209*) 30, 66, 198, 207
Fahmy, M., (*1171 a*) S; (*1194*) S
Farcasiu, D., (*86*) 16; (*88*) 16, 30, 85; (*890*) 197; (*1069*) 244; (*1073*) 246
Farcasiu, M., (*890*) 197
Farid, S., (*252*) 37
Faust, J., (*200*) 30
Fechting, B., (*679*) 136
Feist, F., (*445*) 93; (*447*) 93
Feldman, M., (*382*) 66, 207; (*845*) 182, 206
Fellenberg, T., (*287*) 43
Fellion, Y., (*393*) 71, 73, 241, 243
Fenner, H., (*1053*) 235
Fico, S., (*986*) 215
Filleux-Blanchard, M.-L., (*402*) 74, 113
Findlay, R. H., (*947*) 214
Fischer, D. E., (*685*) 138, 139, 362
Fischer, G. W., (*1*) 2, 3, 8, 10, 21, 25, 27, 30, 40, 55, 56, 59, 83, 169, 215, 217, 221; (*67*) 13, 52, 59, 232–234; (*68*) 13, 52, 323, 233; (*108*) 19, 20; (*122*) 20, 60, 233; (*128*) 21, 59, 128, 231, 280; (*129 a*) 21, 45, 128; (*130*) 22, 111, 256, 257, 259; (*301*) 45, 46, 147; (*307*) 47, 147; (*322*) 52, 147; (*345*) 59, 231–233; (*346*) 59; (*347*) 59; (*349*) 59, 227–229

Fischer, J., (*107*) 19, 31, 266, 267, 270, 271; (*210*) 31
Fischer, O., (*646*) 126
Fleischmann, K., (*775*) 169
Flexser, L. A., (*24*) 9
Floret, E., (*1054*) 237, 241, 263, 265
Flygare, W. H., (*908*) 203
Fochi, R., (*338*) 56, 210; (*448*) 95; (*803*) 174
Föhlisch, B., (*34*) 9, 42, 61, 236, 239
Follweiler, J., (*492*) 105, 282
Fornarini, S., (*294*) 44, 170; (*300*) 45, 170
Förster, D., (*84*) 16, 23; (*143*) 23, 117
Foster, R., (*842*) 182
Fournier, F., (*699*) 143
Fowles, G. W. A., (*903*) 201
Fox, C. J., (*966*) 215; (*790*) 215; (*971*) 215; (*975*) 215; (*981*) 215; (*1005*) 216
Frangopol, M., (*437*) 90, 108, 142, 144, 146, 156, 218; (*832*) 181; (*948*) 214
Frangopol, P. T., (*362*) 61, 212, 236, 239, 267, 278, 288; (*419*) 78, 80, 97; (*437*) 90, 108, 142, 144, 146, 156, 218; (*477*) 103, 218, 244; (*569*) 113, 189, 235; (*826*) 177; (*832*) 181; (*948*) 214; (*949*) 214
Frank, F., (*775*) 169
Freund, H. J., (*1203*) S
Friedel, R. A., (*806*) 174
Fries, F. A., (*606*) 121
Fröde, G., (*1064*) 241, 242, 263, 265, 266, 268, 270, 344
Fugnitto, R., (*202*) 30, 39, 254–256, 263, 264, 273, 276; (*209*) 30, 66, 198, 207; (*266*) 39, 254, 255
Fujii, K., (*671*) 134
Fujino, T., (*995*) 216
Fujita, K., (*194*) 29
Fukui, K., (*490*) 105
Furmanova, N. G., (*397*) 73, 95
Futaki, K., (*982*) 215

G

Gallice, J., (*894*) 198
Gallois, D., (*992*) 216
Garbuglio, C., (*456*) 95; (*457*) 95
Gârd, E., (*178*) 27, 63; (*217*) 32, 62, 63, 70; (*364*) 62, 218; (*365*) 63; (*366*) 63; (*368*) 63; (*370*) 64; (*371*) 64, 70, 279,

Klatte, G., (*375*) 64
Kleimenova, V. I., (*199*) 30; (*201*) 30
Kleipool, R. J. C., (*636*) 126; (*638*) 126
Klevens, H. B., (*815*) 176
Klimenko, S. K., (*90*) 16, 17, 30; (*91*) 16, 17
Klingsberg, E., (*720*) 154
Klostermans, D., (*637*) 126
Knabeschuh, L. H., (*550*) 113; (*551*) 113; (*645*) 126
Knox, W. R., (*533*) 111
Knyazhanskii, M. I., (*21*) 9, 208, 209; (*274*) 39, 208; (*604*) 120; (*833*) 181; (*1031*) 217
Koblik, A. V., (*46*) 12; (*47*) 12, 97, 99, 103, 108, 218, 219, 251, 299, 312, 315; (*48*) 12, 311; (*59*) 13; (*71*) 14; (*72*) 14; (*163*) 26; (*164*) 26; (*166*) 26, 323; (*170*) 27; (*191*) 29, 108; (*731*) 156
Köbrich, G., (*412*) 76, 244; (*423*) 80, 140; (*424*) 80, 140; (*639*) 126
Kodak, Soc., (*983*) 215
Koenen, H., (*1064*) 241, 242, 263, 265, 266, 268, 270, 344
Konovalova, T. M., (*17*) 3, 106, 216
Kopainsky, B., (*1204*) S
Köppel, H., (*910*) 204
Körber, I., (*359*) 59, 255, 263–265
Kornilov, M. Yu., (*357*) 59, 60; (*877*) 187; (*878*) 187
Kornilov, V. I., (*529*) 110
Kornilova, E. N., (*124*) 21, 22, 25, 39, 253, 254, 258, 269, 271
Korobkova, V. G., (*522*) 110; (*524*) 110; (*530*) 110; (*531*) 110; (*568*) 113
Korolchenko, G. A., (*502*) 108, 256, 258, 259, 267, 269, 271, 276, 277
Koslov, E. S., (*682*) 137, 360, 361
Kosower, E. M., (*843*) 182
Kostanecki, S., (*828*) 180
Koutecky, J., (*926*) 211
Koutrakis, H. N., (*589*) 117, 187, 294, 296, 297, 303–306, 309, 311, 315; (*1110*) 294, 296, 309, 315, 318, 322, 327, 330, 331, 338, 346, 348
Kovalenko, N. V., (*541*) 111, 288, 289; (*1072*) 246
Kovalevskii, A. S., (*58*) 13, 108
Koyma, K., (*1000*) 216; (*1003*) 216

Kozhevnikova, N. I., (*336*) 56; (*337*) 56, 339
Kozima, A., (*980*) 215
Kraatz, A., (*299*) 45
Krafft, W., (*146*) 24, 39, 110, 144, 255; (*324*) 53, 144, 236, 237, 364; (*1157*) 363, 364
Krainov, I. P., (*1019*) 216
Krasnikov, V. V., (*51*) 12, 25, 108; (*140*) 23, 108, 172; (*141*) 23, 108, 172; (*785*) 172; (*790*) 172
Kress, A., (*406*) 74, 76, 78, 97, 263, 321, 322, 324, 326, 331, 334, 336, 338
Krivun, S. V., (*5*) 3, 216 (*14*) 3, 26, 207, 216; (*39*) 11, 235, 236, 239; (*40*) 11; (*41*) 11, 12; (*52*) 12; (*53*) 12, 260, 267; (*57*) 13, 56; (*58*) 13, 108; (*61*) 13, 43; (*63*) 13; (*64*) 13; (*65*) 13, 48, 110; (*66*) 13, 15, 39, 46, 66, 207, 254; (*109*) 19, 244, 245, 247, 248, 250, 254, 256; (*168*) 27, 261, 272, 266, 268; (*211*) 31, 32, 286; (*221*) 32, 109, 278; (*260*) 38; (*265*) 39; (*275*) 40; (*279*) 42; (*280*) 42; (*281*) 42; (*282*) 42; (*283*) 42; (*284*) 43; (*285*) 43, 236, 237, 239; (*291*) 44, 72; (*309*) 48; (*310*) 49; (*312*) 49; (*313*) 49; (*316*) 53, 239, 267; (*327*) 53; (*331*) 53, 62; (*479*) 103, 111; (*502*) 108, 256, 258, 259, 267, 269, 271, 276, 277; (*503*) 109, 251, 258, 260, 278; (*504*) 109, 278; (*505*) 109; (*516*) 110; (*517*) 110; (*518*) 110; (*519*) 110; (*522*) 110; (*527*) 110; (*543*) 111, 259, 280; (*544*) 111, 259, 280; (*723*) 154; (*1057*) 239; (*1082*) 255–257, 260, 261, 266, 269, 274; (*1095*) 262, 267–279, 274; (*1096*) 262, 267, 268
Krockenberger, D., (*34*) 9, 42, 61, 236, 239
Kröhnke, F., (*56*) 13, 53, 144
Kroke, H., (*315*) 49, 50, 58, 80, 142; (*316*) 49, 50; (*694*) 142
Kroning, E., (*33*) 9
Krumbholds, E., (*892*) 197
Krutošiková, A., (*1139*) 334, 336, 338, 358
Kryman, F. J., (*1004*) 216
Kubota, T., (*1000*) 216; (*1003*) 216
Kudinova, M. A., (*227*) 32; (*355*) 59; (*358*) 59; (*1143*) 350–353

S

Subject Index*

A

Acetoacetates, 146
Acetylacetone, 81, 146
Acetylenes, 6, 153
Acidity constant, pK_a 58, 69, 204–205
Active methyl(ene) groups, 10, 52–54,
  81–82, 143–150
Acylfurans, 90–92, 101, 161
Aldehydes, 31, 122, 131, 133
Alkanes, bisacylation of, 29
Alkenes (*see also* Olefins)
  bisacylation of, 27–30
  trisacylation of, 27
  from RNH_2, 124, S
Alkoxy anions, 44, S
Alkoxypyrylium, 13, 40, 44
Alkyl substituents (*see* Side chains)
Amidines, 135
Amines, 47, 74–77
  primary, 114–127
  secondary, 127–129
Amino acids, 116
Aminopyrylium, 10, 16, 21, 42
Ammonia, 73–74, 106–114
Ammonium RNR_3^+ from RNH_3, 123
Analytical chemistry, applications in, 217
Anhydrobases (*see also* Methylenepyrans),
  31, 56–64, 71, 132, 225–239
Anilines (*see also* Xylidines), 128
Anions (*see also* Charge transfer), 3, 30
Anthracene as donor in CT complexes,
  183, 184
Anticorrosive agents, 216
Applications (*see* Practical applications)
Aromaticity, 7, 86, 90, 174, 184, 187, 193,
  203, 213, 219, S
Aryl substituents, reactions of, 38
Atropisomerism, 29, 117
Azatetracyanopropenide, 183
Azide, 123, 155

Azirines, 155
Azlactones, 82, 149
Azomethines, 116
Azulenes, 156–158, 366–367, S

B

Bases, 58, 108
Benzene derivatives (*see also* Benzophen-
  ones, Biphenyls, Anilines, Phenols,
  Nitrobenzenes), 102–105, 142
Benzophenones, 104–105, 129, 148
Benzoquinolizinium, 120
Benzylmagnesium halides, 142–143
Betaines, 132
Biological effects, 218–219
Biphenyls, 140, 144
Bipyrans (*see also* Radicals, ESR), 52, 65,
  198, 207–208
Bipyridines, 278–279
Bipyridinium, 116, 355–358
Bipyrylium, 15, 24, 42, 109, 112, 198,
  207–208
Bisacylation, (*see* Olefins, diacylation of)
Bispyridines, bispyridinium (*see* Bipyri-
  dines, Bipyridinium)
Bispyrylium (*see* Bipyrylium)
Bond angles, 201–202
Bond lengths, 201–202, 213
Borohydride (*see also* Hydride), 7, 55
Boron derivatives (*see also* Carboranyl-
  pyrylium), 14, 171
Bromo derivatives from RNH_2, 122
Brönsted relationship, 69

C

Carbenoid pyrans, 39, 40
Carbon isotopes, 103, 108, 142, 144, 218
Carbon nucleophiles, 49–55, 78–82, 101,
  140–153, 156–158

*For an explanation of conventions see p. 405.